Electromagnetic Signals

Reflection, Focusing, Distortion, and Their Practical Applications

Henning F. Harmuth

Formerly of the Catholic University of America
Washington, D.C.

Raouf N. Boules

Towson University
Towson, Maryland

and

Malek G. M. Hussain

Kuwait University
State of Kuwait

SPRINGER SCIENCE+BUSINESS MEDIA, LLC

Library of Congress Cataloging-in-Publication Data

Harmuth, Henning F.
Electromagnetic signals : reflection, focusing, distortion, and their practical applications / Henning F. Harmuth, Raouf N. Boules, and Malek G.M. Hussain.
p. cm.
Includes bibliographical references and index.
ISBN 978-1-4613-7199-1 ISBN 978-1-4615-4849-2 (eBook)
DOI 10.1007/978-1-4615-4849-2
1. Signal processing--Mathematics. 2. Electromagnetic waves--Mathematics. 3. Ultra-wideband devices. 4. Maxwell equations. I. Boules, Raouf N. II. Hussain, Malek G. M. III. Title.
TK5102.9.H385 1999
621.382'2--dc21 98-48687
CIP

ISBN 978-1-4613-7199-1

Originally published by Kluwer Academic / Plenum Publishers, New York in 1999
Softcover reprint of the hardcover 1st edition 1999

10 9 8 7 6 5 4 3 2 1

A C.I.P. record for this book is available from the Library of Congress.

Electromagnetic Signals

Reflection, Focusing, Distortion, and Their Practical Applications

Henning F. Harmuth
Formerly of the Catholic University of America
Washington, D.C.

Raouf N. Boules
Towson University
Towson, Maryland

and

Malek G. M. Hussain
Kuwait University
State of Kuwait

KLUWER ACADEMIC / PLENUM PUBLISHERS
NEW YORK, BOSTON, DORDRECHT, LONDON, MOSCOW

To Rexford M. Morey and Alan E. Schutz,
two pioneers of the carrier-free radar technology

Preface

Maxwell's equations had always an electric current density term that stood for the usual electric *monopole currents* flowing in conductors but also for electric *dipole currents* that flow in insulators and even in vacuum. They were called *polarization currents* by Maxwell since today's atomistic thinking did not exist in his time. If the term dipole current had been used one would have realized immediately that there are magnetic dipoles just like electric ones and that there should be magnetic dipole currents as well as electric dipole currents. This calls for a magnetic dipole current density term in Maxwell's equations even if the controversial magnetic monopoles and magnetic monopole currents should not exist.

An electromagnetic wave cannot exist before a finite time, it must have finite energy, and it must satisfy the causality law. Such waves are represented mathematically by *signal solutions*. The absence of a magnetic dipole current density term has the strange effect that signal solutions do generally not exist for Maxwell's equations. Our theory did not allow the only waves that nature permits. The usual solutions of Maxwell's equations have periodic sinusoidal time variation, which implies that the represented waves started an infinite time ago and have infinite energy, unless their average power is zero. The infinite energy makes the conservation law of energy meaningless while "an infinite time ago" makes the causality law meaningless.

The addition of a magnetic dipole current density term to Maxwell's equations produced equations that permit signal solutions generally. Such solutions have been investigated and elaborated on the academic level for the last 10 years. We have reached the level where this academic work can be applied to practical problems. For instance, this book shows that Snell's law of reflection does not hold generally and that signals can easily be designed that yield significant deviations. Since the *stealth technology* is heavily based on Snell's law we readily find how anti-stealth radars can be designed. Another result is an answer to the classical problem of the airborne anti-submarine radar. A primarily civilian application is the use of focused signals for the improvement of penetration depth and resolution of the ground-probing radar. This civilian application is particularly gratifying since it is always easier to find military applications for electromagnetic waves.

Considerably more mathematical skill and effort are required to find and use signal solutions compared with the usual periodic sinusoidal solutions. We hope this book will convince some readers that the extra effort is worth making.

We want to take this opportunity to commemorate the late Chang Tong, Professor in the Department of Automation, Tsinghua University, Beijing, China. He was one of the early and strongest supporters of what is now variously called non-sinusoidal, large-relative-bandwidth, carrier-free, or ultra-wideband technology.

Contents

Equations are numbered consecutively within each of Sections 1.1 to 5.5. Reference to an equation in a different section is made by writing the number of the section in front of the number of the equation, e.g., Eq.(2.1-50) for Eq.(50) in Section 2.1.

Illustrations and tables are numbered consecutively within each section, with the number of the section given first, e.g., Fig.1.2-3, Table 1.7-1.

References are listed by the name of the author(s), the year of publication, and a lowercase Latin letter if more than one reference by the same author(s) is listed for that year.

List of Frequently Used Symbols

$\mathbf{B}$	$\mathrm{Vs/m^2}$	magnetic flux density
$c = 2.9979 \times 10^8$	m/s	vacuum velocity of light
$\mathbf{D}$	$\mathrm{As/m^2}$	electric flux density
$\mathbf{E}$, E	V/m	electric field strength
E_E	V/m	electric field strength due to electric excitation
E_H	V/m	electric field strength due to magnetic excitation
E_i, E_r, E_t	V/m	electric field strength of incident, reflected, and transmitted wave
e	As	electric charge
f	$\mathrm{s^{-1}}$	frequency
$\mathbf{g}_\mathrm{e}$	$\mathrm{A/m^2}$	electric current density
$\mathbf{g}_\mathrm{m}$	$\mathrm{V/m^2}$	magnetic current density
$\mathbf{H}$, H	A/m	magnetic field strength
H_E	A/m	magnetic field strength due to electric excitation
H_H	A/m	magnetic field strength due to magnetic excitation
H_i, H_r, H_t	A/m	magnetic field strength of incident, reflected, and transmitted wave
J	$\mathrm{kg\,m^2}$	inertial moment of rotation
$K_\mathrm{h,h}$, $K_\mathrm{s,h}$	-	Eqs.(5.3-3), (5.3-4)
m	kg	mass
m_0, m_e, m_p	kg	rest mass, electron mass, proton mass
N_0	$\mathrm{m^{-3}}$	particles per unit volume
$\mathbf{p}$	Asm	electric dipole moment
$p = \tau_\mathrm{mp}/\tau_\mathrm{p}$	-	
$q = \tau_\mathrm{p}/\tau$	-	
q_m	Vs	hypothetical magnetic charge
r	m	Fig.3.1-1
R	m	half length of a bar magnet, Fig.1.4-5
$R(t)$	-	rectangular pulse; Eqs.(2.3-9), (5.1-9)
$\mathbf{s}$, s	m	distance
s	V/Am	conductivity for magnetic currents, including a hypothetical monopole current
s_p	V/Am	conductivity for magnetic dipole currents
$S(t)$, $S(\theta)$	-	unit step function
t	s	time variable
T, ΔT	s	time interval
T_ir	s	Eq.(2.4-1
T_p	s	Fig.(2.4-2a)
$Z = \sqrt{\mu/\epsilon}$	V/A	wave impedance
Z_1, Z_2	V/A	Eqs.(3.1-35), (3.1-45)

$\alpha = \sigma/2\epsilon$	s^{-1}	inverse time constant
α_0	-	angle, Eq.(3.4-24)
$\alpha(y, \Delta T, \Delta T)$	-	attenuation, Fig.5.5-6
$\beta = v/c = g_e/g_c$	-	normalized velocity or current density,
γ_e	-	Eq.(1.5-10)
γ_m	-	Eq.(1.6-2)
$\gamma_s,\ \gamma_\sigma,\ \gamma_\epsilon,\ \gamma_\mu$	-	Eq.(3.1-45)
$\epsilon = 8.854 \times 10^{-12}$	As/Vm	permittivity
ζ	-	normalized distance, Eq.(5.1-18)
η	-	normalized wave number
θ	-	normalized time, Eq.(3.1-32)
$\Theta_0,\ \Theta_{max}$	-	Fig.5.3-1
$\Delta\Theta = \alpha T$	-	normalized time interval, Eq.(5.2-1)
$\Delta\Theta_f$	-	Fig.5.3-1
ϑ	-	angle
$\vartheta_i,\ \vartheta_r,\ \vartheta_t$	-	angle of incidence, reflection, and transmission
ι	-	Eq.(3.1-45)
κ	m^{-1}	wave number
$\mu = 4\pi \times 10^{-7}$	Vs/Am	permeability
μ_r	-	relative permeability
ξ_e	kg/s	constant referring to electric losses, Eq.(1.2-3)
ξ_m	kg m/s	magnetic friction constant for rotation, Eq.(1.4-12)
$\pi = 3.14159$	-	-
ρ	-	normalized spatial variable, Eq.(3.2-6)
ρ_k	m	Eq.(2.2-15)
ρ_r	-	Eq.(3.1-32)
ρ_t	-	Eq.(3.1-42)
σ^2	-	mean-square-deviation, Eq.(5.4-3)
σ	A/Vm	conductivity for electric currents
$\sigma_p = N_0 e^2 \tau_{mp}/m$	A/Vm	conductivity for electric dipole currents
τ_{mp}	s	time constant related to losses, Eq.(1.2-4)
τ_p	s	time constant related to dipole generation, Eq.(1.3-4)
φ	-	angle, Fig.3.1-1
χ	-	Eq.(3.2-8)
$\omega = 2\pi f$	s^{-1}	circular frequency

1 Introduction

1.1 MAXWELL'S EQUATIONS

Using the international system of units, we may write Maxwell's equations in a coordinate system at rest in the following form[1, 2]:

$$\operatorname{curl}\mathbf{H} = \frac{\partial \mathbf{D}}{\partial t} + \mathbf{g}_{\mathrm{e}} \tag{1}$$

$$-\operatorname{curl}\mathbf{E} = \frac{\partial \mathbf{B}}{\partial t} \tag{2}$$

$$\operatorname{div}\mathbf{D} = \rho_{\mathrm{e}} \tag{3}$$

$$\operatorname{div}\mathbf{B} = 0 \tag{4}$$

$$\mathbf{D} = \epsilon\mathbf{E} \tag{5}$$

$$\mathbf{B} = \mu\mathbf{H} \tag{6}$$

$$\mathbf{g}_{\mathrm{e}} = \sigma\mathbf{E} \tag{7}$$

Here $\mathbf{E}$ and $\mathbf{H}$ stand for the electric and magnetic field strength, $\mathbf{D}$ and $\mathbf{B}$ for the electric and magnetic flux density, $\mathbf{g}_{\mathrm{e}}$ and ρ_{e} for the electric current and charge density, while ϵ, μ, and σ represent permittivity, permeability, and conductivity. Maxwell's equations are usually written in more compact and elegant forms than used here. But we want to emphasize lucidity over compactness and elegance.

We have written ϵ, μ, and σ as constants. These parameters may be functions $\epsilon(\mathbf{r},t)$, $\mu(\mathbf{r},t)$, and $\sigma(\mathbf{r},t)$ of location $\mathbf{r}$ and time t since $\mathbf{r}$ and t are the independent variables of Maxwell's equations[3]. We call a medium inhomogeneous and time-variable if ϵ, μ, and σ are functions of location and time. In addition, in anisotropic media ϵ, μ, and σ may have different values for different directions, which means they can be represented mathematically by tensors $\boldsymbol{\epsilon}(\mathbf{r},t)$, $\boldsymbol{\mu}(\mathbf{r},t)$, and $\boldsymbol{\sigma}(\mathbf{r},t)$ that

---

[1] We use the notation of the books by Abraham and Becker, which through 18 editions and for more than half a century have been a definitive standard for electromagnetic theory (Becker, 1964, 1982). These books never assumed that electromagnetic waves had to have a sinusoidal time variation. A term representing an impressed electric field strength is left out in Eq.(7) since it is not needed here. See § 41 of Becker (1964, 1982).

[2] Equations (1)–(4) are usually called Maxwell's equations while the less fundamental Eqs.(5)–(7) are called *constitutive equations*. It is generally understood that the constitutive equations depend on the features of the medium and that Eqs.(5)–(7) are the simplest possible ones if ϵ, μ, and σ are constant scalars. Equation (7) is Ohm's law

[3] For instance, the conductivity of the atmosphere depends on altitude, latitude, and longitude; the rotation of the Earth makes it also a function of time.

are functions of location and time[4]. One thing one cannot do is to write ϵ, μ, and σ as functions of frequency $\epsilon(\omega)$, $\mu(\omega)$, and $\sigma(\omega)$ since a frequency does not occur in Eqs.(1)–(7). Only by giving up the generality of Maxwell's equations and writing equations applying strictly to the *steady state* can one make ϵ, μ, and σ functions of frequency. No *transient* or *signal solutions* can be derived from steady state equations.

Steady state equations and their solutions are always outside the causality law since the concept of cause and effect has no meaning in the steady state[5]. Such solutions are perfectly useful for applications like power transmission where we are interested in power and energy but not in causality. The exact opposite applies in signal transmission. The energy of a signal is of little interest as long as there is enough energy to make it detectable. But different signals should cause different effects and the propagation velocity of the signals determines the time of an effect. A serious study of signal transmission requires equations and solutions that satisfy the causality law.

We define an *electromagnetic signal* as a propagating electromagnetic wave that starts at a certain time and has finite energy. All observed or produced propagating waves are of this type. They satisfy both the causality law and the law of conservation of energy. It is usual to think of a signal as a field strength, a voltage, or a current at a certain location as function of time; but a signal could also be something observable at a certain time as function of one or more spatial variables.

The incomplete symmetry of Maxwell's equations has always attracted interest. Avoiding a philosophical discussion we note that the difference between Eqs.(3) and (4) implies that there are electric charges but not magnetic ones, while the difference between Eqs.(1) and (2) implies that there are electric but not magnetic currents. Authors investigating steady state solutions of Maxwell's equations have often found these two differences to cause difficulties that could be removed by the introduction of magnetic charges and current densities. We cite the very rigorous book by Müller (1967) as an example.

The investigation of signal solutions led to the conclusion that the modification of Maxwell's equations was not a matter of convenience but of mathematical necessity. Two independent proofs of this statement have been published. Apparently they were stated in too complicated a form to be readily understandable, judging from published opposing papers and other sources. We shall try to explain the two proofs here without using any mathematical formalism.

If a physical problem is stated in terms of a partial differential equation in a coordinate system at rest, one must find a function that meets three requirements:

1. The function satisfies the partial differential equation(s).

[4] Anisotropic media are usually crystals, but the laminated iron core of a transformer is anisotropic too since the conductivity σ is large in two directions and small in the third direction, if a Cartesian coordinate system is used for the space variables.

[5] Despite its importance for physics the causality law is usually stated in philosophical books only. The following form is well suited for information transmission: *Every effect has a sufficient cause that occurred a finite time earlier.* The causality law is a physical law and not a mathematical axiom, which implies that it must be specifically introduced into any mathematical model of a physical process subject to causality. See Harmuth (1994e, Section 1.11) for a more detailed discussion of the necessity to introduce physical assumptions or physical laws to derive physical results and some of the startling results obtained by ignoring this necessity.

2. The function satisfies an initial condition that holds at a certain time t_0 for all values of the spatial variable(s).
3. The function satisfies a boundary condition that holds at all times t for certain values of the spatial variable(s).

For a signal solution the boundary condition must be zero for $t < t_0$ and quadratically integrable. In terms of physics the boundary condition is a force—such as an electric or magnetic field strength—with finite energy, since 'quadratically integrable' is the mathematical concept for finite energy.

A second requirement for a signal solution is that the initial condition at the time $t = t_0$ must be independent of the boundary condition at $t > t_0$. Without this independence a cause at a time $t > t_0$ could have an effect at the time $t = t_0$. The requirement of independence of boundary and initial condition in a coordinate system at rest introduces the physical causality law into the mathematical model of the physical process.

The causality law is the cause for the universally observed effect of a distinguished direction of time: the effect comes after the cause. Nothing equivalent exists for spatial coordinates since there is no law that demands that the effect is above, to the right, or in front of the cause. One sometimes reads statements that time has no distinguished direction in mathematics. This is technically correct since neither the time variable nor spatial variables exist in pure mathematics. Instead, pure mathematics has complex variables, real variables, rational variables, integer variables, even and odd variables, prime variables, random variables, etc. Time and space are concepts of physics and subject to the laws of physics. Infinitely large values of t or x or infinitely small values of Δt and Δx are outside the realm of observability and can thus exist only as limiting values for computational convenience. Time has a distinguished direction in all cases subject to the causality law.

The first mathematical proof that Maxwell's equations did not have signal solutions in certain cases was derived for an electric field strength as boundary condition in a coordinate system at rest. It was assumed that a solution for the electric field strength as function of time and space had been derived from the partial differential equation, without any restriction on how this solution was obtained. Then it was shown that the associated magnetic field strength could not be derived from the electric field strength (Harmuth 1986a, b and 1986c, Section 2.5, last paragraph).

Only the determination of the associated magnetic field strength was shown to be impossible, nothing was claimed about the electric field strength due to electric excitation at the boundary. Hence, the proof could not possibly be shown to be wrong by the derivation of an electric field strength caused by an electric excitation, but this fact proved to be an insufficient deterrent.

The second mathematical proof is based on the observation that certain partial differential equations do not permit independent initial and boundary conditions. The mathematician P. Hillion (1990, 1991, 1992a, b, 1993) showed that Maxwell's equations belong to this type. Hillion's proof is made more difficult to understand by *not* assuming a coordinate system at rest but a general, moving one. The simple and lucid distinction between initial and boundary conditions does not apply to moving coordinate systems and one must use the concept of initial-boundary conditions.

The important fact is that two scientists working independently and using different approaches arrived at the same conclusion[6].

A good number of claims and counter claims on Maxwell's equations and the causality law were published in *IEEE Transactions on Electromagnetic Compatibility* from 1987 on. This exchange became irrelevant after 1991 when it was realized that magnetic currents do not require magnetic charges or *monopoles* since rotating magnetic dipoles create magnetic dipole currents (Harmuth 1992a, 1993a). The existence of rotating magnetic dipoles is hard to dispute since most of our electricity is produced by rotating magnetic dipoles. The same cannot be said about magnetic monopoles. There is no direct, generally accepted proof for their existence but there are a number of effects—such as the quantization of electric charges—that we can currently not explain without magnetic monopoles. A comprehensive discussion of these effects was published by Barrett (1993). Hence, one *must* add a magnetic current density term $\mathbf{g}_{\mathrm{m}}$ on the right side of Eq.(2) strictly for physical reasons. That the need for such a term was originally derived from mathematical investigations is perhaps of historical and philosophical interest but has no further bearing on the need for a term $\mathbf{g}_{\mathrm{m}}$.

The electric current density term in Eq.(1) has always stood for the usual monopole current carried by ions, electrons, or other charged particles, as well as for dipole currents, even though this fact is usually not mentioned in text books. Maxwell wrote about electric currents and *electric polarization*, which he needed to explain how a current could flow through a capacitor since the dielectric of a capacitor is an insulator for monopole currents (Maxwell 1891, Art. 111). Equations (1)–(4) do not contain a term specifically for electric polarization current densities. Hence, we must conclude that the term $\mathbf{g}_{\mathrm{e}}$ applies to monopole, dipole, and higher order multipole current densities. Maxwell wrote before today's atomistic thinking developed. As a result he did not know that the macroscopic electric polarization currents were caused by microscopic electric dipoles, and he could not make the connection between electric and magnetic dipoles.

Today we will demand a physical explanation if currents due to electric dipoles are included in electromagnetic theory but currents due to magnetic dipoles are excluded. Hence, the correct way to write Eqs.(1)–(4) in the absence of magnetic monopoles is

$$\operatorname{curl}\mathbf{H} = \frac{\partial\mathbf{D}}{\partial t} + \mathbf{g}_{\mathrm{e}} \tag{8}$$

$$-\operatorname{curl}\mathbf{E} = \frac{\partial\mathbf{B}}{\partial t} + \mathbf{g}_{\mathrm{m}} \tag{9}$$

$$\operatorname{div}\mathbf{D} = \rho_{\mathrm{e}} \tag{10}$$

$$\operatorname{div}\mathbf{B} = 0 \tag{11}$$

[6] Hillion obtained his results earlier than indicated by the dates of his publications but solving the publishing problem was more difficult than solving the scientific problem. It was almost impossible to find a journal that would accept a paper claiming that Maxwell's equations did generally not produce solutions that satisfied the causality law. The editors Richard B. Schulz of *IEEE Transactions on Electromagnetic Compatibility* and Peter W. Hawkes of *Advances in Electronics and Electron Physics* (Academic Press) deserve the credit for having had the courage to publish.

while the inclusion of magnetic monopoles would require Eq.(11) to be rewritten

$$\operatorname{div} \mathbf{B} = \rho_{\mathrm{m}} \tag{12}$$

where the dimensions of ρ_{m} is $\mathrm{Vs/m^3}$. The term $\mathbf{g}_{\mathrm{e}}$ stands for electric monopole, dipole, or higher order multipole current densities while the term $\mathbf{g}_{\mathrm{m}}$ stands for magnetic dipole or higher order multipole current densities in the absence of magnetic monopole currents.

A constitutive equation equivalent to Eq.(7) must be added to connect the magnetic current density $\mathbf{g}_{\mathrm{m}}$ with the magnetic field strength $\mathbf{H}$. We refer to this equation as the *magnetic Ohm's law*. The parameter s is the magnetic (dipole) conductivity:

$$\mathbf{g}_{\mathrm{m}} = s\mathbf{H} \qquad \left[\frac{\mathrm{V}}{\mathrm{m}^2} = \frac{\mathrm{V}}{\mathrm{Am}}\frac{\mathrm{A}}{\mathrm{m}}\right] \tag{13}$$

In terms of group theory Maxwell's equations have the symmetry $U(1)$, while Maxwell's equations with a magnetic current density added have the symmetry $SU(2)$. Both $U(1)$ and $SU(2)$ are gauge groups with distinctive underlying Lie algebras. The $SU(2)$ group formulation of Maxwell's equations with magnetic charges are derived from Yang-Mills theory (Barrett 1993, 1995). No further reference to group theory will be made in this book.

To show the similarity between electric and magnetic dipole currents consider a capacitor with a dielectric such as barium-titanate that has a large relative permittivity. The molecules of such dielectrics are inherent electric dipoles. As long as these dipoles have random orientation we do not notice any dipole effects. An electric field strength rotates the dipoles to make them line up in the direction of the field strength[7]. While this rotation is in progress we observe an electric (dipole) current flowing through the capacitor that is large compared with the current flowing through an equal capacitor but with air as dielectric[8].

Consider next a material made up of (inherent) magnetic dipoles with random orientation. Most materials, from hydrogen gas to ferromagnetic materials, consist of magnetic dipoles. The random orientation of the dipoles hides any dipole effects. A magnetic field strength rotates the dipoles to make them line up in the direction of the field strength. While this rotation is in progress we have a magnetic dipole current flowing. The observation of the magnetic dipole current is more difficult than that of the electric dipole current since practically all of our measurement instruments are based on electric monopole currents, and we need a transducer that transforms a magnetic dipole current into an electric monopole current. The problem of the transducer does not occur with electric dipole currents, since the dipole current in the dielectric of a capacitor becomes automatically a monopole current in the wires leading to the capacitor. Hence, we can use an amperemeter for monopole currents to measure the dipole current. The difference between electric

[7] In addition to the rotation of the inherent dipoles we also get an induced dipole effect from the atoms making up the molecules, since the electric field strength pulls the positive nuclei and the negative electrons slightly apart.

[8] Polarized nitrogen and oxygen atoms like any other polarized atoms are not inherent dipoles but induced dipoles. This applies to the molecules N_2 and O_2 too.

monopole and dipole currents shows up in their time variation or in the phase shift between voltage and current for sinusoidal time variation of voltage and current.

1.2 Electric Monopole Currents

An electric current is usually thought of to consist of charged particles with negligible mass that move with a certain velocity $\mathbf{v}$ in response to an applied electric field strength $\mathbf{E}$. If there are N_0 charged particles per unit volume each with the charge e, we obtain the following relation between their velocity $\mathbf{v}$ and the electric current density $\mathbf{g}_e$:

$$N_0 e\mathbf{v} = \mathbf{g}_e \tag{1}$$

Ohm's law connects the current density $\mathbf{g}_e$ with the electric field strength $\mathbf{E}$ via the conductivity σ, which is in the simplest case a scalar constant:

$$\mathbf{g}_e = \sigma\mathbf{E} \tag{2}$$

In the iron core of a transformer σ is a scalar tensor with large values in two directions (x, y) and a very small value in the third direction (z). Generally, σ can be a tensor function of the time and space variables, $\sigma = \boldsymbol{\sigma}(t, \mathbf{r})$.

We call an electric current described by Eqs.(1) and (2) a *monopole current with negligible mass of the current carriers.* The assumption of a negligible mass usually works well for electrons, but the books by Becker (1964, 1982, § 58) show that there are cases in which it is worthwhile to consider the finite mass of the electron. The finite mass is much more important if ions are the current carriers. The mass m_p of the hydrogen ion equals already 1836 electron masses m_e, while the masses of sodium and chlorine ions, which are the current carriers in seawater, equal about 42000 and 65000 electron masses. Following Becker we note that a current carrier with mass m_0, velocity $\mathbf{v}$, and charge e is pulled by an electric field strength $\mathbf{E}$ with the force $e\mathbf{E}$. Using Newton's mechanic we obtain the following equation of motion:

$$m_0 \frac{d\mathbf{v}}{dt} = e\mathbf{E} - \xi_e\mathbf{v} \tag{3}$$

The term $\xi_e\mathbf{v}$ represents losses proportionate to the velocity. The constant ξ_e is usually referred to as Stokes' friction constant due to its original use in fluid mechanics. In electrodynamics any losses proportionate to $\mathbf{v}$ are more likely to come from near zone radiation that is absorbed by surrounding matter. The term $\xi_e\mathbf{v}$ is clearly the simplest term that can account for losses and we do not have to decide what causes these losses unless we want to represent losses by more complicated terms than $\xi_e\mathbf{v}$.

If we obtain the derivative $d\mathbf{v}/dt$ from Eq.(1) and substitute it into Eq.(3) we obtain an extension of Ohm's law to electric monopole currents having current carriers with finite, constant mass:

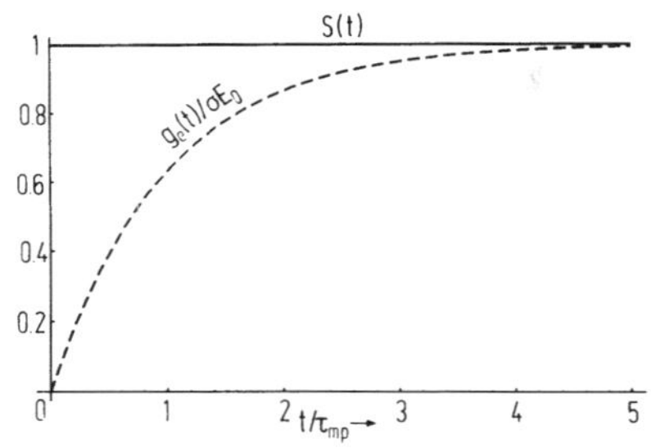

FIG.1.2-1. The normalized electric field strength represented by the step function $S(t)$ and the lagging current density $\mathbf{g}_e(t)/\sigma\mathbf{E}_0$ due to a finite mass of the current carriers.

$$\mathbf{g}_e + \tau_{mp}\frac{d\mathbf{g}_e}{dt} = \sigma\mathbf{E}$$
$$\tau_{mp} = \frac{m_0}{\xi_e}, \quad \sigma = \frac{N_0e^2}{\xi_e} = \frac{N_0e^2\tau_{mp}}{m_0} \tag{4}$$

If the term $\tau_{mp}d\mathbf{g}_e/dt$ can be neglected we obtain the usual Ohm's law with conductivity σ. To see the effect of Eq.(4) consider an electric field strength $\mathbf{E}$ with the time variation of a step function $\mathbf{E}(t) = \mathbf{E}_0S(t)$ as shown in Fig.1.2-1. In order to avoid the point $t = 0$ for which the step function $S(t)$ is not differentiable we consider the infinitesimally larger time $t = +0$. If we require a current density $\mathbf{g}_e(+0)$ to be zero we obtain from Eq.(4) the solution

$$\mathbf{g}_e(t) = \sigma\mathbf{E}_0\left(1 - e^{-t/\tau_{mp}}\right), \quad t > 0, \quad \mathbf{g}_e(+0) = 0 \tag{5}$$

which shows the current density lagging behind the electric field strength $\mathbf{E}_0S(t)$. The normalized form $g_e(t)/\sigma E_0$ of Eq.(5) is plotted in Fig.1.2-1.

In order to extend our results to charge carriers with relativistically variable mass we make in Eq.(3) the standard transition

$$m_0\frac{d\mathbf{v}}{dt} \to \frac{d(m\mathbf{v})}{dt} \tag{6}$$

If the term $\xi_e\mathbf{v}$ is left as it is one obtains clearly wrong results. The simplest change yielding acceptable results is the replacement

$$m_0\mathbf{v} \to m\mathbf{v}, \quad \mathbf{v} \to m\mathbf{v}/m_0 \tag{7}$$

with

$$m = \frac{m_0}{(1 - v^2/c^2)^{1/2}} \tag{8}$$

Hence, we use the following equation as the relativistic generalization of Eq.(3):

$$\frac{d(m\mathbf{v})}{dt} = e\mathbf{E} - \frac{\xi_e m\mathbf{v}}{m_0} \tag{9}$$

Using Eq.(8), the term $d(m\mathbf{v})/dt$ is separated into the velocity $\mathbf{v}$ and the mass m varying with velocity:

$$\begin{aligned}\frac{d(m\mathbf{v})}{dt} &= \frac{dm}{dt}\mathbf{v} + m\frac{d\mathbf{v}}{dt}\\ \frac{dm}{dt} &= \frac{dm}{dv}\frac{dv}{dt} = \frac{v}{c^2}\frac{m_0}{(1-v^2/c^2)^{3/2}}\frac{dv}{dt}\\ \frac{d(m\mathbf{v})}{dt} &= \frac{m_0}{(1-v^2/c^2)^{3/2}}\frac{d\mathbf{v}}{dt}\end{aligned} \tag{10}$$

Equation (9) assumes the form

$$m_0\left(1-\frac{v^2}{c^2}\right)^{-1/2}\left(\frac{1}{1-v^2/c^2}\frac{d\mathbf{v}}{dt} + \frac{\xi_e \mathbf{v}}{m_0}\right) = e\mathbf{E} \tag{11}$$

We observe that $\mathbf{v}$ and $\mathbf{E}$ always have the same direction. Hence, Eq.(11) can be written for the magnitudes v and E. Equation (11) assumes then the following normalized form:

$$(1-\beta^2)^{-1/2}\left(\frac{1}{1-\beta^2}\frac{d\beta}{d\theta} + \frac{1}{pq}\beta\right) = \gamma_e$$

$$\begin{gathered}\beta = v/c = g_e/g_c = N_0 ev/N_0 ec,\ g_e = N_0 ev,\ g_c = N_0 ec,\ \tau_{mp} = m_0/\gamma_e,\\ \theta = t/\tau,\ q = \tau_p/\tau,\ p = \tau_{mp}/\tau_p,\ pq = \tau_{mp}/\tau,\ \gamma_e = \tau_{mp} eE/m_0 c\end{gathered} \tag{12}$$

We observe that β is either the normalized velocity v/c or the normalized current density g_e/g_c, where g_c is the largest possible current density with N_0 charge carriers per unit volume and the charge e per carrier. For $\beta \to 0$ we get the nonrelativistic limit

$$\frac{d\beta}{d\theta} + \frac{1}{pq}\beta = \gamma_e \tag{13}$$

with the solution

$$\beta(\theta) = pq\gamma_e\left(1-e^{-\theta/pq}\right),\quad \beta(0) = 0,\ \theta \geq 0 \tag{14}$$

Plots of $\beta(\theta)$ for $pq = 1$ and $\gamma_e = 0.5$, 1, 2 are shown in Fig.1.2-2.

Very different plots are obtained for $\beta(\theta)$ by numerical integration of Eq.(12) with the initial condition $\beta(0) = 0$. Figure 1.2-3 shows the relativistic plots for $pq = 1$ and various values of γ_e. The limit $\beta(\theta) = 1$ is evident.

One can solve Eq.(12) by means of a series expansion of $(1-\beta^2)^{-1/2}$ and $(1-\beta^2)^{-3/2}$:

$$(1-\beta^2)^{-1/2} \approx 1 + \frac{1}{2}\beta^2,\quad (1-\beta^2)^{-3/2} \approx 1 + \frac{3}{2}\beta^2,\quad \beta^2 \ll 1 \tag{15}$$

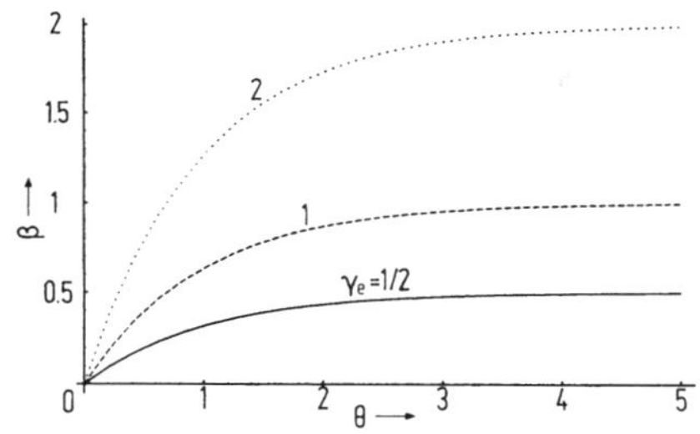

FIG.1.2-2. Nonrelativistic current density $\beta(\theta)$ according to Eq.(14) for $pq = 1$ and $\gamma_e = 0.5, 1, 2$ in the time interval $0 \leq \theta \leq 5$.

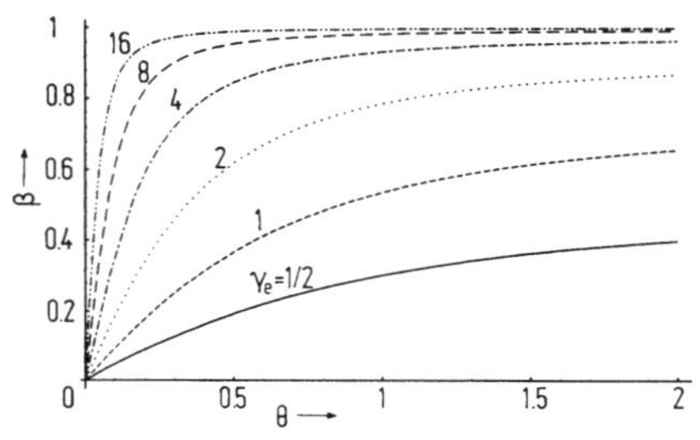

FIG.1.2-3. Relativistic current density $\beta(\theta)$ according to the numerical integration of Eq.(12) with the initial condition $\beta(0) = 0$ for $pq = 1$ and $\gamma_e = 0.5, 1, 2, 4, 8, 16$ in the interval $0 \leq \theta \leq 2$.

Substitution into Eq.(12) yields with the changed notation $\beta \to \beta_s$:

$$\left(1 + \frac{3}{2}\beta_s^2\right)\frac{d\beta_s}{d\theta} + \frac{1}{pq}\left(1 + \frac{1}{2}\beta_s^2\right)\beta_s = \gamma_e \tag{16}$$

A further series expansion of $\beta_s(\theta)$ is made:

$$\beta_s(\theta) = \beta_0(\theta) + \beta_1(\theta) + \ldots, \qquad \beta_1 \ll \beta_0 \ll 1$$

$$\left(1 + \frac{3}{2}\beta_0^2\right)\frac{d}{d\theta}(\beta_0 + \beta_1) + \frac{1}{pq}\left(1 + \frac{1}{2}\beta_0^2\right)(\beta_0 + \beta_1) = \gamma_e \tag{17}$$

This equation is broken into two equations for β_0 and β_1:

$$\frac{d\beta_0}{d\theta} + \frac{1}{pq}\beta_0 = \gamma_e, \qquad \beta_0(0) = 0 \tag{18}$$

$$\frac{d\beta_1}{d\theta} + \frac{1}{pq}\beta_1 = -\frac{3}{2}\beta_0^2\frac{d\beta_0}{d\theta}, \quad \beta_1(0) = 0 \tag{19}$$

Equation (18) yields

$$\beta_0(\theta) = pq\gamma_e\left(1 - e^{-\theta/pq}\right), \quad \theta \geq 0 \tag{20}$$

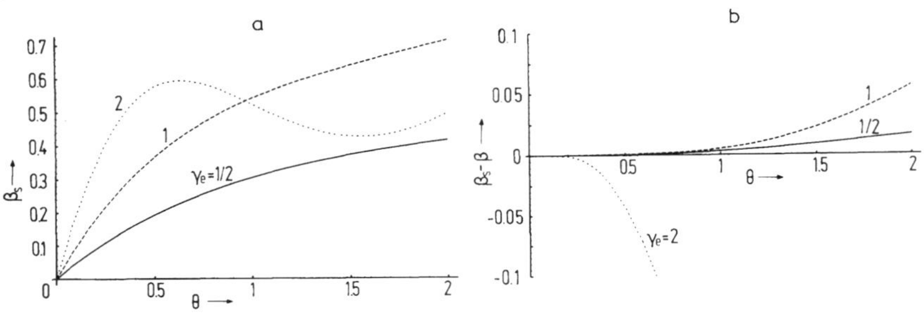

FIG.1.2-4. (a) Plots of $\beta_s(\theta)$ according to Eq.(23) for $pq = 1$ and $\gamma_e = 0.5$, 1, 2 in the interval $0 \leq \theta \leq 2$. (b) The difference $\beta(\theta) - \beta_s(\theta)$ between the plots of Fig.1.2-3 and 1.2-4a.

and Eq.(19) becomes:

$$\frac{d\beta_1}{d\theta} + \frac{1}{pq}\beta_1 = -\frac{3}{2}p^2q^2\gamma_e^3\left(1 - e^{-\theta/pq}\right)^2 e^{-\theta/pq} \tag{21}$$

Variation of the constant yields:

$$\beta_1(\theta) = \frac{3}{2}p^2q^2\gamma_e^3\left(\frac{3}{2}pq - \theta - 2pqe^{-\theta/pq} + \frac{pq}{2}e^{-2\theta/pq}\right)e^{-\theta/pq}$$

$$\beta_s(\theta) = pq\gamma_e\left(1 - e^{\theta/pq}\right) + \beta_1(\theta) \tag{23}$$

Figure 1.2-4a shows plots of $\beta_s(\theta)$ for $pq = 1$ and various values of γ_e while Fig.1.2-4b shows the difference $\beta(\theta) - \beta_s(\theta)$, where $\beta(\theta)$ is obtained by numerical integration of Eq.(12).

1.3 ELECTRIC DIPOLE CURRENTS

An electric insulator is supposed to prevent the flow of a current. This is correct only if the current is a monopole current. The insulators of any power line or the insulating dielectric between the plates of a capacitor can carry a current, but it must be a dipole or higher order multipole current. In the case of the insulators of a power line the dipole current means an undesirable loss that can be prevented by using DC rather than AC transmission. For a capacitor, the dipole current flowing through the dielectric is a necessity to make the capacitor work. We do not often read about dipole currents and they are usually called *polarization currents*, which obscures their physical significance. But there is at least one book that discusses them thoroughly (Reitz, Milford, Christy 1980).

Electric dipole currents are caused either by *induced dipoles* or by rotating *inherent dipoles*. As example of an induced dipole consider atomic hydrogen. Initially the positive proton and the negative electron are close together as shown in Fig.1.3-1 on the left. If a voltage is applied to the metal plates on top and bottom of Fig.1.3-1, the electron will move toward the positive voltage and the proton toward the negative one as shown on the right of Fig.1.3-1. A current flows as long as these two charge carriers are moving. If the applied electric field strength is less than

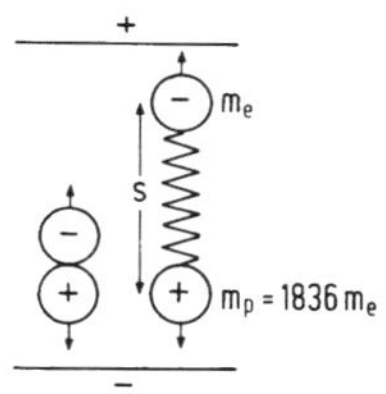

FIG.1.3-1. Polarization current produced by pulling the proton and the electron of a hydrogen atom apart by an electric field strength without breaking the bond. The large mass m_p of a proton relative to the mass m_e of an electron implies that the current is almost exclusively due to the electron.

required for ionization, the binding force between proton and electron will eventually halt any further movement and the current will stop flowing. This is a dipole current. On the other hand, if the electric field strength is larger than required for ionization, the binding force between proton and electron will be overcome and the two charged particles will keep moving independently. Now we have a monopole current. We note that one cannot tell at the beginning of the current flow whether the current will eventually be a dipole or a monopole current. For this reason the two types of currents must be represented mathematically equally in Eqs.(1.1-1) or (1.1-8) and (1.1-9).

We see from this example that a dipole current can become a monopole current if a sufficiently large electric field strength is applied sufficiently long. Whenever an insulator breaks down due to excessive voltage, a dipole current changes to a monopole current. Monopole and dipole currents usually exist together. A partly ionized gas permits a monopole current carried by the separated ions and electrons, but also a dipole current by neutral, polarized atoms. Even a metallic conductor permits both currents. The free electrons in the conduction band carry a monopole current, but the finite conductivity of most conductors permits an electric field strength that pulls bound electrons and their positive nuclei slightly apart. Only superconductors with infinite conductivity will permit pure monopole currents. Vice versa, a perfect insulator would permit dipole or higher order multipole currents only.

The two charge carriers held together by a spring in Fig.1.3-1 are a sufficiently good model for our purposes. In terms of Bohr's atomic model the circular orbit of the electron is stretched by the electric field to produce an electric dipole in the direction of the electric field strength as a time average. In terms of current quantum mechanics, a spherically symmetric probability density function of the location of the electron is stretched into the shape of an ellipsoid or an American football by the electric field strength. We will not need such fine details since our simple model is sufficient to permit the use of Maxwell's equations and their modification well into the visible light region in terms of sinusoidal waves (Becker 1964, 1982, § 58).

The electric field strength **E** applied to a medium does generally not imply the same field strength at the location of a polarized atom or molecule since the surrounding polarized molecules produce a field strength of their own. For instance, the *effective field strength* **F**

$$\mathbf{F} = \frac{\mathbf{E}}{1 - N_0 \alpha_e / 3\epsilon} \tag{1}$$

applies to a cubic crystal or a liquid without any permanent dipole moment, where N_0 is the number of atoms or molecules per unit volume, α_e the electric polarizability, and ϵ the dielectric constant (Becker 1964, 1982, § 26). For a gas one may use the simpler relation

$$\mathbf{F} = \mathbf{E} \tag{2}$$

For a description of the movement of the electron in our two-spheres-with-a-spring model of the hydrogen atom we follow Becker (1964, 1982, § 58) and use the equation

$$m_0 \left(\frac{d^2\mathbf{s}}{dt^2} + \frac{1}{\tau_{mp}} \frac{d\mathbf{s}}{dt} + \frac{\mathbf{s}}{\tau_p^2} \right) = e\mathbf{F} = e\mathbf{E} \tag{3}$$

To make this formula plausible consider a charge at a distance s from its rest position. A force pulling it back to its rest position shall be proportionate to s and have the direction $-\mathbf{s}$. We use Newton's mechanic and ignore the change of mass with velocity until later. The following equation of motion for a particle with constant mass m_0 is obtained

$$m_0 \frac{d\mathbf{v}}{dt} = m_0 \frac{d^2\mathbf{s}}{dt^2} = -\frac{m_0 \mathbf{s}}{\tau_p^2} \tag{4}$$

where τ_p is a constant with the dimension of time. We add the force $e\mathbf{F} = e\mathbf{E}$ due to the (effective) field strength and rewrite the equation:

$$\frac{d^2\mathbf{s}}{dt^2} + \frac{\mathbf{s}}{\tau_p^2} = \frac{e}{m_0} \mathbf{E} \tag{5}$$

For a certain field strength $\mathbf{E}$ a large mass m_0 implies a small acceleration $d^2\mathbf{s}/dt^2$ and distance $\mathbf{s}$. To clarify the implication consider Fig.1.3-1 once more. The spring shown symbolizes a force proportionate to the distance s that works against the electric force $e\mathbf{E}$. The center of mass of the proton-electron system is not moved by the electric field strength due to the conservation of momentum. Since the mass m_p of the proton equals 1836 electron masses m_e, the electron must move 1836 times as fast as the proton to conserve momentum. The movement of a charge creates an electric current. According to Fig.1.3-1 the current due to the electron is 1836 times the current due to the proton. Hence, m_0 in Eqs.(3) and (5) will stand for the mass of the electron and the contribution of the proton to the dipole current will be ignored.

We must add a term $\tau_{mp}^{-1} d\mathbf{s}/dt$ to Eq.(5) to obtain Eq.(3). Such a term yields an attenuation proportionate to the velocity of the electron. A physical explanation for such a term is the near-zone radiation which produces field strengths $\mathbf{E}$ and $\mathbf{H}$ that decrease with the distance r like $1/r^2$. The losses due to far-field radiation caused by the acceleration $d^2\mathbf{s}/dt^2$ of a charge are ignored. We have to keep in mind that Eq.(3) is a successful extension of classical physics to a problem of quantum mechanics, using the simplest possible equation that yields results in accordance with observation.

We multiply Eq.(3) with $e\tau_{\mathrm{mp}}/m_0$ as well as with the number N_0 of atoms or molecules in a unit volume. Furthermore, we write $\mathbf{v} = d\mathbf{s}/dt$:

$$N_0 e\mathbf{v} + \tau_{\mathrm{mp}} N_0 e \frac{d\mathbf{v}}{dt} + \frac{\tau_{\mathrm{mp}}}{\tau_{\mathrm{p}}^2} \int N_0 e\mathbf{v}\, dt = \frac{N_0 e^2 \tau_{\mathrm{mp}}}{m_0} \mathbf{E} \tag{6}$$

For $N_0 e\mathbf{v}$ we write the current density $\mathbf{g}_{\mathrm{e}}$ of Eq.(1.1-1):

$$\mathbf{g}_{\mathrm{e}} + \tau_{\mathrm{mp}} \frac{d\mathbf{g}_{\mathrm{e}}}{dt} + \frac{\tau_{\mathrm{mp}}}{\tau_{\mathrm{p}}^2} \int \mathbf{g}_{\mathrm{e}}\, dt = \sigma_{\mathrm{p}} \mathbf{E}$$

$$\tau_{\mathrm{mp}} = \frac{m_0}{\xi_{\mathrm{e}}}, \qquad \sigma_{\mathrm{p}} = \frac{N_0 e^2 \tau_{\mathrm{mp}}}{m_0} = \frac{N_0 e^2}{\xi_{\mathrm{e}}} \tag{7}$$

Here σ_{p} with the dimension A/Vm is the *electric polarization current conductivity* or the *electric dipole current conductivity*. Equation (7) is Ohm's law extended to electric *polarization* or *dipole currents* with constant mass of the charge or current carriers. A comparison with Eq.(1.2-4) shows that the integral is characteristic for dipole currents, while a term $d\mathbf{g}_{\mathrm{e}}/dt$ occurs in Ohm's law for monopole as well as for dipole currents if a delay caused by the need to accelerate the charge carriers to give them a velocity is taken into account; the term $d\mathbf{g}_{\mathrm{e}}/dt$ in Eq.(7) evidently comes from the term $d\mathbf{v}/dt$ in Eq.(6).

Let $\mathbf{E}$ in Eq.(7) be a step function

$$\mathbf{E} = \frac{1}{q} \mathbf{E}_0 S(t) \tag{8}$$

where q is a factor to be explained shortly. For the solution of Eq.(7) we may then do away with the inhomogeneous term by differentiation:

$$\mathbf{g}_{\mathrm{e}} + \frac{\tau_{\mathrm{p}}^2}{\tau_{\mathrm{mp}}} \frac{d\mathbf{g}_{\mathrm{e}}}{dt} + \tau_{\mathrm{p}}^2 \frac{d^2 \mathbf{g}_{\mathrm{e}}}{dt^2} = 0 \qquad \text{for } t > 0 \tag{9}$$

Substitution of

$$\mathbf{g}_{\mathrm{e}} = \mathbf{g}_0 e^{-t/\tau_{\mathrm{e}}} \tag{10}$$

yields an equation for τ_{e}

$$\tau_{\mathrm{e}}^2 - \frac{\tau_{\mathrm{p}}^2}{\tau_{\mathrm{mp}}} \tau_{\mathrm{e}} + \tau_{\mathrm{p}}^2 = 0$$

$$\begin{aligned}
\tau_{\mathrm{e1,e2}} &= \frac{1}{2}\frac{\tau_{\mathrm{p}}^2}{\tau_{\mathrm{mp}}}\left[1 \pm \left(1 - \frac{4\tau_{\mathrm{mp}}^2}{\tau_{\mathrm{p}}^2}\right)^{1/2}\right] && \text{for } \frac{\tau_{\mathrm{mp}}}{\tau_{\mathrm{p}}} < \frac{1}{2} \\
&= \frac{1}{2}\frac{\tau_{\mathrm{p}}^2}{\tau_{\mathrm{mp}}}\left[1 \pm j\left(\frac{4\tau_{\mathrm{mp}}^2}{\tau_{\mathrm{p}}^2} - 1\right)^{1/2}\right] && \text{for } \frac{\tau_{\mathrm{mp}}}{\tau_{\mathrm{p}}} > \frac{1}{2} \\
\tau_{\mathrm{e}} &= \frac{1}{2}\frac{\tau_{\mathrm{p}}^2}{\tau_{\mathrm{mp}}} = \tau_{\mathrm{p}} && \text{for } \frac{\tau_{\mathrm{mp}}}{\tau_{\mathrm{p}}} = \frac{1}{2}
\end{aligned} \tag{11}$$

and we obtain the following solutions of Eq.(9):

$$\begin{aligned}
\mathbf{g}_{\mathrm{e}} &= \mathbf{g}_{\mathrm{e1}}e^{-t/\tau_{\mathrm{e1}}} + \mathbf{g}_{\mathrm{e2}}e^{-t/\tau_{\mathrm{e2}}} && \text{for } \tau_{\mathrm{mp}}/\tau_{\mathrm{p}} \neq 1/2 \\
\mathbf{g}_{\mathrm{e}} &= (\mathbf{g}_{\mathrm{e3}} + \mathbf{g}_{\mathrm{e4}}t)e^{-t/\tau_{\mathrm{p}}} && \text{for } \tau_{\mathrm{mp}}/\tau_{\mathrm{p}} = 1/2
\end{aligned} \tag{12}$$

Consider the solution for $\tau_{\mathrm{mp}}/\tau_{\mathrm{p}} = 1/2$ first. A current density equal to zero at $t = 0$ requires $\mathbf{g}_{\mathrm{e3}} = 0$. Differentiation of the remaining part of $\mathbf{g}_{\mathrm{e}}$ yields:

$$\frac{d\mathbf{g}_{\mathrm{e}}}{dt} = \mathbf{g}_{\mathrm{e4}}\left(1 - \frac{t}{\tau_{\mathrm{p}}}\right)e^{-t/\tau_{\mathrm{p}}} \tag{13}$$

If $\mathbf{g}_{\mathrm{e}}$ for $t = +0$ is zero then the integral $\int \mathbf{g}_{\mathrm{e}}\,dt$ must be zero too. Only the derivative $d\mathbf{g}_{\mathrm{e}}/dt$ is not zero. Substitution of Eq.(13) into Eq.(7) yields for $t = +0$

$$\mathbf{g}_{\mathrm{e4}} = \frac{\sigma_{\mathrm{p}}\mathbf{E}}{\tau_{\mathrm{mp}}} = \frac{\sigma_{\mathrm{p}}\mathbf{E}_0}{q\tau_{\mathrm{mp}}} \qquad \text{for } t = +0 \tag{14}$$

and we obtain:

$$\mathbf{g}_{\mathrm{e}} = 2\sigma_{\mathrm{p}}\mathbf{E}_0\frac{1}{q}\frac{t}{\tau_{\mathrm{p}}}e^{-t/\tau_{\mathrm{p}}} \qquad \text{for } \frac{\tau_{\mathrm{mp}}}{\tau_{\mathrm{p}}} = \frac{1}{2} \tag{15}$$

With the definitions

$$\tau_{\mathrm{p}} = q\tau, \quad t/\tau = \theta \tag{16}$$

we rewrite Eq.(15) in normalized form:

$$\mathbf{g}_{\mathrm{e}} = 2\sigma_{\mathrm{p}}\mathbf{E}_0\frac{1}{q^2}\theta e^{-\theta/q} \qquad \text{for } \frac{\tau_{\mathrm{mp}}}{\tau_{\mathrm{p}}} = p = \frac{1}{2}, \quad \theta > 0 \tag{17}$$

The integral

$$\int_0^\infty \frac{1}{q^2}\theta e^{-\theta/q}d\theta = 1 = \frac{1}{2p} \tag{18}$$

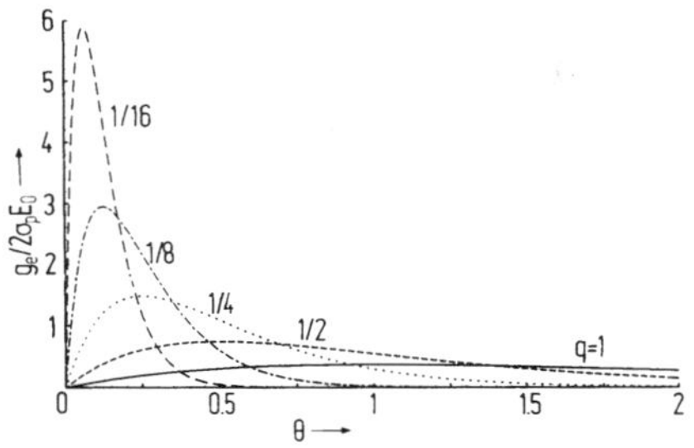

FIG.1.3-2. Time variation of dipole current densities according to Eq.(17) that transport equal charges through a certain cross section of the current path for $p = 1/2$ and $q = 1$, 1/2, 1/4, 1/8, 1/16. The areas under the plots in the interval $0 < \theta < \infty$ are equal.

explains the use of the factor $1/q$ in Eq.(8). The same charge will pass through a certain cross section of the path for the current density $\mathbf{g}_e$ during the time $0 < t < \infty$. The function $\mathbf{g}_e/2\sigma_p \mathbf{E}_0$ is plotted in Fig.1.3-2 for various values of q. The areas under these plots are all equal; they represent the constant charge passing through a certain cross section of the current path. As q decreases, the current density increases for an ever shorter time and approaches for $q \to 0$ an infinitely large and infinitely short needle pulse similar to Dirac's delta function[1].

We turn to the case $\tau_{mp}/\tau_p \neq 1/2$ in Eq.(12). The condition $\mathbf{g}_e(+0) = 0$ yields

$$\mathbf{g}_{e1} = -\mathbf{g}_{e2} \tag{19}$$

and

$$\mathbf{g}_e = \mathbf{g}_{e1}\left(e^{-t/\tau_{e1}} - e^{-t/\tau_{e2}}\right) \tag{20}$$

The derivative

$$\frac{d\mathbf{g}_e(+0)}{dt} = \mathbf{g}_{e1}\left(\frac{1}{\tau_{e2}} - \frac{1}{\tau_{e1}}\right) \tag{21}$$

substituted into Eq.(7) with $\mathbf{g}_e(+0) = 0$ and $\int \mathbf{g}_e(+0)dt = 0$ yields $\mathbf{g}_{e1}$:

$$\mathbf{g}_{e1} = \sigma_p \mathbf{E}_0 \frac{1}{q(1 - 4\tau_{mp}^2/\tau_p^2)^{1/2}} \qquad \text{for } \frac{\tau_{mp}}{\tau_p} < \frac{1}{2} \tag{22}$$

Using the definitions

$$\begin{gathered}\tau_p = q\tau, \quad t/\tau = \theta, \quad \tau_{mp}/\tau_p = p < 1/2 \\ \theta_1 = q\left[1 + \left(1 - 4p^2\right)^{1/2}\right]/2p, \quad \theta_2 = q\left[1 - \left(1 - 4p^2\right)^{1/2}\right]/2p\end{gathered} \tag{23}$$

we obtain the solution:

[1]It is *not* a Dirac delta function since the integral of Eq.(18) is over the interval $0 < \theta < \infty$ for which Dirac's delta function has no defined value. Dirac's delta function yields 1 if integrated over the intervals $-\infty < \theta < +\infty$ or $-\epsilon < \theta < +\epsilon$.

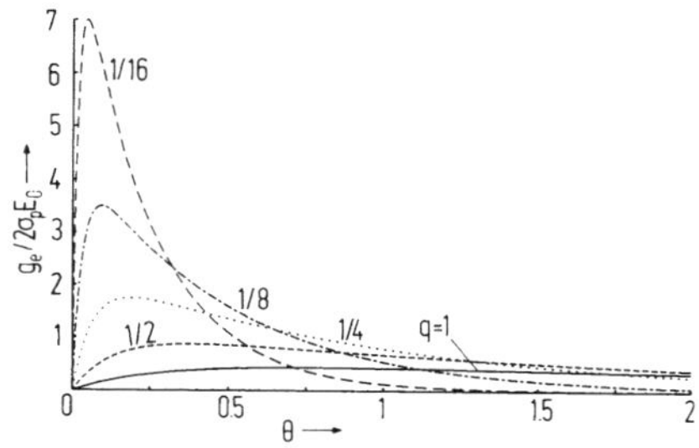

FIG.1.3-3. Time variation of dipole current densities according to Eq.(24) that transport equal charges through a certain cross section of the current path for $p = 1/4$ and $q = 1$, $1/2$, $1/4$, $1/8$, $1/16$. The areas under the plots in the interval $0 < \theta < \infty$ are equal. They are also equal to the corresponding areas in Fig.1.3-2.

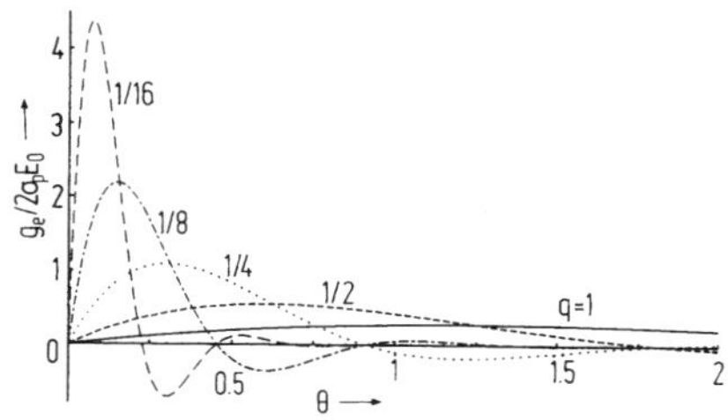

FIG.1.3-4. Time variation of dipole current densities according to Eq.(28) that transport equal charges through a certain cross section of the current path for $p = 1$ and $q = 1$, $1/2$, $1/4$, $1/8$, $1/16$. The integrals over the plots in the interval $0 < \theta < \infty$ are equal. They equal $1/2p = 1/2$ the value of the corresponding integrals in Figs.1.3-2 and 1.3-3.

$$\mathbf{g}_e = 2\sigma_p \mathbf{E}_0 \frac{e^{-\theta/\theta_1} - e^{-\theta/\theta_2}}{2q\,(1-4p^2)^{1/2}}, \qquad p < \frac{1}{2} \tag{24}$$

Plots of $\mathbf{g}_e/2\sigma_p\mathbf{E}_0$ are shown in Fig.1.3-3 for $p = 1/4$ and various values of q. The areas under these plots are all equal and also equal to those in Fig.1.3-2:

$$\int_0^\infty \frac{e^{-\theta/\theta_1} - e^{-\theta/\theta_2}}{2q\,(1-4p^2)^{1/2}}\,d\theta = \frac{1}{2p} \tag{25}$$

For $\tau_{mp}/\tau_p = p > 1/2$ we obtain from Eq.(11)

$$\frac{t}{\tau_{e1}} = \frac{1 - j(4p^2-1)^{1/2}}{2pq}\theta, \quad \frac{t}{\tau_{e2}} = \frac{1 + j(4p^2-1)^{1/2}}{2pq}\theta$$
$$\tau_p = q\tau, \quad t/\tau = \theta, \quad \tau_{mp}/\tau_p = p > 1/2 \tag{26}$$

while Eq.(22) is replaced by

$$\mathbf{g}_{e1} = -j\frac{\sigma_p \mathbf{E}_0}{q(4p^2-1)^{1/2}} \tag{27}$$

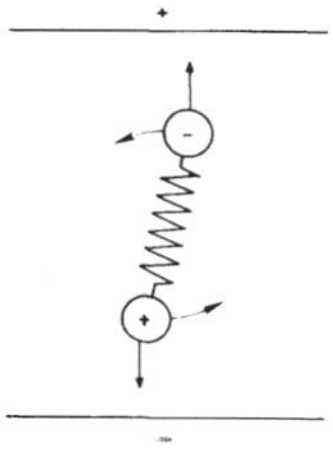

FIG.1.3-5. Superposition of induced electric polarization that is typical for atoms and electric orientation polarization that is typical for molecules.

and Eq.(24) by:

$$\mathbf{g}_{\mathrm{e}} = 2\sigma_{\mathrm{p}}\mathbf{E}_0 \frac{e^{-\theta/2pq}}{q(4p^2-1)^{1/2}} \sin\frac{(4p^2-1)^{1/2}\theta}{2pq} \tag{28}$$

Plots of $\mathbf{g}_{\mathrm{e}}/2\sigma_{\mathrm{p}}\mathbf{E}_0$ are shown in Fig.1.3-4 for $p = 1$ and various values of q. We note the value of the integral

$$\frac{1}{q(4p^2-1)^{1/2}} \int_0^{\infty} e^{-\theta/2pq} \sin\frac{(4p^2-1)^{1/2}\theta}{2pq}\, d\theta = \frac{1}{2p} \tag{29}$$

According to Eqs.(18) and (25) the charge flowing through a certain cross section of the current path is constant for $p \leq 1/2$ but drops like $1/2p$ for $p > 1/2$ according to Eq.(29).

For the experimental determination of τ_{p} and τ_{mp} in Eq.(7) consider a capacitor with two metal plates and hydrogen gas between the plates. This should be atomic hydrogen H rather than molecular hydrogen H_2. The value of σ_{p} is not needed directly if an electric voltage V_0 with the time variation of a step function $q^{-1}E_0S(t)$ is applied to the capacitor plates since the derivative of $S(t)$ is zero for $t > 0$. The voltage $V_0S(t)$ drives a current with a time variation as those of the functions in Figs.1.3-2 to 1.3-4. Matching the observed time variation of the current with plots according to these illustrations determines τ_{p} and τ_{mp} directly from observation.

In principle one should be able to calculate τ_{p} and τ_{mp} from one of the models of the hydrogen atom, but this calls for a transient solution for a problem of quantum mechanics. The solution applying to the Stark effect is of no help since this is a steady state effect rather than a transient effect.

Indirectly the value of σ_{p} in Eq.(7) enters our experimental determination of τ_{p} and τ_{mp} even though the right side of Eq.(9) is zero. The number N_0 of the hydrogen atoms per unit volume depends on the pressure and temperature of the hydrogen gas between the capacitor plates. By changing pressure and temperature one can measure τ_{p} and τ_{mp} for various values of N_0.

Atoms are polarized by an electric field strength but most molecules are inherent dipoles. Examples are H_2O, HCl, NH_3, and barium-titanate $BaTiO_3$. Only a few molecules with symmetric structure, such as CO_2 or CH_4, have no permanent dipole moment. Inherent electric dipoles are rotated in an electric field to line up with the field as shown in Fig.1.3-5. Since the molecules consist of atoms that are polarized by pulling apart the positive nuclei and the negative electrons, we get a superposition of orientation and induced polarization. We will deal with orientation polarization in connection with magnetic dipoles. An analytical derivation of dipole currents is generally not possible for orientation polarization, but plots can be derived numerically.

The dielectric of a capacitor can be polarized either by induced of by orientation polarization. This explains how a current can pass through a capacitor with a dielectric material between its plates. But how does a current pass through a capacitor whose dielectric material is replaced by vacuum or near-vacuum? Within Maxwell's original theory—which does not contain the concept of mass and charges carried by particles with mass—the answer is vacuum polarization. An electric dipole consisting of two equal charges with opposite polarity, held together by some elastic force, can be generated from *nothing* by an electric force. This idea of creating "something" from "nothing" may not be readily accepted but it does not contradict any basic laws of classical physics. The conservation of charge is satisfied since the total charge is always zero. Conservation of mass or momentum is not an issue when there is no mass. Causality is observed, since the application of an electric force is sufficient cause for the creation of dipoles after the application of the force. Conservation of energy is satisfied since the increased energy is accounted for by the applied electric force and the dipole current produced by the vacuum polarization.

A problem not resolvable within classical physics appears when we accept that observed charges have always been connected with particles that have a mass. The creation of a dipole from nothing implies that at least an electron and a positron are created, since these are the charged particles with lowest mass in low-energy physics. The rest mass of the electron is $m_e = 9.109 \times 10^{-31}$ kg. Within quantum mechanics the temporary existence of particles with mass $2m_e$ is permitted by Heisenberg's uncertainty relation in the form $\Delta E \Delta t \geq \hbar$, which states a mass $2m_e = \Delta E/c^2$ can exist during a time

$$\begin{aligned}\Delta t &\leq \hbar/\Delta E = \hbar/2m_e c^2\\ &\leq 1.055\times10^{-34}\ [\mathrm{Nms}]/(2\times9.109\times10^{-31}\ [\mathrm{kg}]\times8.988\times10^{16}\ [\mathrm{m^2/s^2}])\\ &\leq 6.440\times10^{-22}\ [\mathrm{s}]\end{aligned} \tag{30}$$

We have to think that dipoles are constantly created and annihilated within 6.44×10^{-22} s. There must be enough dipoles on the average to carry the dipole current. Furthermore, there must be on the average a certain mass of the dipoles that represents an energy supplied by the dipole current and the applied electric force. All this becomes more acceptable if we think of vacuum without dipole current and electric force not as "nothing" but as "nothing observable".

1.4 Magnetic Dipole Currents

The magnetic compass needle or the ferromagnetic bar magnet are typical representatives of magnetic dipoles. In the presence of a magnetic field strength they try to rotate so as to line up with the field strength. The atoms or molecules of a gas

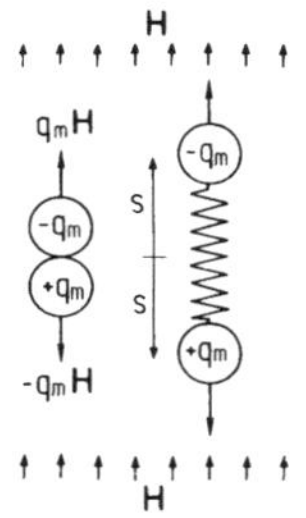

FIG.1.4-1. Two magnetic charges $\pm q_{\mathrm{m}}$ with equal mass m_0 are pulled with the force $\pm q_{\mathrm{m}}\mathbf{H}$ the distance $\pm s$ from their rest position.

usually have a magnetic dipole moment which makes them like small bar magnets. These dipole moments are the sum of a number of components that we do not need to discuss here. In solids the atoms or molecules interact to yield the effects of para-, dia-, or ferromagnetism but there is no such differentiation for a gas. In addition to these inherent, experimentally well documented dipoles we have the hypothetical magnetic charge dipole that is produced by pulling a positive and a negative charge $\pm q_{\mathrm{m}}$ apart by a magnetic field strength just like the electric dipole discussed in Section 1.3 was produced by pulling apart a positive and negative electric charge by an electric field strength. This hypothetical dipole is easy to analyze and we will use it as a standard of comparison for inherent dipoles which we will be able to investigate generally by numerical methods only.

We start with the model of Fig.1.4-1 that shows two magnetic charges $\pm q_{\mathrm{m}}$ with equal mass m_0 pulled with the force $\pm q_{\mathrm{m}}\mathbf{H}$ the distance $\pm\mathbf{s}$ from the rest position. The magnetic charge q_{m} has the dimension Vs. We may use the left side of Eq.(1.3-3) and keep in mind that the total magnetic dipole current density will be twice as large as that calculated for charges with one polarity according to Fig.1.4-1:

$$\frac{d^2\mathbf{s}}{dt^2}+\frac{1}{\tau_{\mathrm{mp}}}\frac{d\mathbf{s}}{dt}+\frac{\mathbf{s}}{\tau_{\mathrm{p}}^2}=\frac{q_{\mathrm{m}}\mathbf{H}}{m_0} \tag{1}$$

Deviating from the course of calculation in Section 1.3 we introduce the definitions $\tau_{\mathrm{p}}=q\tau$, $\tau_{\mathrm{mp}}/\tau_{\mathrm{p}}=p$, and $t/\tau=\theta$. Furthermore we use the magnitudes s and H since $\mathbf{s}$ and $\mathbf{H}$ have always the same direction:

$$\frac{d^2s}{d\theta^2}+\frac{1}{pq}\frac{ds}{d\theta}+\frac{1}{q^2}s=\frac{q_{\mathrm{m}}\tau_{\mathrm{p}}^2H}{q^2m_0}$$
$$\theta=t/\tau,\quad q=\tau_{\mathrm{p}}/\tau,\quad p=\tau_{\mathrm{mp}}/\tau_{\mathrm{p}} \tag{2}$$

For a magnetic field strength

$$H=\frac{1}{q}H_0S(\theta) \tag{3}$$

having the time variation of a step function and satisfying the initial conditions $s(+0)=ds(+0)/d\theta=0$ we obtain the solution:

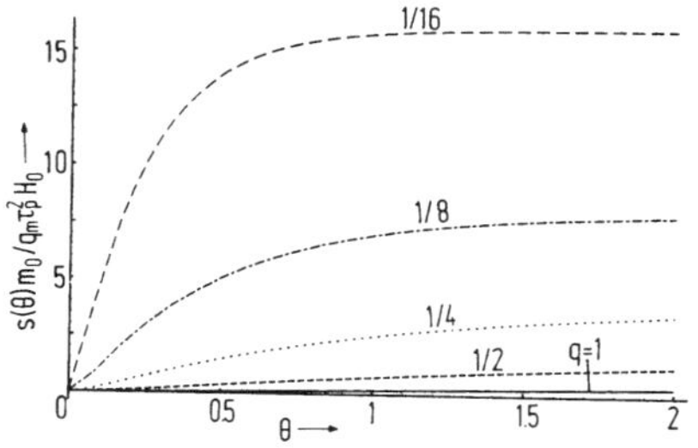

FIG.1.4-2. The distance $s(\theta)$ from the rest position $s = 0$ of a magnetic charge $q_{\rm m}$ according to Eq.(4) for $p = 1/4$ and $q = 1,\ 1/2,\ 1/4,\ 1/8,\ 1/16$.

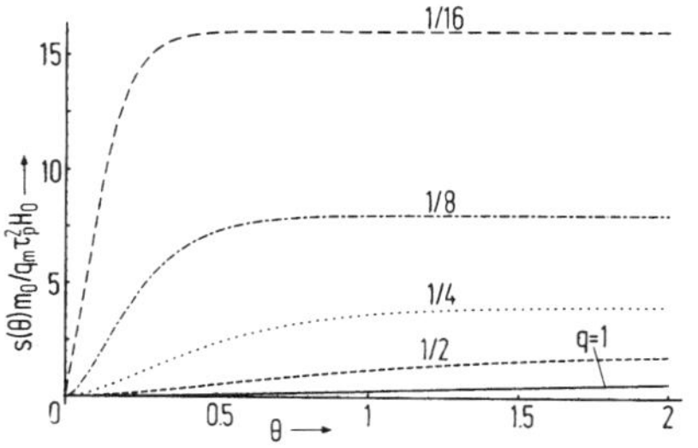

FIG.1.4-3. The distance $s(\theta)$ from the rest position $s = 0$ of a magnetic charge $q_{\rm m}$ according to Eq.(5) for $p = 1/2$ and $q = 1,\ 1/2,\ 1/4,\ 1/8,\ 1/16$,

$$\begin{aligned} s(\theta) &= \frac{q_{\rm m}\tau_{\rm p}^2 H_0}{qm_0}\left(\frac{\theta_1}{\theta_2-\theta_1}e^{-\theta/\theta_1} + \frac{\theta_2}{\theta_1-\theta_2}e^{-\theta/\theta_2} + 1\right) \\ \theta_1 &= \frac{q^2}{\theta_2} = \frac{q}{2p}\left[1 + (1-4p^2)^{1/2}\right] \\ \theta_2 &= \frac{q^2}{\theta_1} = \frac{q}{2p}\left[1 - (1-4p^2)^{1/2}\right], \quad p \neq \frac{1}{2} \end{aligned} \tag{4}$$

There are oscillations for $p > 1/2$ and no oscillations for $p < 1/2$. Plots of $s(\theta)m_0/q_{\rm m}\tau_{\rm p}^2 H_0$ for $p = 1/4$ and various values of q are shown in Fig.1.4-2.

The aperiodic limit $p = 1/2$ yields the special solution

$$\begin{aligned} s(\theta) &= \frac{q_{\rm m}\tau_{\rm p}^2 H_0}{qm_0}\left[1 - \left(1 + \frac{\theta}{q}\right)e^{-\theta/q}\right] \\ \theta_1 = \theta_2 = q, &\quad \theta = t/\tau, \quad q = \tau_{\rm p}/\tau, \quad p = 1/2 \end{aligned} \tag{5}$$

Plots of $s(\theta)m_0/q_{\rm m}\tau_{\rm p}^2 H_0$ for various values of q are shown in Fig.1.4-3. These functions show the distance s the magnetic charges have been pulled from their original rest position $s = 0$ at $\theta = 0$.

For $p > 1/2$ we obtain from Eq.(4):

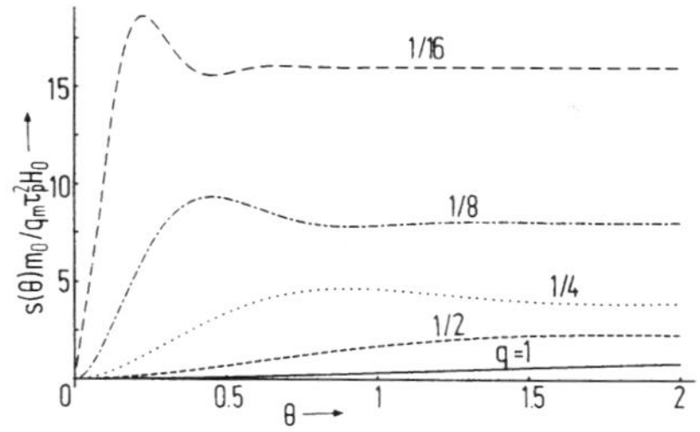

FIG.1.4-4. The distance $s(\theta)$ from the rest position $s=0$ of a magnetic charge $q_{\rm m}$ according to Eq.(6) for $p=1$ and $q = 1, 1/2, 1/4, 1/8, 1/16$.

$$s(\theta) = \frac{q_{\rm m}\tau_{\rm p}^2 H_0}{q m_0}\left[1 - \left(\cos\frac{(4p^2-1)^{1/2}\theta}{2pq} + \frac{1}{(4p^2-1)^{1/2}}\sin\frac{(4p^2-1)^{1/2}\theta}{2pq}\right)e^{-\theta/2pq}\right], \quad p > \frac{1}{2} \tag{6}$$

Plots of $s(\theta)m_0/q_{\rm m}\tau_{\rm p}^2 H_0$ for various values of q are shown in Fig.1.4-4.

Differentiation of Eqs.(4), (5), and (6) yields the velocities

$$v(\theta) = \frac{ds(\theta)}{d\theta}\frac{d\theta}{dt} = \frac{1}{\tau}\frac{ds(\theta)}{d\theta} \tag{7}$$

and with the substitution

$$g_{\rm m}(\theta) = 2N_0 q_{\rm m} v \ [{\rm V/m^2}] \tag{8}$$

we obtain:

$$\begin{aligned} g_{\rm m}(\theta) &= 2s_{\rm p}H_0\frac{e^{-\theta/\theta_1} - e^{\theta/\theta_2}}{2q(1-4p^2)^{1/2}}, & \frac{\tau_{\rm mp}}{\tau_{\rm p}} = p < \frac{1}{2} \\ &= 2s_{\rm p}H_0\frac{1}{q^2}e^{-\theta/q}, & p = \frac{1}{2} \\ &= 2s_{\rm p}H_0\frac{e^{-\theta/2pq}}{q(4p^2-1)^{1/2}}\sin\frac{(4p^2-1)^{1/2}}{2pq}, & p > \frac{1}{2} \end{aligned}$$

$$s_{\rm p} = 2N_0 q_{\rm m}^2\tau_{\rm mp}/m_0 \ [{\rm V/Am}] \tag{9}$$

These equations correspond to Eqs.(1.3-24), (1.3-17), and (1.3-28). The time variation of $g_{\rm m}/2s_{\rm p}H_0$ is thus as shown in Figs.1.3-3, 1.3-2, and 1.3-4.

We turn from the hypothetical induced magnetic dipoles to the well-established inherent magnetic dipoles. Consider a ferromagnetic bar magnet of length $2R$ in a homogeneous magnetic field of strength $\mathbf{H}$ and flux density $\mathbf{B}$ as shown in Fig.1.4-5a. Let $m_{\rm mo}$ with dimension[1] $\rm Am^2$ denote the magnetic dipole moment and J the

[1] If we write $m_{\rm mo}\mathbf{B} = m_{\rm mo}\mu\mathbf{H}$, the term $m_{\rm mo}\mu$ has the dimension Vsm and the symmetry with the electric dipole moment in Eq.1.3-3 with dimension Asm is obtained, if we multiply that equation with the distance s and obtain on the right side $(es)\mathbf{E}$ with the dimension Asm×V/m.

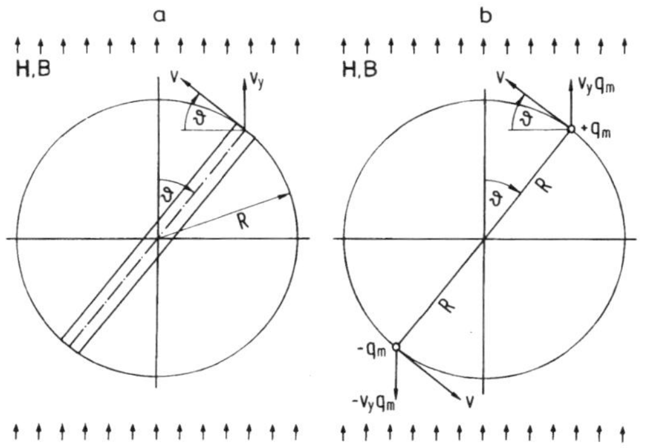

FIG.1.4-5. Ferromagnetic bar magnet in a homogeneous field (a) and its replacement by a thin rod with magnetic charges $\pm q_{\rm m}$ at its ends (b).

moment of inertia with dimension $\mathrm{Nms^2} = \mathrm{kg\,m^2}$ of the bar magnet. The equation of motion equals

$$J\frac{d^2\vartheta}{dt^2} = -m_{\rm mo}B\sin\vartheta \tag{10}$$

where ϑ is the angle between the direction of the field strength and the bar magnet. The velocity of the end points of the bar has the value

$$v(t) = -R\frac{d\vartheta}{dt} \tag{11}$$

which suggests to introduce a friction or generally an attenuation term $\xi_{\rm m}$ into Eq.(10):

$$J\frac{d^2\vartheta}{dt^2} + \xi_{\rm m}R\frac{d\vartheta}{dt} + m_{\rm mo}B\sin\vartheta = 0 \tag{12}$$

An analytical solution of this differential equation is generally impossible due to the term $\sin\vartheta$, but it succeeds for small values $\vartheta \approx \sin\vartheta$. We substitute $\vartheta = Ae^{-t/\tau}$ into the equation

$$J\frac{d^2\vartheta}{dt^2} + \xi_{\rm m}R\frac{d\vartheta}{dt} + m_{\rm mo}B\vartheta = 0 \tag{13}$$

and obtain:

$$\tau^2 - \frac{\xi_{\rm m}R}{m_{\rm mo}B}\tau + \frac{J}{m_{\rm mo}B} = 0$$

$$\tau_{3,4} = \frac{\xi_{\rm m}R}{2m_{\rm mo}B}\left[1 \pm \left(1 - \frac{4Jm_{\rm mo}B}{\xi_{\rm m}^2R^2}\right)^{1/2}\right] \tag{14}$$

For

$$\xi_{\rm m}^2 R^2 > 4Jm_{\rm mo}B \tag{15}$$

we get non-oscillating solutions

$$\vartheta = A_1 e^{-t/\tau_3} + A_2 e^{-t/\tau_4} \tag{16}$$

corresponding to the solutions of Figs.1.4-2 and 1.3-3, while

$$\xi_{\rm m}^2 R^2 < 4Jm_{\rm mo}B \tag{17}$$

yields oscillating solutions corresponding to the ones of Figs.1.4-4 and 1.3-4 with complex time constants τ_3 and τ_4:

$$\begin{aligned}\tau_3 &= \frac{\xi_{\rm m} R}{2m_{\rm mo}B}\left[1 - j\left(\frac{4Jm_{\rm mo}B}{\xi_{\rm m}^2 R^2} - 1\right)^{1/2}\right]\\ \tau_4 &= \frac{\xi_{\rm m} R}{2m_{\rm mo}B}\left[1 + j\left(\frac{4Jm_{\rm mo}B}{\xi_{\rm m}^2 R^2} - 1\right)^{1/2}\right]\end{aligned} \tag{18}$$

We will analyze first the aperiodic limit case:

$$\tau_3 = \tau_4 = \tau_{\rm p} = \frac{\xi_{\rm m} R}{2m_{\rm mo}B} = \sqrt{\frac{J}{m_{\rm mo}B}}, \qquad \xi_{\rm m} R = 2\sqrt{Jm_{\rm mo}B}$$

$$\vartheta(t) = (c_1 + c_2 t)e^{-t/\tau_{\rm p}} \tag{19}$$

The velocity $v(t)$ of the end points of the bar magnet equals:

$$\begin{aligned}v(t) &= -R\frac{d\vartheta}{dt} = -Rc_2 e^{-t/\tau_{\rm p}} + \frac{R}{\tau_{\rm p}}(c_1 + c_2 t)e^{-t/\tau_{\rm p}}\\ v(0) &= -Rc_2 + \frac{Rc_1}{\tau_{\rm p}}\end{aligned} \tag{20}$$

We choose c_1 and c_2 in Eq.(19) so that the following initial conditions are satisfied at the time $t = 0$

$$\begin{aligned}\vartheta(0) &= \vartheta_0 \quad \rightarrow \quad c_1 = \vartheta_0\\ v(0) &= 0 \quad \rightarrow \quad c_2 = \vartheta_0/\tau_{\rm p}\end{aligned} \tag{21}$$

and obtain:

$$\vartheta(t) = \vartheta_0\left(1 - \frac{t}{\tau_{\rm p}}\right)e^{-t/\tau_{\rm p}}, \quad \vartheta_0 \approx \sin\vartheta_0 \tag{22}$$

A series expansion yields for small values of t/τ_{p}:

$$\vartheta(t) = \vartheta_0 \left(1 - \frac{t^2}{2\tau_{\mathrm{p}}^2}\right) \tag{23}$$

The velocities $v(t)$ and $v_y(t)$ have the values:

$$v(t) = -R\frac{d\vartheta}{dt} = \frac{R\vartheta_0 t}{\tau_{\mathrm{p}}^2} e^{-t/\tau_{\mathrm{p}}} \tag{24}$$

$$v_y(t) = v(t)\sin\vartheta = \frac{R\vartheta_0 t}{\tau_{\mathrm{p}}^2} e^{-t/\tau_{\mathrm{p}}} \sin\left[\vartheta_0 \left(1 + \frac{t}{\tau_{\mathrm{p}}}\right) e^{-t/\tau_{\mathrm{p}}}\right] \tag{25}$$

Division of Eq.(9) for $p = 1/2$ by $N_0 q_{\mathrm{m}}$ yields

$$\frac{q_{\mathrm{m}}(\theta)}{N_0 q_{\mathrm{m}}} = v(\theta) = \frac{2q_{\mathrm{m}}\tau_{\mathrm{mp}} H_0}{m_0} \frac{1}{q^2}\theta e^{-\theta/q} \tag{26}$$

while Eqs.(22) and (24) assume for $\tau_{\mathrm{p}} = q\tau$ and $t/\tau = \theta$ the form

$$\vartheta(\theta) = \vartheta_0 \left(1 - \frac{\theta}{q}\right) e^{-\theta/q} \tag{27}$$

$$v(\theta) = \frac{R\vartheta_0}{\tau} \frac{1}{q^2} \theta e^{-\theta/q} \tag{28}$$

The two equations are equal for $2q_{\mathrm{m}}\tau_{\mathrm{mp}}H_0/m_0 = R\vartheta_0/\tau$.

In order to simplify the notation we substitute p^2, q^2, and $q\tau/p$ for certain more complicated terms

$$\frac{Jm_{\mathrm{mo}}B}{\xi_{\mathrm{m}}^2 R^2} = p^2, \quad \frac{J}{m_{\mathrm{mo}}B\tau^2} = q^2, \quad \frac{\xi_{\mathrm{m}}R}{m_{\mathrm{mo}}B} = \frac{q\tau}{p} \tag{29}$$

and rewrite τ_3 and τ_4 of Eq.(14) for $p < 1/2$:

$$\frac{\tau_3}{\tau} = \theta_1 = \frac{q}{2p}\left[1 + (1 - 4p^2)^{1/2}\right], \quad \frac{\tau_4}{\tau} = \theta_2 = \frac{q}{2p}\left[1 - (1 - 4p^2)^{1/2}\right] \tag{30}$$

For the initial conditions $\vartheta(0) = \theta_0$ and $v(0) = 0$ we obtain from Eq.(13):

$$\vartheta(\theta) = \vartheta_0 \left(\frac{\theta_1}{\theta_1 - \theta_2} e^{-\theta/\theta_1} + \frac{\theta_2}{\theta_2 - \theta_1} e^{-\theta/\theta_2}\right), \quad p < 1/2 \tag{31}$$

$$v(\theta) = -\frac{R}{\tau}\frac{d\vartheta(\theta)}{d\theta} = \frac{2R\vartheta_0 p}{\tau} \frac{e^{-\theta/\theta_1} - e^{-\theta/\theta_2}}{2q(1 - 4p^2)^{/2}} \tag{32}$$

From Eq.(8) we obtain for $p < 1/2$ the velocity $v(\theta)$:

$$v(\theta) = \frac{g_{\mathrm{m}}(\theta)}{N_0 q_{\mathrm{m}}} = \frac{2q_{\mathrm{m}}\tau_{\mathrm{mp}}H_0}{m_0}\frac{e^{-\theta/\theta_1} - e^{-\theta/\theta_2}}{2q(1-4p^2)^{1/2}}, \quad p < \frac{1}{2} \tag{33}$$

Equations (32) and (33) are equal for $2q_{\mathrm{m}}\tau_{\mathrm{mp}}H_0/m_0 = 2R\vartheta_0 p/\tau$.

We turn to the case $p > 1/2$ and obtain τ_3/τ as well as τ_4/τ from Eqs.(18) and (29):

$$\frac{\tau_3}{\tau} = \theta_1 = \frac{q}{2p}\left[1 + j(4p^2-1)^{1/2}\right], \quad \frac{\tau_4}{\tau} = \theta_2 = \frac{q}{2p}\left[1 - j(4p^2-1)^{1/2}\right] \tag{34}$$

The initial conditions $\vartheta(0) = \vartheta_0$ and $v(0) = 0$ yield the following solutions of Eq.(13):

$$\vartheta(\theta) = \vartheta_0\left(\cos\frac{(4p^2-1)^{1/2}\theta}{2pq} + \frac{1}{(4p^2-1)^{1/2}}\sin\frac{(4p^2-1)^{1/2}\theta}{2pq}\right)e^{-\theta/2pq}, \quad p > \frac{1}{2} \tag{35}$$

$$v(\theta) = \frac{2R\vartheta_0 p}{\tau}\frac{e^{-\theta/2pq}}{q(4p^2-1)^{1/2}}\sin\frac{(4p^2-1)^{1/2}\theta}{2pq} \tag{36}$$

Equation (9) yields for $p > 1/2$ the velocity

$$v(\theta) = \frac{g_{\mathrm{m}}(\theta)}{N_0 q_{\mathrm{m}}} = \frac{2q_{\mathrm{m}}\tau_{\mathrm{mp}}H_0}{m_0}\frac{e^{-\theta/2pq}}{q(4p^2-1)^{1/2}}\sin\frac{(2p^2-1)^{1/2}\theta}{2pq} \tag{37}$$

Equations (36) and (37) are equal for $2q_{\mathrm{m}}\tau_{\mathrm{mp}}H_0/m_0 = 2R\vartheta_0 p/\tau$.

The velocity $v(\theta)$ of the end points of the bar magnet in Fig.1.4-5a and of the magnetic charges of the induced dipole in Fig.1.4-1 is the same but the velocity determining the magnetic dipole current is $v_y(\theta)$

$$v_y(\theta) = v(\theta)\sin\vartheta(\theta) \tag{38}$$

according to Eq.(25). Plots of $v_y(\theta)\tau/R$ are shown in Figs.1.4-6a and b for $p = 1/2$ and $p = 1$, using the initial value $\vartheta_0 = \pi/8$ and various values of q. Figure 1.4-6a should be compared with Fig.1.3-2 and Fig.1.4-6b with Fig.1.3-4.

For larger values of ϑ_0 we rewrite Eq.(12) with the help of Eqs.(2) and (29)

$$\frac{d^2\vartheta}{d\theta^2} + \frac{1}{pq}\frac{d\vartheta}{d\theta} + \frac{1}{q^2}\sin\vartheta = 0$$

$$\theta = t/\tau, \quad p^2 = \frac{Jm_{\mathrm{mo}}B}{\xi_{\mathrm{m}}^2 R^2}, \quad q^2 = \frac{J}{m_{\mathrm{mo}}B\tau^2}, \quad \frac{1}{pq} = \frac{\xi_{\mathrm{m}}R\tau}{J} \tag{39}$$

solve it numerically for the initial conditions $\vartheta(0) = n\vartheta_0 = \vartheta_n$ and $d\vartheta(0)/d\vartheta = 0$ by computer and plot

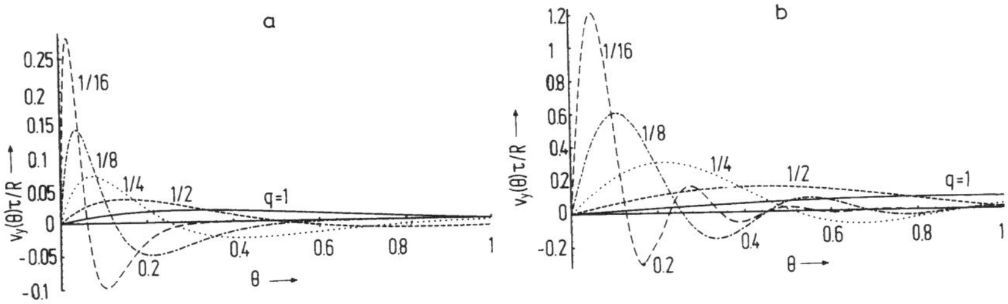

FIG.1.4-6. The function $v_y(\theta)\tau/R = v(\theta)\sin\vartheta(\theta)\tau/R$ for $p = 1/2$ according to Eqs.(28) and (38) for $q = 1$, $1/2$, $1/4$, $1/8$, $1/16$ and $\vartheta_0 = \pi/8$ (a) and for $p = 1$ according to Eqs.(36) and (38) (b).

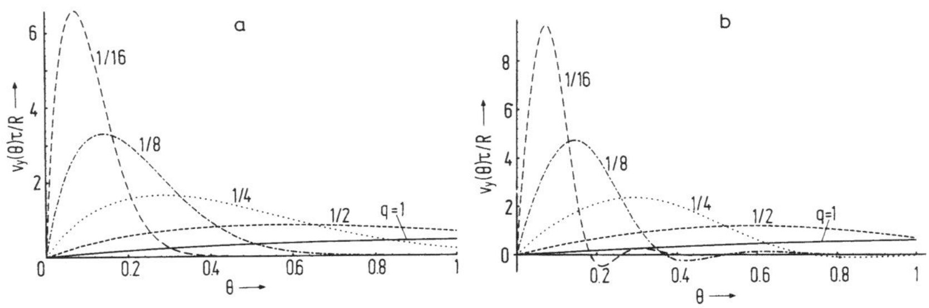

FIG.1.4-7. The function $v_y(\theta)\tau/R$ according to Eq.(40) for the initial angle $\vartheta_n = \pi/2$, initial velocity $v_y(0) = 0$, and $q = 1$, $1/2$, $1/4$, $1/8$, $1/16$. The left illustration (a) holds for $p = 1/2$, the right (b) for $p = 1$.

$$\frac{v_y(\theta)\tau}{R} = \frac{v(\theta)\tau}{R}\sin\vartheta(\theta) = -\frac{d\vartheta(\theta)}{d\theta}\sin\vartheta(\theta) \tag{40}$$

as shown for $\vartheta_n = \pi/2$ and various values of q in Fig.1.4-7a for $p = 1/2$ and in Fig.1.4-7b for $p = 1$.

Let us connect the velocities $v_y(\theta)$ with the current density $\mathbf{g}_\mathrm{m}(\theta)$ of a magnetic dipole current. First we replace the bar magnet in Fig.1.4-5a by a thin rod with fictitious magnetic charges $+q_\mathrm{m}$ and $-q_\mathrm{m}$ at its ends as shown in Fig.1.4-5b. The magnetic dipole moment m_mo equals $2Rq_\mathrm{m}$. The charge q_m must be connected with the magnetic dipole moment by the relation

$$q_\mathrm{m}\ [\mathrm{Vs}] = \mu\frac{m_\mathrm{mo}}{2R}\ \left[\frac{\mathrm{Vs}}{\mathrm{Am}}\frac{\mathrm{Am}^2}{\mathrm{m}}\right] \tag{41}$$

where μ is the magnetic permeability, to obtain the dimension Vs for q_m. The magnetic dipole current produced by such a bar magnet is $2q_\mathrm{m}v_y(\theta)$. If there are N_0 bar magnets in a unit volume, all having the direction ϑ_0 and velocity $v_y(\theta) = 0$ at $t = 0$, we would get the dipole current density $\mathbf{g}_\mathrm{m}(\theta) = 2N_0q_\mathrm{m}v_y(\theta)\mathbf{y}/y$ for the current flowing in the direction of the y-axis. For many randomly oriented bar magnets we must average all velocities $v_y(\theta)$ in the sector $0 \le \vartheta \le \pi$ as well as in the sector $\pi \le \vartheta \le 2\pi$, which yields the same result:

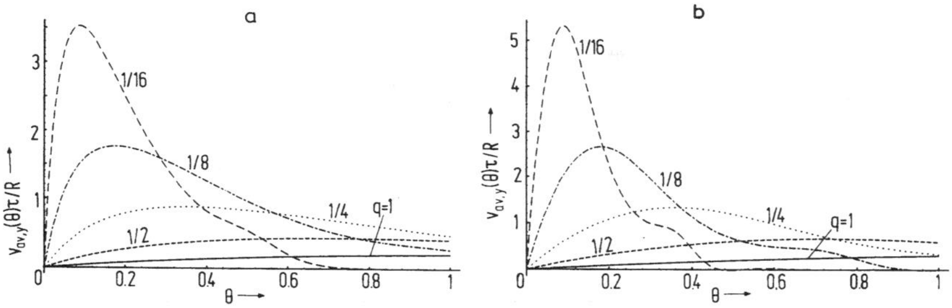

FIG.1.4-8. The average velocity $v_{av,y}(\theta)\tau/R$ according to Eqs.(39), (40), and (42) for $q = 1$, $1/2$, $1/4$, $1/8$, $1/16$. The left illustration (a) holds for $p = 1/2$, the right (b) for $p = 1$.

$$v_{av,y}(\theta) = \frac{1}{N}\sum_{n=1}^{N} v_y(\vartheta_n, \theta), \quad 0 \le \vartheta \le \pi \tag{42}$$

Figures 1.4-8a and b show the average velocity $v_{av,y}(\theta)\tau/R$ for $p = 1/2$ and 1, and for various values of q. Multiplication of $v_{av,y}(\theta)$ with $2N_0q_m$ yields the magnitude of the current density $\mathbf{g}_m$ as function of time. Decreasing values of q imply according to Eq.(39) a vanishing moment of inertia J. The plots in Figs.1.4-8a and b approach in this case a time variation like that of the Dirac delta function[2].

The magnetic dipole current density in the direction of the y-axis becomes with the help of Eq.(8):

$$\mathbf{g}_m = 2N_0q_mv_{av,y}(\theta)\frac{\mathbf{y}}{y} \tag{43}$$

With $\mathbf{v} = d\mathbf{s}/dt$ and $\mathbf{g}_m = 2N_0q_m\mathbf{v}$ one may rewrite Eq.(1) as Ohm's law for magnetic current densities due to a hypothetical, polarizable magnetic dipole; we emphasize that N_0 is the number of dipoles per unit volume, each having a charge $+q_m$ and $-q_m$:

$$\begin{aligned} \mathbf{v} + \tau_{mp}\frac{d\mathbf{v}}{dt} + +\frac{\tau_{mp}}{\tau_p^2}\int \mathbf{v}\,dt &= \frac{q_m\tau_{mp}}{m_0}\mathbf{H} \\ \mathbf{g}_m + \tau_{mp}\frac{d\mathbf{g}_m}{dt} + \frac{\tau_{mp}}{\tau_p^2}\int \mathbf{g}_m\,dt &= 2s_p\mathbf{H} \\ s_p = N_0q_m^2\tau_{mp}/m_0 \quad &[\mathrm{V/Am}] \end{aligned} \tag{44}$$

The attempt to write the extension of Ohm's law for rotating magnetic dipoles in a form comparable to Eq.(44) runs into difficulties. First, we obtain from Eq.(24)

$$\frac{d\vartheta}{dt} = -\frac{v(t)}{R}, \quad \frac{d^2\vartheta}{dt^2} = -\frac{1}{R}\frac{dv}{dt}, \quad \vartheta = -\frac{1}{R}\int v\,dt \tag{45}$$

and we rewrite Eq.(12):

[2]The integral over the interval $0 < \theta < \varepsilon$ of the limit function $q \to 0$ is finite while for the Dirac delta function the integration interval is $-\varepsilon < \theta < \varepsilon$.

$$\frac{J}{R}\frac{dv}{dt} + \xi_{\mathrm{m}} v + m_{\mathrm{mo}} B \sin\left(\frac{1}{R}\int v\,dt\right) = 0 \tag{46}$$

Then we substitute from Eq.(25)

$$v(t) = \frac{v_y(t)}{\sin\vartheta(t)} \tag{47}$$

to obtain

$$\frac{J}{R}\frac{d}{dt}\left(\frac{v_y}{\sin\vartheta}\right) + \xi_{\mathrm{m}}\frac{v_y}{\sin\vartheta} + m_{\mathrm{mo}} B \sin\left(\frac{1}{R}\int \frac{v_y}{\sin\vartheta}dt\right) = 0 \tag{48}$$

Next we substitute

$$v_y = \frac{g_{\mathrm{m},\vartheta}}{2N_0 q_{\mathrm{m}}} \tag{49}$$

and multiply with $2N_0 q_{\mathrm{m}}$:

$$\frac{J}{R}\frac{d}{dt}\left(\frac{g_{\mathrm{m},\vartheta}}{\sin\vartheta}\right) + \xi_{\mathrm{m}}\frac{g_{\mathrm{m},\vartheta}}{\sin\vartheta} + 2N_0 q_{\mathrm{m}} m_{\mathrm{mo}} B \sin\left(\frac{1}{2N_0 q_{\mathrm{m}} R}\int \frac{g_{\mathrm{m},\vartheta}}{\sin\vartheta}dt\right) = 0 \tag{50}$$

This differential equation yields the current density $g_{\mathrm{m},\vartheta}$ for a certain initial angle $\vartheta = \vartheta_0$ of the dipoles as function of the magnetic flux density $\mathbf{B}$ or field strength $\mathbf{H}$. The direction of $g_{\mathrm{m},\vartheta}$ is that of the positive y-axis:

$$\mathbf{g}_{\mathrm{m},\vartheta} = g_{\mathrm{m},\vartheta}\frac{\mathbf{y}}{y} \tag{51}$$

Equations (50) and (51) are the extension of Ohm's law to magnetic dipole currents caused by inherent dipoles with a certain initial orientation angle ϑ_0. If we add to the electric Ohm's law $\mathbf{g}_{\mathrm{e}} = \sigma\mathbf{E}$ in Maxwell's equations the magnetic Ohm's law defined by Eqs.(50) and (51) we can calculate—at least in principle—an electric field strength $\mathbf{E}$ and a magnetic field strength $\mathbf{H}$ for the initial angle ϑ_0. One must calculate $\mathbf{E}$ and $\mathbf{H}$ for many possible angles ϑ_0 and average the obtained field strengths to get the field strengths $\mathbf{E}$ and $\mathbf{H}$ for dipoles with initially random orientation.

1.5 Relativistic Electric Dipole Currents

For nonrelativistic velocities we had deduced from Fig.1.3-1 that the contribution of the proton to the dipole current of the hydrogen atom or any other atoms as well was negligible. This does not hold for velocities v_{e} of the electron or v_{p} of the proton close to the velocity of light, since the greater velocity of the electron implies that its mass m_{e} is going to increase faster than the mass m_{p} of the proton. Conservation of momentum demands

$$m_{\mathrm{p}}\mathbf{v}_{\mathrm{p}} = m_{\mathrm{e}}\mathbf{v}_{\mathrm{e}} \tag{1}$$

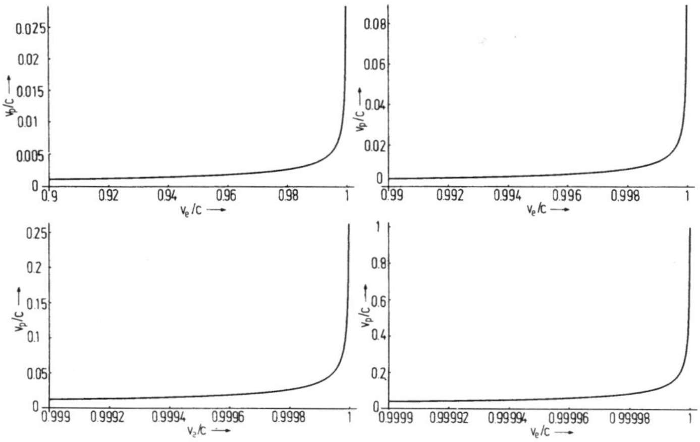

FIG.1.5-1. The ratio v_p/c of the velocity of the proton as function of the ratio v_e/c of the electron of a hydrogen atom being polarized by an electric field strength according to Eq.(3); $m_{0p}/m_{0e} = 1836$.

Denoting the rest mass of proton and electron by m_{0p} and m_{0e}, and observing that the direction of $\mathbf{v}_e$ and $\mathbf{v}_p$ is always opposite, we obtain from Eq.(1):

$$\frac{m_{0p}v_p}{(1-v_p^2/c^2)^{1/2}} = \frac{m_{0e}v_e}{(1-v_e^2/c^2)^{1/2}} \tag{2}$$

The velocity v_p of the proton is expressed as a function of the velocity v_e of the electron:

$$v_p = \frac{m_{0e}v_e}{m_{0p}[1-(1-m_{0e}^2/m_{0p}^2)v_e^2/c^2]^{1/2}} \tag{3}$$

For $v_e \to c$ we get $v_p \to v_e$. In this case the proton contributes as much as the electron to the dipole current. Figure 1.5-1 shows a plot of Eq.(3).

We may see from Fig.1.5-1 that the velocity v_e of the electron must be very close to the velocity c of light before the proton contributes significantly to the current density. For instance, a proton velocity $v_p/c = 0.01$ requires an electron velocity $v_e/c = 0.99852$, which implies a ratio of the current densities $g_e/g_p = 0.99852/0.01 = 99.85$; hence, the proton contributes about 0.99% to the current density. For $v_p/c = 0.1$ we get $v_e/c = 0.9999853$ and $g_e/g_p = 0.9999853/0.1 = 9.999853$ and the proton contributes about 9.1% to the current density.

We may extend Eq.(1.3-3) to charge carriers with relativistically varying mass by replacing m_0 with m and $d\mathbf{s}/dt$ with $\mathbf{v}$:

$$\frac{d(m\mathbf{v})}{dt} + \frac{(m\mathbf{v})}{\tau_{mp}} + \frac{1}{\tau_p^2}\int (m\mathbf{v})\,dt = e\mathbf{E}$$
$$m = \frac{m_0}{(1-v^2/c^2)^{1/2}} \tag{4}$$

The integral is eliminated by differentiation with respect to t. Since $\mathbf{E}$ and $\mathbf{v}$ have always the same direction we may use their magnitudes E and v:

$$\frac{d^2(mv)}{dt^2} + \frac{1}{\tau_{\text{mp}}}\frac{d(mv)}{dt} + \frac{1}{\tau_{\text{p}}^2}(mv) = e\frac{dE}{dt} \tag{5}$$

The term $d(mv)/dt$ was evaluated in Eq.(1.2-10). We still must calculate the term $d^2(mv)/dt^2$:

$$\frac{d^2(mv)}{dt^2} = \frac{d^2m}{dt^2}v + 2\frac{dm}{dt}\frac{dv}{dt} + m\frac{d^2v}{dt^2} \tag{6}$$

The term dm/dt may be taken from Eq.(1.2-10), which leaves only d^2m/dt^2 to be calculated:

$$\begin{aligned}\frac{d^2m}{dt^2} &= \frac{d}{dt}\left(\frac{dm}{dt}\right) = \frac{d}{dt}\left(\frac{dm}{dv}\right)\frac{dv}{dt} + \frac{dm}{dv}\frac{d^2v}{dt^2} = \frac{d^2m}{dv^2}\left(\frac{dv}{dt}\right)^2 + \frac{dm}{dv}\frac{d^2v}{dt^2} \\ &= \frac{m_0}{c^2\,(1-v^2/c^2)^{3/2}}\left[\frac{1+2v^2/c^2}{1-v^2/c^2}\left(\frac{dv}{dt}\right) + v\frac{d^2v}{dt^2}\right]\end{aligned} \tag{7}$$

We obtain from Eqs.(6) and (7):

$$\frac{d^2(mv)}{dt^2} = \frac{3m_0}{c^2\,(1-v^2/c^2)^{5/2}}v\left(\frac{dv}{dt}\right)^2 + \frac{m_0}{(1-v^2/c^2)^{3/2}}\frac{d^2v}{dt^2} \tag{8}$$

Equation (5) can now be written for the velocity v as variable and the rest mass m_0 as a constant:

$$\begin{aligned}\frac{m_0}{(1-v^2/c^2)^{1/2}}\bigg[\frac{3}{c^2\,(1-v^2/c^2)^2}v\left(\frac{dv}{dt}\right)^2 + \frac{1}{1-v^2/c^2}\frac{d^2v}{dt^2} \\ + \frac{1}{\tau_{\text{mp}}}\frac{1}{1-v^2/c^2}\frac{dv}{dt} + \frac{1}{\tau_{\text{p}}^2}v\bigg] = e\frac{dE}{dt}\end{aligned} \tag{9}$$

Multiplication of Eq.(9) with τ^2/c and introduction of normalized variables brings it into the following form:

$$\begin{aligned}\frac{1}{(1-\beta^2)^{1/2}}\left[\frac{1}{1-\beta^2}\left(\frac{d^2\beta}{d\theta^2} + \frac{1}{pq}\frac{d\beta}{d\theta}\right) + \frac{3}{(1-\beta^2)^2}\beta\left(\frac{d\beta}{d\theta}\right)^2 + \frac{1}{q^2}\beta\right] \\ = \frac{\gamma_{\text{e}}}{pq}\frac{1}{E_0}\frac{dE}{d\theta}\end{aligned}$$

$$\begin{gathered}\beta = v/c = N_0ev/N_0ec,\ \theta = t/\tau,\ q = \tau_{\text{p}}/\tau \\ p = \tau_{\text{mp}}/\tau_{\text{p}},\ pq = \tau_{\text{mp}}/\tau,\ \gamma_{\text{e}} = \tau_{\text{mp}}eE_0/m_0c\end{gathered} \tag{10}$$

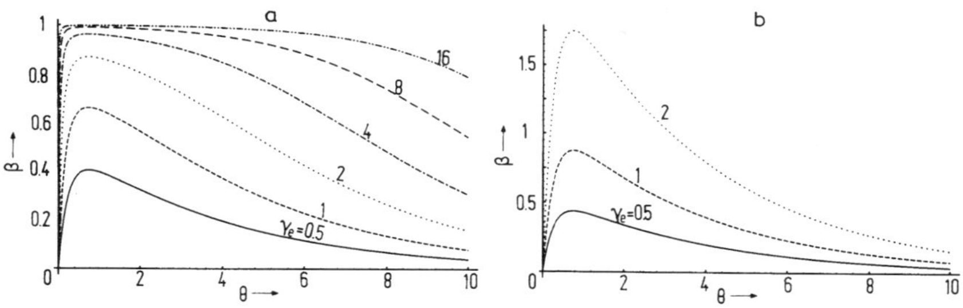

FIG.1.5-2. Plots of the normalized electric dipole current density $\beta(\theta)$ for $q = 1$, $p = 1/4$, and various values of γ_e according to Eqs.(10) to (16). (a) Relativistic plots for $\gamma_e = 0.5$, 1, 2, 4, 8, 16 using Eq.(10). (b) Nonrelativistic plots for $\gamma_e = 0.5$, 1, 2 using Eq.(11).

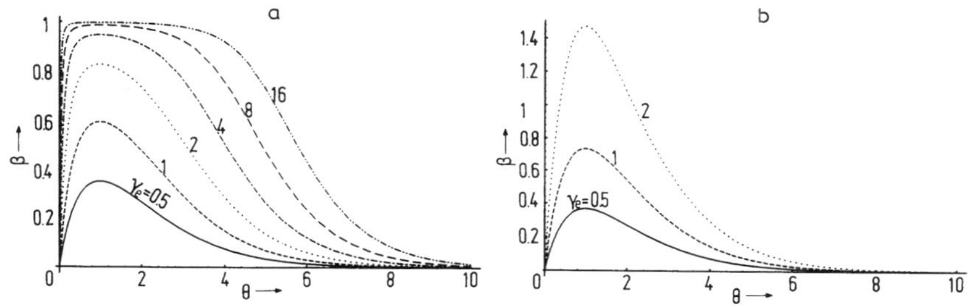

FIG.1.5-3. Plots of the normalized electric dipole current density $\beta(\theta)$ for $q = 1$, $p = 1/2$, and various values of γ_e according to Eqs.(10) to (16). (a) Relativistic plots for $\gamma_e = 0.5$, 1, 2, 4, 8, 16 using Eq.(10). (b) Nonrelativistic plots for $\gamma_e = 0.5$, 1, 2 using Eq.(11).

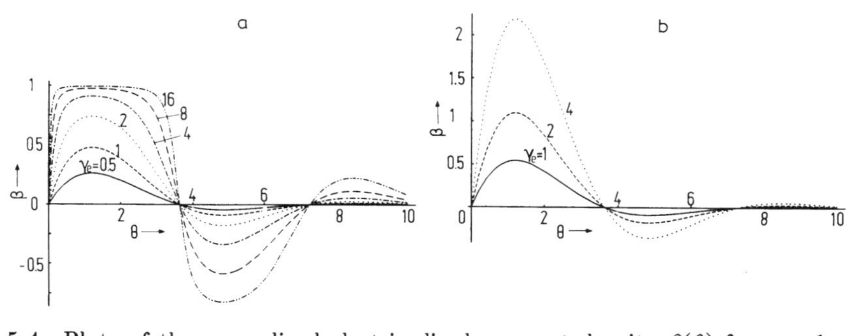

FIG.1.5-4. Plots of the normalized electric dipole current density $\beta(\theta)$ for $q = 1$, $p = 1$, and various values of γ_e according to Eqs.(10) to (16). (a) Relativistic plots for $\gamma_e = 0.5$, 1, 2, 4, 8, 16 using Eq.(10). (b) Nonrelativistic plots for $\gamma_e = 1$, 2, 4 using Eq.(11).

The classical limit of this equation is obtained for $\beta^2 \to 0$ and $\beta(d\beta/d\theta)^2 \to 0$:

$$\frac{d^2\beta}{d\theta^2} + \frac{1}{pq}\frac{d\beta}{d\theta} + \frac{1}{q^2}\beta = \frac{\gamma_{\mathrm{e}}}{pq}\frac{1}{E_0}\frac{dE}{d\theta} = \frac{\sigma_{\mathrm{p}}}{N_0 ecpq}\frac{dE}{d\theta} \qquad (11)$$

Two initial conditions are required for Eqs.(10) and (11). The first one is

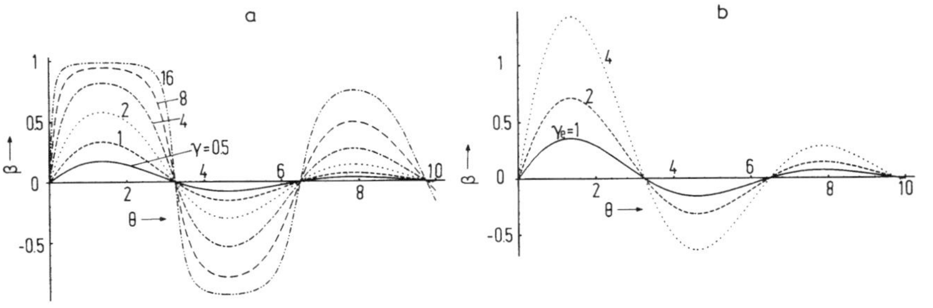

FIG.1.5-5. Plots of the normalized electric dipole current density $\beta(\theta)$ for $q = 1$, $p = 2$, and various values of γ_e according to Eqs.(10) to (16). (a) Relativistic plots for $\gamma_e = 0.5$, 1, 2, 4, 8, 16 using Eq.(10). (b) Nonrelativistic plots for $\gamma_e = 1$, 2, 4 using Eq.(11).

$$\beta(0) = 0 \tag{12}$$

since the current or the current density should start from zero at $\theta = 0$. A second initial condition for $\theta = 0$ is obtained by rewriting Eq.(10) for $\beta = 0$:

$$\frac{d^2\beta}{d\theta^2} + \frac{1}{pq}\frac{d\beta}{d\theta} = \frac{\gamma_e}{pq}\frac{1}{E_0}\frac{dE}{d\theta} \tag{13}$$

Integration of this equation yields:

$$\begin{aligned} \frac{d\beta}{d\theta} + \frac{1}{pq}\beta &= \frac{\gamma_e}{pq}\frac{1}{E_0}E(\theta) \\ \frac{d\beta(0)}{d\theta} &= \frac{\gamma_e}{pq}\frac{1}{E_0}E(0), \quad \theta = 0,\ \beta(0) = 0 \end{aligned} \tag{14}$$

Let $E(\theta)$ be an electric field strength with the time variation of a step function:

$$E(\theta) = \frac{1}{q}E_0 S(\theta) \tag{15}$$

Substitution into Eq.(14) yields the second initial condition:

$$\frac{d\beta}{d\theta} = \frac{\gamma_e}{pq^2} \tag{16}$$

To recognize the difference between the relativistic and the nonrelativistic differential equation we show in Figs.1.5-2 to 1.5-5 various plots of the normalized velocity or current density β. Figure 1.5-2 shows relativistic plots for $p = 1/4$ on the left and nonrelativistic plots on the right. The amplitudes of the nonrelativistic plots vary proportionately to the parameter γ_e. For the relativistic plots this holds true only for small values of γ_e, for large values the plots approach the limit $\beta(\theta) = 1$. Figures 1.5-3 to 1.5-5 are repetitions of Fig.1.5-2 with the value of p increased from 1/4 to 1/2, 1, and 2.

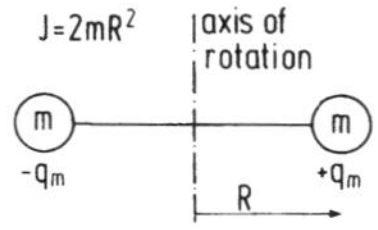

FIG.1.6-1. Dumb-bell model of a rotating bar magnet with two masses m at the ends of a thin rod of length $2R$.

1.6 Relativistic Magnetic Dipole Currents

We consider first the hypothetical magnetic dipole of Fig.1.4-1 and obtain the relativistic version of Eq.(1.4-44):

$$m\mathbf{v} + \tau_{\mathrm{mp}}\frac{d(m\mathbf{v})}{dt} + \frac{\tau_{\mathrm{mp}}}{\tau_{\mathrm{p}}^2}\int (m\mathbf{v})\,dt = q_{\mathrm{m}}\tau_{\mathrm{mp}}\mathbf{H}$$
$$m = \frac{m_0}{(1 - v^2/c^2)^{1/2}} \tag{1}$$

This is the same equation as Eq.(1.5-4) if we make the replacements $e \to q_{\mathrm{m}}$ and $\mathbf{E} \to \mathbf{H}$. Instead of Eq.(1.5-10) we obtain the following equation for the normalized magnetic current density due to one magnetic charge q_{m}:

$$\frac{1}{(1-\beta^2)^{1/2}}\left[\frac{1}{1-\beta^2}\left(\frac{d^2\beta}{d\theta^2} + \frac{1}{pq}\frac{d\beta}{d\theta}\right) + \frac{3}{(1-\beta^2)^2}\beta\left(\frac{d\beta}{d\theta}\right)^2 + \frac{1}{q^2}\beta\right] = \frac{\gamma_{\mathrm{m}}}{pq}\frac{1}{H_0}\frac{dH}{d\theta}$$

$$\beta = v/c = N_0 q_{\mathrm{m}} v / N_0 q_{\mathrm{m}} c,\ \theta = t/\tau,\ q = \tau_{\mathrm{p}}/\tau$$
$$p = \tau_{\mathrm{mp}}/\tau_{\mathrm{p}},\ pq = \tau_{\mathrm{mp}}/\tau,\ \gamma_{\mathrm{m}} = \tau_{\mathrm{mp}} q_{\mathrm{m}} H_0/m_0 c \tag{2}$$

For a pair of magnetic charges $\pm q_{\mathrm{m}}$ we get the normalized magnetic current density $2\beta(\theta)$.

We turn to the rotating magnetic dipole described by Eq.(1.4-12). In order to obtain the relativistic form of the moment of inertia J we must specify in more detail where the mass of the bar magnet in Fig.1.4-5 is located. We assume the bar magnet can be represented by the dumb-bell model shown in Fig.1.6-1 with the masses m at the end of a thin rod of length $2R$, just as was assumed in Fig.1.4-5b for the magnetic charges $\pm q_{\mathrm{m}}$, and obtain the momentum

$$J = 2mR^2 \tag{3}$$

Introduction of Eq.(3) into Eq.(1.4-13) yields:

$$2mR^2\frac{d^2\vartheta}{dt^2} + \xi_{\mathrm{m}} R\frac{d\vartheta}{dt} + m_{\mathrm{mo}} B \sin\vartheta = 0 \tag{4}$$

For the transition to the relativistic version of this equation we consider the velocity $v(t)$ of Eq.(1.4-11)

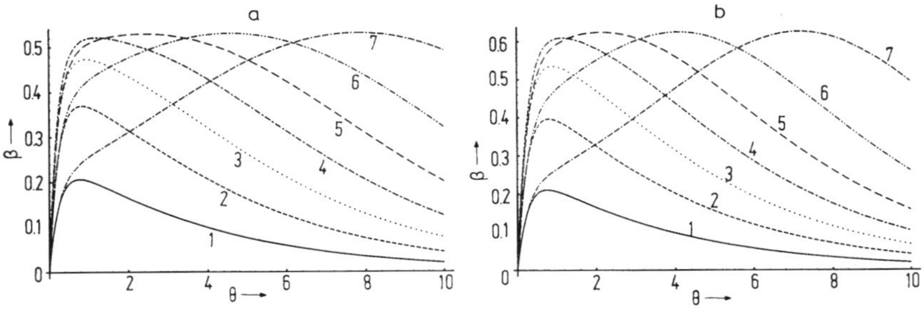

FIG.1.6-2. (a) Relativistic plots of $\beta = v/c$ obtained from Eqs.(10) and (13), and (b) nonrelativistic plots obtained from Eqs.(11) and (13) for $q = 1$, $p = 1/4$, $\rho = 2.5$, and initial angles $\vartheta(0) = n\pi/8$, with $n = 1, 2, \ldots, 7$; $0 \leq \theta \leq 10$.

$$R\frac{d\vartheta}{dt} = -v, \quad R\frac{d^2\vartheta}{dt^2} = -\frac{dv}{dt} \tag{5}$$

and rewrite Eq.(4):

$$2mR\frac{dv}{dt} + \xi_{\mathrm{m}}v = m_{\mathrm{mo}}B\sin\vartheta \tag{6}$$

The term $m_{\mathrm{mo}}B\sin\vartheta$ represents the force due to the flux density B which is acting on the magnetic dipole. This force is not affected by any relativistic change of the mass of the dipole. The terms mdv/dt and $\xi_{\mathrm{m}}v$ are affected. Their relativistic form was derived in Eqs.(1.2-7) and (1.2-10). In analogy, Eq.(6) is rewritten as follows:

$$2R\frac{d(mv)}{dt} + \frac{\xi_{\mathrm{m}}}{m_0}(mv) = m_{\mathrm{mo}}B\sin\vartheta \tag{7}$$

With m and $d(mv)/dt$ from Eqs.(1.2-8) and (1.2-10) we obtain:

$$\frac{2Rm_0}{(1-v^2/c^2)^{3/2}}\frac{dv}{dt} + \frac{\xi_{\mathrm{m}}}{(1-v^2/c^2)^{1/2}}v = m_{\mathrm{mo}}B\sin\vartheta \tag{8}$$

We re-substitute for v and dv/dt from Eq.(5):

$$\frac{2R^2m_0}{[1-(R/c)^2(d\vartheta/dt)^2]^{3/2}}\frac{d^2\vartheta}{dt^2} + \frac{R\xi_{\mathrm{m}}}{[1-(R/c)^2(d\vartheta/dt)^2]^{1/2}}\frac{d\vartheta}{dt} = -m_{\mathrm{mo}}B\sin\vartheta \tag{9}$$

With the notation $\theta = t/\tau$ and $\rho = R/c\tau$ one can bring Eq.(9) into the following form:

$$\left[1-\rho^2\left(\frac{d\vartheta}{d\theta}\right)^2\right]^{-1/2}\left(\frac{1}{1-\rho^2(d\vartheta/d\theta)^2}\frac{d^2\vartheta}{d\theta^2} + \frac{1}{pq}\frac{d\vartheta}{d\theta}\right) + \frac{1}{q^2}\sin\vartheta = 0$$

$$\theta = \frac{t}{\tau},\ \rho = \frac{R}{c\tau},\ q = \frac{R}{\tau}\sqrt{\frac{2m_0}{m_{\mathrm{mo}}B}} = \frac{\tau_{\mathrm{p}}}{\tau},\ p = \frac{\sqrt{2m_0m_{\mathrm{mo}}B}}{\xi_{\mathrm{m}}} = \frac{\tau_{\mathrm{mp}}}{\tau_{\mathrm{p}}}$$

$$pq = \frac{2Rm_0}{\tau\xi_{\mathrm{m}}},\ \tau_{\mathrm{p}} = R\sqrt{\frac{2m_0}{m_{\mathrm{mo}}B}},\ \tau_{\mathrm{mp}} = \frac{2m_0R}{\xi_{\mathrm{m}}} \tag{10}$$

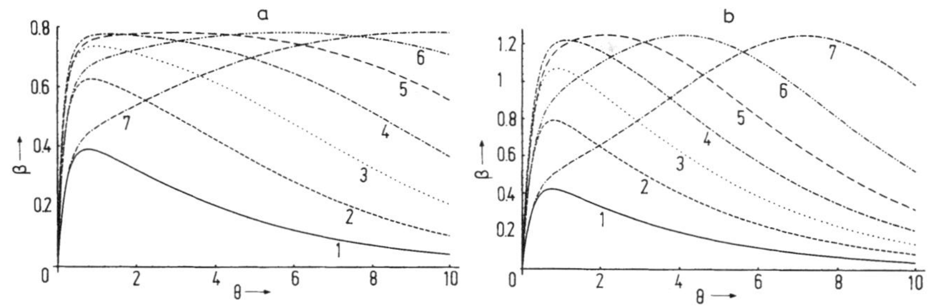

FIG.1.6-3. (a) Relativistic plots of $\beta = v/c$ obtained from Eqs.(10) and (13), and (b) nonrelativistic plots obtained from Eqs.(11) and (13) for $q = 1$, $p = 1/4$, $\rho = 5$, and initial angles $\vartheta(0) = n\pi/8$, with $n = 1, 2, \ldots, 7$; $0 \leq \theta \leq 10$.

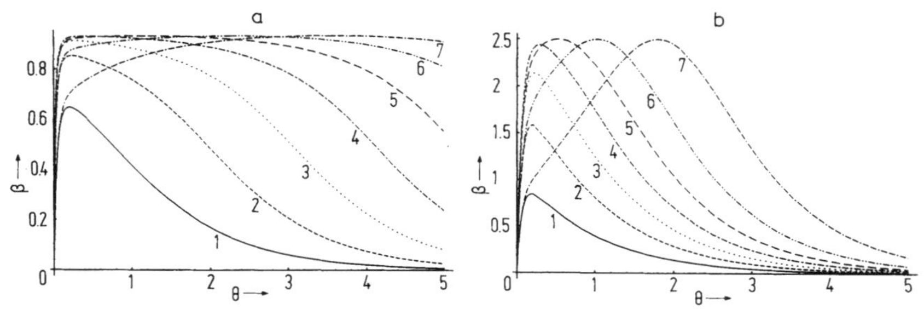

FIG.1.6-4. (a) Relativistic plots of $\beta = v/c$ obtained from Eqs.(10) and (13), and (b) nonrelativistic plots obtained from Eqs.(11) and (13) for $q = 1/4$, $p = 1/4$, $\rho = 2.5$, and initial angles $\vartheta(0) = n\pi/8$, with $n = 1, 2, \ldots, 7$; $0 \leq \theta \leq 5$.

For $\rho^2(d\vartheta/d\theta)^2 \to 0$ we obtain from Eq.(10) the nonrelativistic limit:

$$\frac{d^2\vartheta}{d\theta^2} + \frac{1}{pq}\frac{d\vartheta}{d\theta} + \frac{1}{q^2}\sin\vartheta = 0 \tag{11}$$

The initial conditions for Eqs.(10) and (11) are

$$\vartheta(0) = \vartheta_0, \quad d\vartheta(0)/d\theta = 0 \tag{12}$$

where ϑ_0 is the initial angle ϑ in Fig.1.4-5 before the flux density B is applied.

For the further calculation we still need the velocities $v(\theta)$ and $v_y(\theta)$ of Eqs.(1.4-24) and (1.4-25):

$$\beta = \frac{v(\theta)}{c} = -\frac{R}{c\tau}\frac{d\vartheta}{d\theta} = -\rho\frac{d\vartheta}{d\theta} \tag{13}$$

$$\beta_y = \frac{v_y(\theta)}{c} = -\rho\frac{d\vartheta}{d\theta}\sin\vartheta \tag{14}$$

The results obtained by numeric integration of Eq.(10) with the initial conditions $\vartheta(0) = n\vartheta_0 = \vartheta_n$ as well as $d\vartheta(0)/d\theta = 0$ are shown in Figs.1.6-2 to 1.6-10. Relativistic plots according to Eq.(10) are shown on the left side and corresponding nonrelativistic plots according to Eq.(11) on the right side.

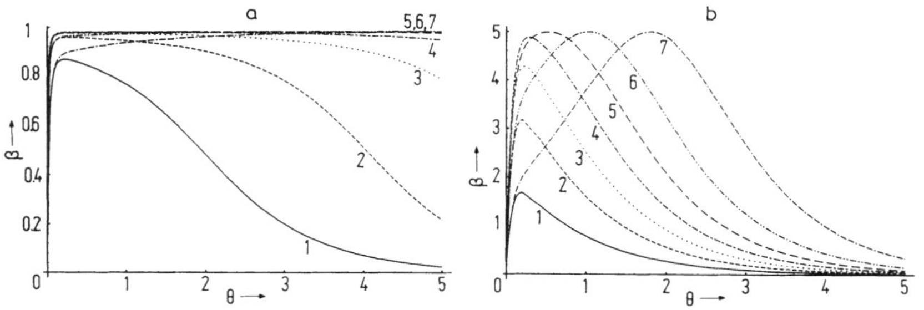

FIG.1.6-5. (a) Relativistic plots of $\beta = v/c$ obtained from Eqs.(10) and (13), and (b) nonrelativistic plots obtained from Eqs.(11) and (13) for $q = 1/4$, $p = 1/4$, $\rho = 5$, and initial angles $\vartheta(0) = n\pi/8$, with $n = 1, 2, \ldots, 7$; $0 \leq \theta \leq 5$.

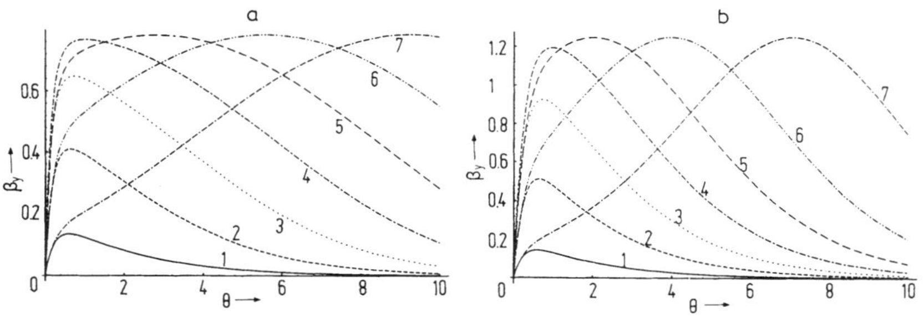

FIG.1.6-6. (a) Relativistic plots of $\beta_y = v_y/c$ obtained from Eqs.(10) and (14), and (b) nonrelativistic plots obtained from Eqs.(11) and (14) for $q = 1$, $p = 1/4$, $\rho = 5$, and initial angles $\vartheta(0) = n\pi/8$, with $n = 1, 2, \ldots, 7$; $0 \leq \theta \leq 10$.

The first four of these illustrations, Figs.1.6-2 to 1.6-5, show the normalized velocity $\beta = v/c$ according to Eq.(13) for various values of the parameters q, p, and ρ, while the initial angles $\vartheta(0)$ have the same seven values. As one would expect, the normalized velocity v/c can exceed 1 in the nonrelativistic plots but not in the relativistic ones.

Figure 1.6-6 shows the normalized velocity $\beta_y = v_y/c$ in the direction of the y-axis for the same values of q, p, and $\vartheta(0)$ as in Fig.1.6-3. The relativistic velocities in Fig.1.6-6a are significantly lower than in Fig.1.6-3a while the peaks of the nonrelativistic velocities for $n = 4, 5, 6, 7$ in Fig.1.6-6b are as large as in Fig.1.6-3b.

The average normalized velocity $\beta_{\mathrm{a,v}y} = v_{\mathrm{av},y}(\theta)/c$ according to Eqs.(1.4-42), (10), and (14)

$$\begin{aligned}\beta_{\mathrm{av},y} &= \frac{v_{\mathrm{av},y}(\theta)}{c} \qquad\qquad 0 \leq \vartheta \leq \pi,\ \vartheta_n = n\vartheta(0)\\ &= -\frac{1}{N}\sum_{n=1}^{N} \rho \frac{d\vartheta(\vartheta_n,\theta)}{d\theta}\sin\vartheta(\vartheta_n,\theta),\end{aligned} \tag{15}$$

is plotted for various values of the parameters ρ and p in Figs.1.6-7 to 1.6-10. The parameter q has the values 1, 1/2, 1/4, 1/8, 1/16 in Figs.1.6-7 to 1.6-9, but the values 1, 1/2, 1/4 in Fig.1.6-10.

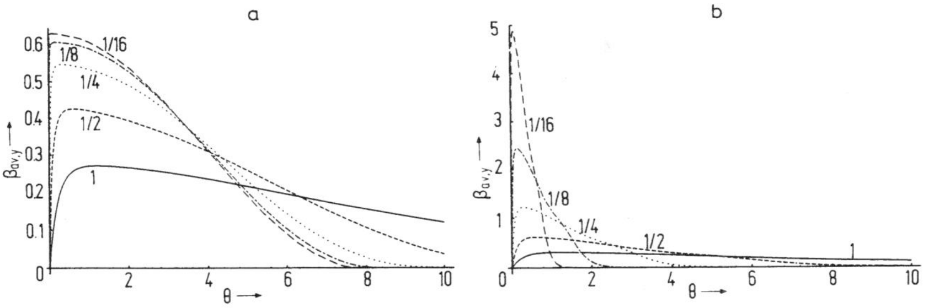

FIG.1.6-7. (a) Relativistic plots of $\beta_{\mathrm{av},y} = v_{\mathrm{av},y}/c$ obtained from Eqs.(10), (14), and (15), and (b) nonrelativistic plots obtained from Eqs.(11), (14), and (15) for $p = 1/4$, $\rho = 2.5$, and $q = 1, 1/2, 1/4, 1/8, 1/16$; $0 \leq \theta \leq 10$.

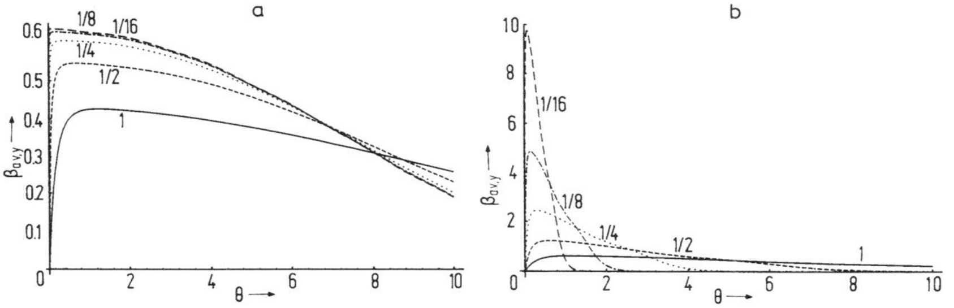

FIG.1.6-8. (a) Relativistic plots of $\beta_{\mathrm{av},y} = v_{\mathrm{av},y}/c$ obtained from Eqs.(10), (14), and (15), and (b) nonrelativistic plots obtained from Eqs.(11), (14), and (15) for $p = 1/4$, $\rho = 5$, and $q = 1, 1/2, 1/4, 1/8, 1/16$; $0 \leq \theta \leq 10$.

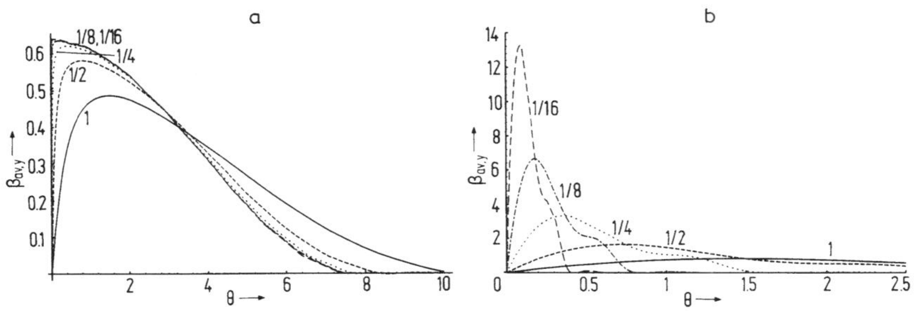

FIG.1.6-9. (a) Relativistic plots of $\beta_{\mathrm{av},y} = v_{\mathrm{av},y}/c$ obtained from Eqs.(10), (14), and (15), and (b) nonrelativistic plots obtained from Eqs.(11), (14), and (15) for $p = 1$, $\rho = 2.5$, and $q = 1, 1/2, 1/4, 1/8, 1/16$.

Recalling from Eq.(1.4-43) that the magnetic current density $\mathbf{g}_{\mathrm{m}}$ is proportionate to the average velocity $v_{\mathrm{av},y}(\theta)$ we see from Figs.1.6-7 to 1.6-10 that the time variation of the relativistic current densities can be radically different from that of the nonrelativistic ones. Let us observe that these plots derived for magnetic dipole current densities also apply to electric dipole current densities caused by the rotation of electrically polarized molecules in the form of a gas, such as water vapor H_2O. Polarized molecules in a solid, such as barium-titanate $BaTiO_3$, act more complicatedly since they interact with neighboring molecules.

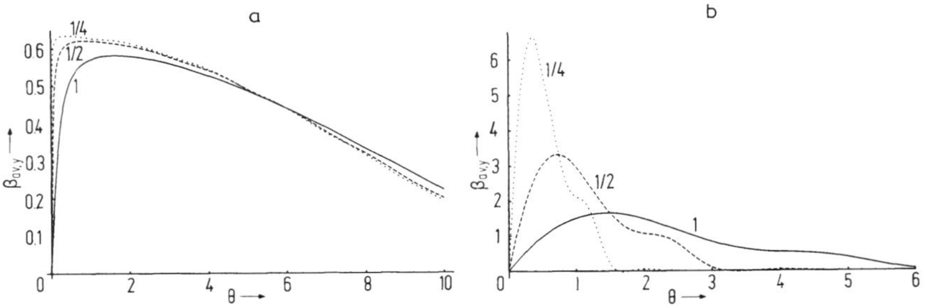

FIG.1.6-10. (a) Relativistic plots of $\beta_{\mathrm{av},y} = v_{\mathrm{av},y}/c$ obtained from Eqs.(10), (14), and (15), and (b) nonrelativistic plots obtained from Eqs.(11), (14), and (15) for $p = 1$, $\rho = 5$, and $q = 1, 1/2, 1/4$.

1.7 Electromagnetic Missiles

Starting with a paper by Wu (1985) the concept of electromagnetic missiles, bullets, or propagating focused waves has attracted considerable attention[1]. The principle can readily be illustrated by means of ray optics. Consider Fig.1.7-1 that shows a point-like light source denoted radiator R at the location y_{R} in the focal plane f_1 of a lens. The focal plane has the distance x_{f} from the center $x = 0$ of the lens. A light beam is formed that has the angle β relative to the lens axis. The value of β is given by the formula

$$\tan\beta = \frac{y_{\mathrm{R}}}{x_{\mathrm{f}}} \tag{1}$$

In terms of ray optics the power or energy density of the radiation propagating through a plane perpendicular to the radiation axis does not depend on the distance r from the radiator. The transition from ray optics to wave optics brings a decrease of the power or energy density proportionate to $1/r^2$. For sinusoidal waves one must use the concept of energy density since a periodic, infinitely extended sine wave has infinite energy. Waves representing signals—which includes sinusoidal pulses with a finite number of cycles—have a finite energy and energy density, their beam patterns differ from those of sinusoidal waves, but their energy density decreases like $1/r^2$ just like the power density of sinusoidal waves.

Consider now Fig.1.7-2 where the radiator R is not located in the focal plane of the lens but somewhat further away in a radiator plane. An image I of the radiator is produced in the image plane, which means all the radiation is concentrated in one point at a certain distance, which is quite different from Fig.1.7-1. For the angles β and γ in Fig.1.7-2 we recognize the relations

$$\tan\beta = \frac{y_{\mathrm{R}}}{x_{\mathrm{R}}} = \frac{y_{\mathrm{I}}}{x_{\mathrm{I}}} \tag{2}$$

$$\tan\gamma = \frac{y_{\mathrm{R}}}{x_{\mathrm{f}}} = \frac{y_{\mathrm{I}}}{x_{\mathrm{I}} + x_{\mathrm{f}}} \tag{3}$$

[1] Carian and Agi (1993); Geyi, Chengli, and Weigan (1991a, b; 1992a, b, c); Wu and Shen (1988, 1989, 1990); Yingzheng and Weigan (1988); see also Pan (1985) and Zeiping (1983).

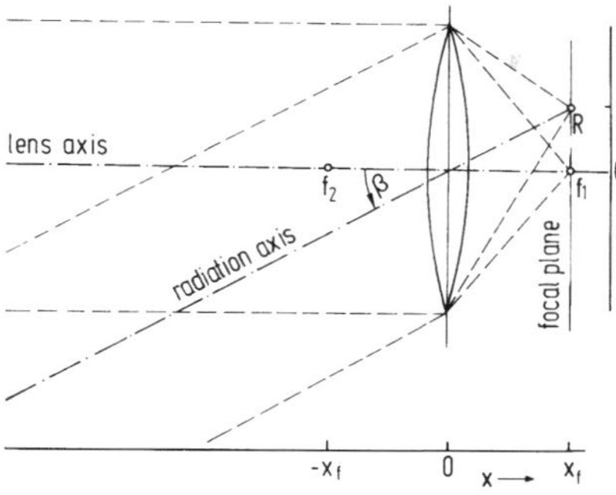

FIG.1.7-1. A point-like radiator R in the focal plane of a lens produces a pencil beam with angle β relative to the lens axis.

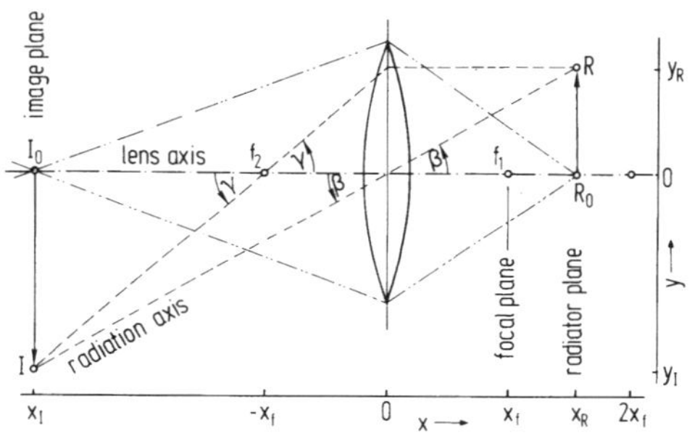

FIG.1.7-2. A point-like radiator R in the radiator plane produces a point-like image I in the image plane for $x_R > x_f$.

from which we get the x- and y-coordinates of the image I or R:

$$x_I = -\frac{x_f}{x_R - x_f} x_R \tag{4}$$

$$y_I = -\frac{x_f}{x_R - x_f} y_R \tag{5}$$

For $x_R \to x_f$ we get $x_I \to -\infty$ and $y_I \to -\infty$, which represents Fig.1.7-1: The image of R is produced at an infinite distance x and it is infinitely far from the lens axis with sign reversal of the location y_R of R.

Let the radiator R in Fig.1.7-2 be moved to the location $y = 0$ in the radiator plane. The new radiator is denoted R_0 and its image I_0. If the radiator R_0 is moved in the interval $2x_f \geq x_R > x_f$ the image I_0 will move in the interval $-2x_f \geq x_I > -\infty$. The velocity dx_I/dt of the image I_0 as function of the velocity dx_R/dt of the radiator R_0 is obtained by differentiation of Eq.(4):

$$\frac{dx_I}{dt} = \frac{x_f^2}{(x_R - x_f)^2} \frac{dx_R}{dt} \tag{6}$$

At $x_R = 2x_f$ the image moves with the velocity of the radiator while for $x_R \to x_f$ the velocity of the image approaches infinity since ray optics is not a relativistic theory. Radiator and image always move in the same direction.

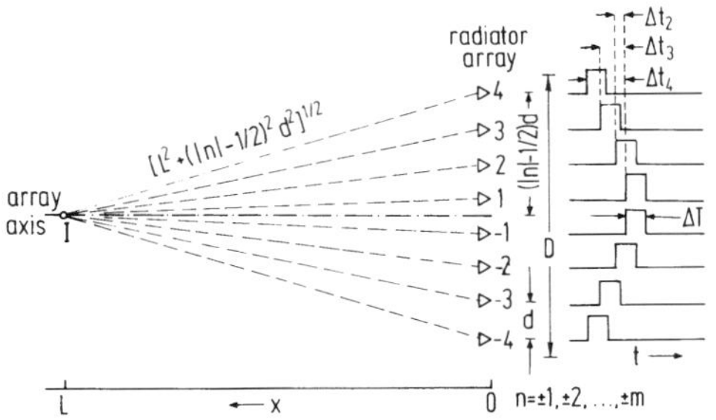

FIG.1.7-3. Replacement of the lens and the radiator in Fig.1.7-2 by a line array of radiators that emit electromagnetic waves whose electric and magnetic field strengths have the time variation of rectangular pulses.

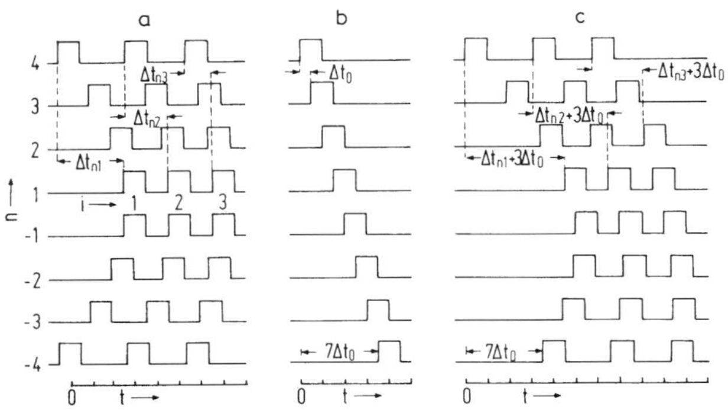

FIG.1.7-4. Generalization of the radiated pulses in Fig.1.7-3 to permit image points I at three successive distances (a). Timing of pulses for off-axis radiation without focusing (b). Combination of off-axis radiation with focusing (c).

The comparison of Figs.1.7-1 and 1.7-2 suggests that variable focusing should provide a narrower beam or angular resolution than fixed focusing at infinity, provided the distances of interest are not too large compared with the diameter of the lens. To investigate this possibility further we replace the lens and the radiator of Fig.1.7-2 by an array of radiators that emit electromagnetic waves with carefully chosen time variation and timing as shown in Fig.1.7-3. A line array with $2m = 8$ radiators with distance $(|n| - 1/2)d$, $n = \pm 1, \pm 2, \pm 3, \pm 4$, from the array axis radiates electromagnetic waves whose electric and magnetic field strengths have the time variation of rectangular pulses. These pulses should arrive at the image point I on the array axis at the distance L from the center of the array at the same time. The distance x_n between a radiator n and the image point I is given by

$$x_n = \left[L^2 + (|n| - 1/2)^2 d^2\right]^{1/2} \tag{7}$$

and the propagation time is x_n/c. In order to have the leading edge of all pulses arrive simultaneously at the image point I at the time $t_0 = L/c$ one must radiate them from radiator n at the time

$$t_n = \frac{L - \left[L^2 + (|n| - 1/2)^2 d^2\right]^{1/2}}{c}$$
$$\approx -\frac{(2|n|-1)^2}{8}\frac{d^2}{Lc} \qquad \text{for } (2|n|-1)^2 d^2/4L^2 \ll 1 \tag{8}$$

The time difference between the pulse radiation from radiator n and from radiator 1 is of practical interest:

$$\Delta t_n = t_n - t_1 = \frac{L}{c}\left\{\left(1 + \frac{d^2}{4L^2}\right)^{1/2} - \left[1 + \left(|n| - \frac{1}{2}\right)^2 \frac{d^2}{L^2}\right]^{1/2}\right\}$$
$$\approx -\frac{(2|n|-1)^2}{8}\frac{d^2}{Lc} \tag{9}$$

If one wants a concentration of energy by focusing not in one image point at the distance L but generally at k points L, $L + \Delta L$, ..., $L + (i-1)\Delta L$, ... for $i = 1, 2, \ldots, k$ one must replace the $k = 1$ set of $2m$ pulses in Fig.1.7-3 by k sets of $2m$ pulses each. An example is given in Fig.1.7-4a for $k = 3$. Let us demand that the image point I shall be produced at the distance $L + \Delta L$ at the time $t_1 = (L + \Delta L)/c + \Delta T$, and generally at the location $L + (i-1)\Delta L$ at the time $t_{i-1} = [L + (i-1)\Delta L]/c + (i-1)\Delta T$. The extra time $(i-1)\Delta T$ is required to permit a chosable minimum time difference between the pulses of the radiators $n = \pm 1$. The generalization of the time t_n in Eq.(8) to $t_{n,i}$ becomes:

$$t_{n,i} = \frac{1}{c}\left\{L + (i-1)\Delta L - \left[[L + (i-1)\Delta L]^2 + \left(|n| - \frac{1}{2}\right)^2 d^2\right]^{1/2}\right\} + (i-1)\Delta T \tag{10}$$

The generalization of Eq.(9) does not contain ΔT:

$$\Delta t_{n,i} = t_{n,i} - t_{1,i} = \frac{L + (i-1)\Delta L}{c}\left\{\left(1 + \frac{d^2}{4[L + (i-1)\Delta L]^2}\right)^{1/2} - \left[1 + \left(|n| - \frac{1}{2}\right)^2 \frac{d^2}{[L + (i-1)\Delta L]^2}\right]^{1/2}\right\} \tag{11}$$

In order to obtain some numerical values to judge the potential applications of focused waves we take from Fig.1.7-3 the relation

$$d = D/2m \tag{12}$$

TABLE 1.7-1

CHARACTERISTIC TIME DELAYS Δt_m REQUIRED TO PRODUCE A FOCUSED WAVE AT THE DISTANCE L WITH A LINE ARRAY OF LENGTH D.

D [m]	L [m]	$-\Delta t_m = D^2/8Lc$ [ps]
1	10	41.7
	100	4.17
	1000	0.417
2	20	83.3
	200	8.33
	2000	0.833
5	50	208.3
	500	20.8
	5000	2.8
10	100	417
	1000	41.7
	10000	4.17
20	200	833.3
	2000	83.3
	20000	8.33
50	500	2083
	5000	208
	50000	20.8
100	1000	4167
	10000	417
	100000	41.7

and substitute it into Eq.(9):

$$\Delta t_m = t_m - t_1 \approx -D^2/8Lc \qquad (13)$$

Table 1.7-1 shows Δt_m as function of D and L. A few values of Δt_m are larger than 1 ns and a few are smaller than 1 ps, but most lie in the range from 1 ps to 1 ns. This determines the typical required accuracy of the beginning of the pulses in Figs.1.7-3 and 1.7-4.

If one wants to produce one focused point I_0 in Fig.1.7-3 one needs one set of pulses ($i = 1, 2 \dots$) according to Fig.1.7-3a, which shall have the combined energy W. For k focused points at k different distances we need k sets of pulses with a combined energy kW. Hence, in principle the required energy increases linearly with distance. The usual beamforming requires that the radiated energy increases with the square of the distance if a certain energy density at the chosen distance is to be achieved. For a numerical comparison of the two methods consider Fig.1.7-5. An array R of radiators radiates the energy W_0 into the spherical angle ϵ. At the distance L we get the energy density w_L

$$w_L = W_0/\epsilon L^2 \qquad (14)$$

while at the distance $L + (k-1)\Delta L$ we get the energy density $w_{L,k}$:

$$w_{L,k} = W_0/\epsilon[L + (k-1)\Delta L]^2 \qquad (15)$$

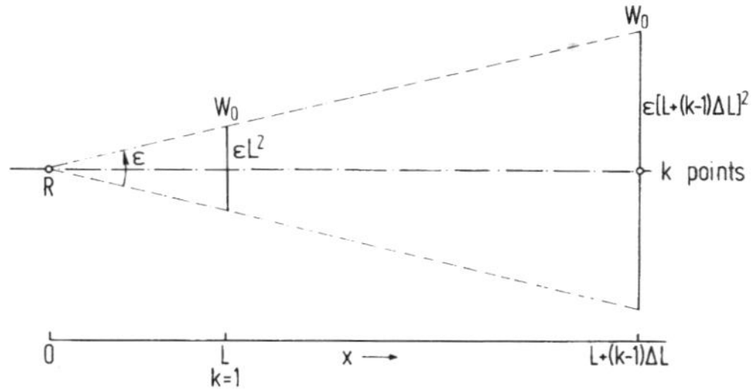

FIG.1.7-5. Derivation of the energy density of a spherical wave with opening angle ϵ at the distances L and $L+(k-1)\Delta L$ for a fixed radiated energy W_0.

If the energy density $w_{L,k}$ is to be increased to the value of w_L one must multiply W_0 by a factor K:

$$K = [L+(k-1)\Delta L]^2/L^2 = [1+(k-1)\Delta L/L]^2 \tag{16}$$

If a focused wave makes the energy W flow through the point at distance L in Fig.1.7-5, one needs the energy kW to make the energy W flow through each one of the k points L, $L+\Delta L$, ... , $L+(k-1)\Delta L$. If the condition

$$kW < LW_0 = [1+(k-1)\Delta L/L]^2 W_0 \tag{17}$$

is satisfied the focused wave will require less radiated energy than the spherical wave. Equation (17) may be rewritten:

$$0 < k^2 - \left[\frac{W}{W_0}\left(\frac{L}{\Delta L}\right)^2 - 2\frac{L}{\Delta L} + 2\right] k + \left(\frac{\Delta L}{L}\right)^2 \left(1 - \frac{\Delta L}{L}\right)^2 \tag{18}$$

If the sign $<$ is replaced by an equality sign we obtain for the larger root k_1 of k the following second order approximation:

$$k_1 \approx \frac{W}{W_0}\left(\frac{L}{\Delta L}\right)^2 - 2\frac{L}{\Delta L} \tag{19}$$

As a first numerical example consider the values $W = W_0$, $L/\Delta L = 10$. From Eq.(19) we get the approximation $k_1 \approx 100 - 20 = 80$. From the exact formula of Eq.(16) we derive the following values:

$k \rightarrow$	79	80	81	82	83
$K \rightarrow$	77.4	79.2	81	82.8	84.6

We see that $k_1 = 81$ rather than 80 is the exact value. For $k > 81$ the focused wave will require less radiated energy. The distance x corresponding to $k = 81$ follows from Fig.1.7-5:

$$x = L + (k-1)\Delta L = L(1 + 80 \times 0.1) = 9L \tag{20}$$

As a second example consider $W = W_0$, $L/\Delta L = 100$. Equation (19) yields $k_1 \approx 9800$. The exact formula (16) yields:

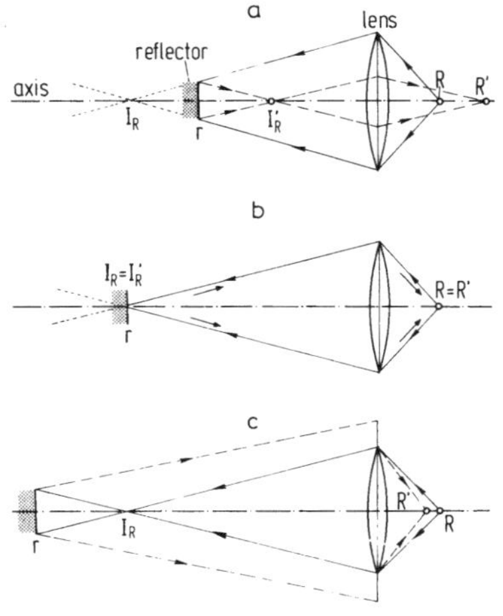

FIG.1.7-6. Explanation in terms of ray-optics why the energy of a reflected focused wave does not decrease with the distance r like $1/r^2$ as would be the case for an unfocused wave.

$k \rightarrow$	9800	9801	9802
$K \rightarrow$	9799.0	9801	9802.9

The exact value of k_1 is 9801 rather than 9800. The focused wave requires less radiated energy for $k > 9801$ and the distance x corresponding to $k = 9801$ becomes:

$$x = L + (k-1)\Delta L = L(1 + 9800 \times 0.01) = 99L \tag{21}$$

Comparison with Eq.(20) shows that the distance has increased by a factor 11 while the focused points increased by a factor $9800/81 = 121$.

One of the best features of focused waves applied to radar is that the energy of a pulse returned by a reflector does not vary with the distance r like $1/r^2$ as would be the case for an unfocused wave. Similarly, the energy of a pulse returned by a point-like scatterer decreases like $1/r^2$ rather than $1/r^4$. The principle of this effect can readily be shown by means of ray optics. Consider Fig.1.7-6a which shows a radiator R, a lens, the image point I_R, and a reflector at the distance r between the lens and the image point. A perfect reflector perpendicular to the optical axis will return all the energy coming from R via the lens to the new image point I'_R and via the lens to R'. If the reflector is located at the image point I_R as in Fig.1.7-6b the points I_R and I'_R as well as R and R' will coincide.

Figure 1.7-6c shows the case of the reflector having a larger distance r from the lens than the image point I_R. Only part of the incident energy is returned and focused in the point R'. The returned energy decreases with the distance r proportionate to $1/r^2$. The use of a larger lens for the returned signal would avoid the decrease.

If the reflector in Fig.1.7-6 is replaced by a point-like scatterer we obtain Fig.1.7-7. The returned energy decreases now like $1/r^2$ since the scatterer produces a spherical wave whose surface increases like r^2. Using Fig.1.7-6 one may readily

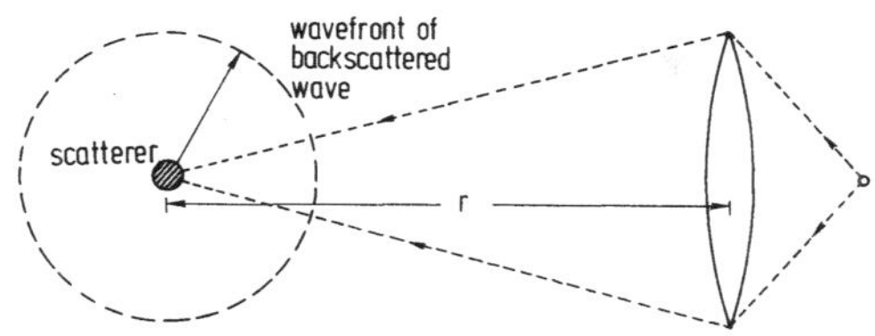

FIG.1.7-7. Scattering of a focused wave by a point-like scatterer. The backscattered energy decreases with the distance r proportionate to $1/r^2$ rather than $1/r^4$ as in unfocused radar.

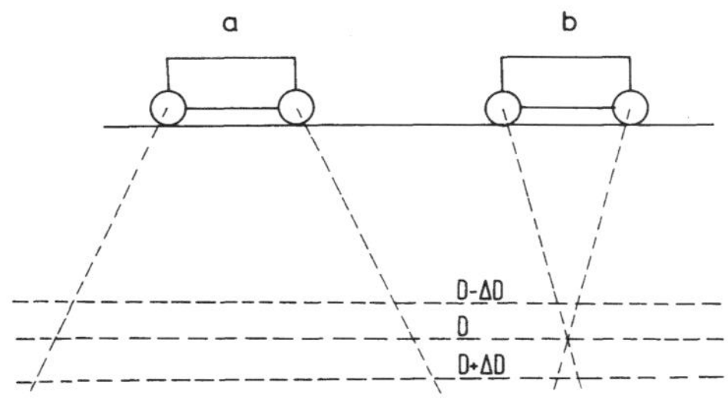

FIG.1.7-8. The angular resolution of an unfocused ground-probing radar (a) is greatly improved by focusing (b).

generalize this illustration to the cases when the scatterer is to the left or right of the image point of the radiation source.

Figure 1.7-8 shows that focusing can greatly improve the angular resolution of a radar. Figure 1.7-8a assumes unfocused waves for a ground-probing radar that is mounted on a cart. The cart is pulled along the surface of the ground. A large area is illuminated at a distance $D - \Delta D \leq d \leq D + \Delta D$. Figure 1.7-8b shows the same radar using focused waves. The improved angular resolution in a layer at the depth $D \pm \Delta D$ is striking.

2 Reflection and Transmission of Incident Signals

2.1 Partial Differential Equations for Perpendicular Polarization

Consider an electromagnetic wave excited in the *plane of excitation* in Fig.2.1-1, beginning at the time $t = t_0$. The 'beginning' or *front (plane)* of the wave reaches the origin of a Cartesian coordinate system with the axes x, y, z at the time $t = 0$. The x-axis is assumed to be in the wave front and to stick out perpendicularly from the paper plane. The plane $z = 0$ is the boundary between two media characterized by the parameters ϵ_1, μ_1, σ_1, s_1 and ϵ_2, μ_2, σ_2, s_2. The angle of incidence between the boundary plane and the wave front is denoted ϑ_i in Fig.2.1-1. All points $y \leq 0$ of the boundary plane have been reached at a time $t \leq 0$ by the wave, while none of the points $y > 0$ have been reached yet at the time $t = 0$. A reflected cylinder wave C_r is excited at the boundary plane $z = 0$ for any value $y \leq 0$ in medium 1 and a transmitted cylinder wave C_t in medium 2. Figure 2.1-1 is the basic

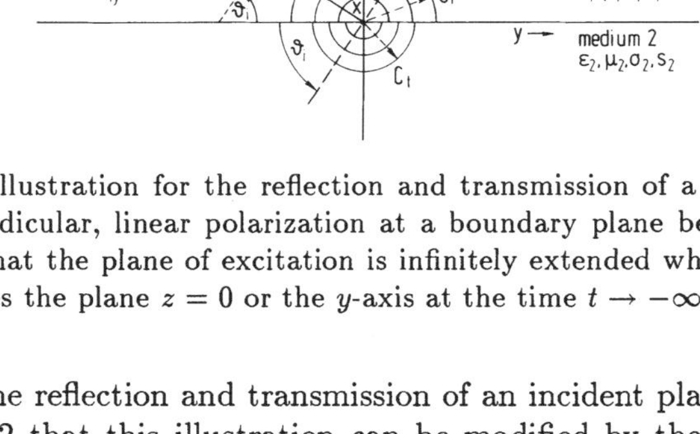

FIG.2.1-1. Basic illustration for the reflection and transmission of a *planar TEM signal wave* with perpendicular, linear polarization at a boundary plane between two media 1 and 2. We note that the plane of excitation is infinitely extended which implies that the signal front reaches the plane $z = 0$ or the y-axis at the time $t \to -\infty$ at $y \to -\infty$.

illustration for the reflection and transmission of an incident planar wave. We will see in Section 2.2 that this illustration can be modified by the introduction of a *signal boundary* to produce a model that is better suited for waves that represent signals.

The usual illustrations for reflection and transmission of planar waves at the boundary plane $z = 0$ assume *steady state waves.* The steady state case is obtained from Fig.2.1-1 by showing the signal front not at the time $t = 0$ but at the time $t \to \infty$. All points $y > 0$ in the boundary plane will then have been reached by the incident wave. Snell's laws of reflection and transmission apply to such steady state waves. We emphasize that Snell's laws do not assume a sinusoidal time variation of the incident wave[1] (Harmuth 1986c, p. 163).

We turn to the solution of the reflection and transmission of electromagnetic signals at a boundary plane according to Fig.2.1-1. We emphasize that this is an approximation that we will improve upon. In Fig.2.1-1 the signal front reaches the plane $z = 0$ at the time $t \to -\infty$. A better approximation will require that the plane $z = 0$ is reached by the signal front at a finite time.

As the first step we write the modified Maxwell equations, Eqs.(1.1-8)–(1.1-11), in component form for a Cartesian coordinate system with axes x, y, z using Eqs.(1.1-5), (1.1-6), (1.1-7), and (1.1-13):

$$\frac{\partial H_z}{\partial y} - \frac{\partial H_y}{\partial z} = \epsilon \frac{\partial E_x}{\partial t} + \sigma E_x \tag{1}$$

$$\frac{\partial H_x}{\partial z} - \frac{\partial H_z}{\partial x} = \epsilon \frac{\partial E_y}{\partial t} + \sigma E_y \tag{2}$$

$$\frac{\partial H_y}{\partial x} - \frac{\partial H_x}{\partial y} = \epsilon \frac{\partial E_z}{\partial t} + \sigma E_z \tag{3}$$

$$-\frac{\partial E_z}{\partial y} + \frac{\partial E_y}{\partial z} = \mu \frac{\partial H_x}{\partial t} + sH_x \tag{4}$$

$$-\frac{\partial E_x}{\partial z} + \frac{\partial E_z}{\partial x} = \mu \frac{\partial H_y}{\partial t} + sH_y \tag{5}$$

$$-\frac{\partial E_y}{\partial x} + \frac{\partial E_x}{\partial y} = \mu \frac{\partial H_z}{\partial t} + sH_z \tag{6}$$

$$\frac{\partial E_x}{\partial x} + \frac{\partial E_y}{\partial y} + \frac{\partial E_z}{\partial z} = 0 \tag{7}$$

$$\frac{\partial H_x}{\partial x} + \frac{\partial H_y}{\partial y} + \frac{\partial H_z}{\partial z} = 0 \tag{8}$$

Consider first the planar wave excited in the *plane of excitation* in Fig.2.1-1. We describe it in the coordinate system x_i, y_i, z_i that is rotated by $-(\pi/2 - \vartheta_\mathrm{i})$ relative to the coordinate system x, y, z. Let us assume a transverse, electromagnetic (TEM) wave propagating in the direction y_i:

$$E_{y_\mathrm{i}} = H_{y_\mathrm{i}} = 0 \tag{9}$$

A planar wave calls for the following additional relations:

[1] Not even the wave theory of light developed by Christian Huygens (1629–1695) existed at the time of Willebrord Snell (1591–1626).

$$\frac{\partial E_{x_i}}{\partial x_i} = \frac{\partial E_{x_i}}{\partial z_i} = \frac{\partial E_{z_i}}{\partial x_i} = \frac{\partial E_{z_i}}{\partial z_i} = 0 \tag{10}$$

$$\frac{\partial H_{x_i}}{\partial x_i} = \frac{\partial H_{x_i}}{\partial z_i} = \frac{\partial H_{z_i}}{\partial x_i} = \frac{\partial H_{z_i}}{\partial z_i} = 0 \tag{11}$$

We can use Eqs.(1)–(8) by writing x_i, y_i, z_i for x, y, z. Equations (7) and (8) are automatically satisfied by a planar TEM wave; the same holds for Eqs.(2) and (5). The remaining Eqs.(1), (3), (4), and (6) assume the following form with ϵ, μ, σ, s replaced by ϵ_1, μ_1, σ_1, s_1:

$$\frac{\partial H_{z_i}}{\partial y_i} = \epsilon_1 \frac{\partial E_{x_i}}{\partial t} + \sigma_1 E_{x_i} \tag{12}$$

$$-\frac{\partial H_{x_i}}{\partial y_i} = \epsilon_1 \frac{\partial E_{z_i}}{\partial t} + \sigma_1 E_{z_i} \tag{13}$$

$$-\frac{\partial E_{z_i}}{\partial y_i} = \mu_1 \frac{\partial H_{x_i}}{\partial t} + s_1 H_{x_i} \tag{14}$$

$$\frac{\partial E_{x_i}}{\partial y_i} = \mu_1 \frac{\partial H_{z_i}}{\partial t} + s_1 H_{z_i} \tag{15}$$

We make the substitutions

$$E_{x_i} = E_i \cos\chi, \quad E_{z_i} = E_i \sin\chi \tag{16}$$

$$H_{x_i} = H_i \sin\chi, \quad H_{z_i} = -H_i \cos\chi \tag{17}$$

and obtain from Eqs.(12)–(15):

$$\frac{\partial E_i}{\partial y_i} + \mu_1 \frac{\partial H_i}{\partial t} + s_1 H_i = 0 \tag{18}$$

$$\frac{\partial H_i}{\partial y_i} + \epsilon_1 \frac{\partial E_i}{\partial t} + \sigma_1 E_i = 0 \tag{19}$$

The variable χ is the polarization angle. For $\chi = 0$ we get a planar TEM wave with field strengths $\mathbf{E}_i(y_i, t)$, $\mathbf{H}_i(y_i, t)$ having perpendicular, linear polarization as shown in Fig.2.1-1

$$\mathbf{E}_i(y_i, t) = E_i \mathbf{e}_{x_i} \tag{20}$$

$$\mathbf{H}_i(y_i, t) = -H_i \mathbf{e}_{z_i} \tag{21}$$

where $\mathbf{e}_{x_i}$, $\mathbf{e}_{y_i}$, $\mathbf{e}_{z_i}$ are unit vectors in the direction of the coordinate axes x_i, y_i, z_i. A planar TEM wave with parallel, linear polarization is obtained for $\chi = \pi/2$:

$$\mathbf{E}_\mathrm{i}(y_\mathrm{i},t) = E_\mathrm{i}\mathbf{e}_{z_\mathrm{i}} \tag{22}$$
$$\mathbf{H}_\mathrm{i}(y_\mathrm{i},t) = H_\mathrm{i}\mathbf{e}_{x_\mathrm{i}} \tag{23}$$

For $\chi = \omega t$ and $\chi = -\omega t$ we get planar TEM waves with left circular polarization

$$\mathbf{E}_\mathrm{i}(y_\mathrm{i},t) = E_\mathrm{i}(\mathbf{e}_{x_\mathrm{i}}\cos\omega t + \mathbf{e}_{z_\mathrm{i}}\sin\omega t) \tag{24}$$
$$\mathbf{H}_\mathrm{i}(y_\mathrm{i},t) = H_\mathrm{i}(\mathbf{e}_{x_\mathrm{i}}\sin\omega t - \mathbf{e}_{z_\mathrm{i}}\cos\omega t) \tag{25}$$

and right circular polarization:

$$\mathbf{E}_\mathrm{i}(y_\mathrm{i},t) = E_\mathrm{i}(\mathbf{e}_{x_\mathrm{i}}\cos\omega t - \mathbf{e}_{z_\mathrm{i}}\sin\omega t) \tag{26}$$
$$\mathbf{H}_\mathrm{i}(y_\mathrm{i},t) = -H_\mathrm{i}(\mathbf{e}_{x_\mathrm{i}}\sin\omega t + \mathbf{e}_{z_\mathrm{i}}\cos\omega t) \tag{27}$$

If χ is a general time function

$$\chi = \chi(t) \tag{28}$$

we get planar TEM waves with *general polarization*. We will discuss polarization in Section 2.6 to avoid deviating from the topic of this section.

Equations (18) and (19) have been solved analytically for the initial conditions $E_\mathrm{i}(y_\mathrm{i},0) = H_\mathrm{i}(y_\mathrm{i},0) = 0$ and boundary conditions given by a step function (Harmuth 1986c; Harmuth and Hussain 1994a)

$$\begin{aligned} E_\mathrm{i}(0,t) = E_0 S(t) &= 0 \quad \text{for } t \le 0 \\ &= E_0 \quad \text{for } t > 0 \end{aligned} \tag{29}$$

or an exponential ramp function:

$$\begin{aligned} E_\mathrm{i}(0,t) = E_0 r(t) &= 0 \quad &&\text{for } t \le 0 \\ &= E_0(1 - e^{-t/\tau}) \quad &&\text{for } t > 0 \end{aligned} \tag{30}$$

Let us assume we have obtained the solution of Eqs.(18) and (19) for perpendicular, linear polarization. The electric and magnetic field strengths of Eqs.(20) and (21) are then transformed from the coordinate system x_i, y_i, z_i to the system x, y, z according to Fig.2.1-1. Using the unit vectors $\mathbf{e}_x$, $\mathbf{e}_y$, $\mathbf{e}_z$ in the direction of the axes x, y, z we get:

$$\mathbf{E}_\mathrm{i}(y_\mathrm{i},t) = \mathbf{E}_\mathrm{i}(y,z,t) = E_{\mathrm{i}x}\mathbf{e}_x$$
$$\mathbf{H}_\mathrm{i}(y_\mathrm{i},t) = H_{\mathrm{i}y}\mathbf{e}_y + H_{\mathrm{i}z}\mathbf{e}_z$$

$$\begin{gathered} E_{\mathrm{i}x} = E_\mathrm{i}(y,z,t), \quad E_{\mathrm{i}y} = 0, \quad E_{\mathrm{i}z} = 0 \\ H_{\mathrm{i}x} = 0, \quad H_{\mathrm{i}y} = -H_\mathrm{i}(y,z,t)\cos\vartheta_\mathrm{i}, \quad H_{\mathrm{i}z} = -H_\mathrm{i}(y,z,t)\sin\vartheta_\mathrm{i} \end{gathered} \tag{31}$$

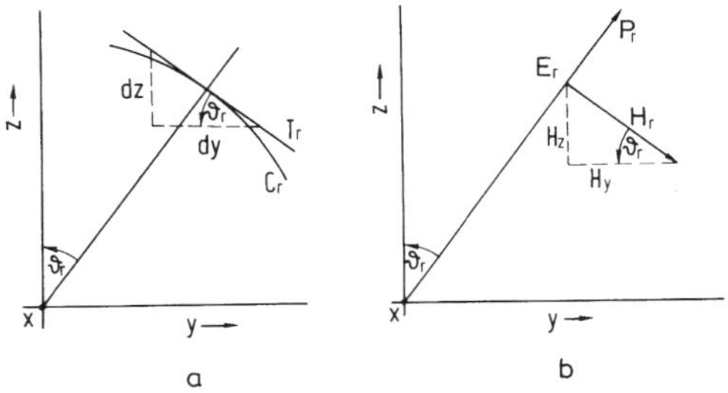

FIG.2.1-2. (a) The relation $dz = \operatorname{tg}\vartheta_r dy$ holds for a cylinder wave for any angle ϑ_r. (b) The relation $H_z = -H_y \operatorname{tg}\vartheta_r$ holds for a cylinder TEM wave for any angle ϑ_r.

Let the wave defined by Eq.(31) reach the boundary plane $z = 0$ in Fig.2.1-1. The media shall be homogeneous, isotropic, and time-invariant. The parameters ϵ_1, μ_1, σ_1, s_1 and ϵ_2, μ_2, σ_2, s_2 are then constant scalars. A cylinder wave with axis parallel to the x-axis will be excited in every point y of the boundary plane. No electric field strengths in the directions of y and z, and no magnetic field strength in the direction of x will be excited due to Eq.(31):

$$E_y = E_y(y,0,t) = 0, \quad E_z = E_z(y,0,t) = 0, \quad H_x = H_x(y,0,t) = 0 \tag{32}$$

With these three conditions we get from Eqs.(1)–(8) the following relations holding in both media 1 and 2; no indices 1 or 2 are written for this reason:

$$\frac{\partial H_z}{\partial y} - \frac{\partial H_y}{\partial z} = \epsilon\frac{\partial E_x}{\partial t} + \sigma E_x \tag{33}$$

$$-\frac{\partial H_z}{\partial x} = 0 \tag{34}$$

$$\frac{\partial H_y}{\partial x} = 0 \tag{35}$$

$$0 = 0 \tag{36}$$

$$-\frac{\partial E_x}{\partial z} = \mu\frac{\partial H_y}{\partial t} + sH_y \tag{37}$$

$$\frac{\partial E_x}{\partial y} = \mu\frac{\partial H_z}{\partial t} + sH_z \tag{38}$$

$$\frac{\partial E_x}{\partial x} = 0 \tag{39}$$

$$\frac{\partial H_y}{\partial y} + \frac{\partial H_z}{\partial z} = 0 \tag{40}$$

From Eqs.(32), (34), (35), and (39) we get the relations

$$\partial E_x/\partial x = \partial E_y/\partial x = \partial E_z/\partial x = 0 \tag{41}$$

$$\partial H_x/\partial x = \partial H_y/\partial x = \partial H_z/\partial x = 0 \tag{42}$$

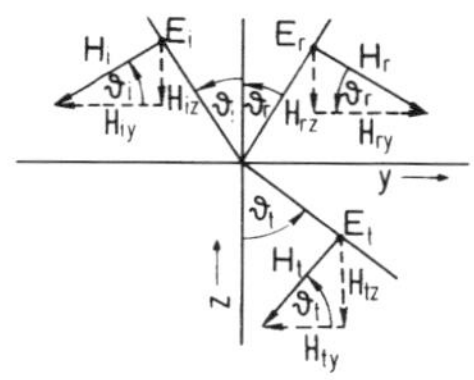

FIG.2.1-3. Relationships between the angles ϑ_i, ϑ_r, ϑ_t and the components H_y, H_z of the magnetic field strengths H_i, H_r, H_t.

which state that the electric and magnetic field strength are independent of x. This is a typical feature of a cylinder wave, including its limit of a planar wave.

Equations (37) and (38) are equal for a TEM wave and one of them can be left out. To see this equality consider Fig.2.1-2a. The cylinder wave C_r of Fig.2.1-1 is shown enlarged. A tangential plane T_r defines a reflection angle ϑ_r and the relation

$$\frac{dz}{dy} = \operatorname{tg}\vartheta_r \tag{43}$$

The vectors $\mathbf{E}_r$ and $\mathbf{H}_r$ of a perpendicularly polarized TEM wave reflected in the direction of the angle ϑ_r are shown in Fig.2.1-2b. The relation[2]

$$\frac{H_z}{H_y} = \frac{H_{rz}}{H_{ry}} = -\operatorname{tg}\vartheta_r \tag{44}$$

is readily recognizable. Substitution of Eqs.(43) and (44) into Eq.(38) yields

$$\frac{\partial E_x}{\partial y} = \frac{\partial E_x}{\partial z}\frac{\partial z}{\partial y} = \frac{\partial E_x}{\partial z}\operatorname{tg}\vartheta_r = -\left(\mu\frac{\partial H_y}{\partial t} + sH_y\right)\operatorname{tg}\vartheta_r \tag{45}$$

which is Eq.(37). Furthermore, substitution of Eqs.(43) and (44) into Eq.(40) shows that this relation is satisfied for a TEM wave too:

$$\frac{\partial H_y}{\partial y} + \frac{\partial H_z}{\partial z} = \frac{\partial H_y}{\partial y} + \frac{\partial H_z}{\partial y}\frac{\partial y}{\partial z} = \frac{\partial H_y}{\partial y} - \operatorname{tg}\vartheta_r\frac{\partial H_y}{\partial y}\frac{1}{\operatorname{tg}\vartheta_r} = 0 \tag{46}$$

Only Eqs.(33) and (37) remain for a planar TEM wave. Using Eq.(44) we can eliminate H_z in Eq.(33). We add the index r to indicate a reflected wave and write ϵ_1, μ_1, σ_1, s_1 to indicate that the wave propagates in medium 1 and use Eq.(43):

$$-\frac{1}{\cos^2\vartheta_r}\frac{\partial H_{ry}}{\partial z} = \epsilon_1\frac{\partial E_{rx}}{\partial t} + \sigma_1 E_{rx} \tag{47}$$

$$-\frac{\partial E_{rx}}{\partial z} = \mu_1\frac{\partial H_{ry}}{\partial t} + s_1 H_{ry} \tag{48}$$

For the transmitted wave in medium 2 in Fig.2.1-1 we get with the index t from Figs.2.1-3 and 2.1-4 the relationships

[2]The index r is added for "reflection" as will be explained in the text following Eq.(46).

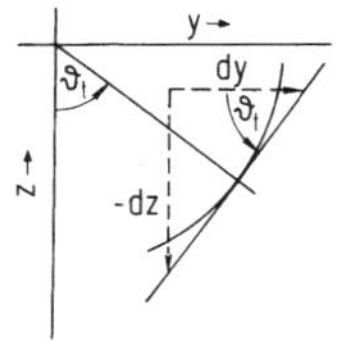

FIG.2.1-4. The relation $-dz = \operatorname{tg}\vartheta_t dy$ holds for a cylinder wave for any angle ϑ_t.

$$\frac{H_{tz}}{H_{ty}} = \operatorname{tg}\vartheta_t \tag{49}$$

$$\frac{dz}{dy} = -\operatorname{tg}\vartheta_t \tag{50}$$

Equations (33) and (37) assume the following form with t and ϵ_2, μ_2, σ_2, s_2:

$$-\frac{1}{\cos^2\vartheta_t}\frac{\partial H_{ty}}{\partial z} = \epsilon_2\frac{\partial E_{tx}}{\partial t} + \sigma_2 E_{tx} \tag{51}$$

$$-\frac{\partial E_{tx}}{\partial z} = \mu_2\frac{\partial H_{ty}}{\partial t} + s_2 H_{ty} \tag{52}$$

The four Eqs.(47), (48), (51), and (52) contain the four unknowns E_{rx}, H_{ry}, E_{tx}, and H_{ty}, provided we obtain the angles ϑ_r and ϑ_t from somewhere else. Two additional equations are obtained from boundary conditions that must be met at the boundary plane $z = 0$ in Fig.2.1-1. These boundary conditions are as much a part of the physical content of Maxwell's theory as Maxwell's equations themselves:

(a) The tangential components ($\mathbf{e}_x$) of the electric field strengths in both media must be equal at $z = 0$:

$$E_i(y, z=0, t) = E_{ix}(y, z=0, t) = E_{rx}(y, z=0, t) + E_{tx}(y, z=0, t) \tag{53}$$

(b) The tangential components ($\mathbf{e}_y$) of the magnetic field strengths in both media must be equal at $z = 0$:

$$\begin{aligned} &-H_i(y, z=0, t)\cos\vartheta_i \\ &\quad = -H_{iy}(y, z=0, t) = H_{ry}(y, z=0, t) - H_{ty}(y, z=0, t) \end{aligned} \tag{54}$$

The solutions obtained from Eqs.(47), (48), (51), and (52) must be general enough to satisfy Eqs.(53) and (54). We get thus E_{rx}, H_{ry}, E_{tx}, H_{ty} from the differential equations, H_{rz} and H_{tz} from Eqs.(44) and (49), while the remaining six field strengths E_{ry}, E_{rz}, H_{rx}, H_{ty}, E_{tz}, and H_{tx} are zero. Hence, all twelve components of the reflected and transmitted electric and magnetic field strengths are determined.

Despite all these results the approach taken so far based on an incident planar wave front is not satisfactory. There is no hint what determines the angles ϑ_r of reflection and ϑ_t of transmission. A further basic problem is that the planar wave in Fig.2.1-1 reaches the plane $z = 0$ first at the location $y \to -\infty$ and the time $t \to -\infty$. Such infinite distances and times make it impossible to introduce the causality law, which in turn makes it impossible to derive any results for the reflection or transmission of signals from Fig.2.1-1. We must develop a more sophisticated model for reflection and transmission than the planar wave front model.

2.2 GENERALIZATION OF SNELL'S LAW FOR SIGNALS

The reflection of infinitely extended electromagnetic waves with arbitrary time variation has been described for close to four centuries by Snell's law of reflection:

$$\text{angle of incidence } \vartheta_i = \text{angle of reflection } \vartheta_r$$

The theory developed in Section 2.1 gave no hint how the angle of incident and the angle of reflection are connected. We want to find the physical cause for the relation $\vartheta_r = \vartheta_i$ and see whether it applies not only for steady state waves but for signals too.

To recognize the physical basis for Snell's law refer to Fig.2.2-1. A planar wave arrives from the left with angle of incidence ϑ_i. Two rays are distinguished by their Poynting vectors $\mathbf{P}_1$ and $\mathbf{P}_2$. At the points A_1 and A_2 the waves of the two rays are *synchronous*[1]. If these two rays are reflected at the points A_1 and B_2 with an angle of reflection $\vartheta_{r1} = \vartheta_i$, they will again by synchronous at the points C_1 and C_2 since the distances $\overline{A_1B_1} + \overline{B_1C_1}$ and $\overline{A_2B_2} + \overline{B_2C_2}$ to the reflected wave front WF_1 are equal. On the other hand, angles of reflection $\vartheta_{r2} > \vartheta_i$ and $\vartheta_{r3} < \vartheta_i$ yield unequal distances to the reflected wave fronts WF_2 and WF_3:

$$\begin{aligned}
&\overline{A_1C_3} > \overline{A_2B_2} + \overline{B_2C_4}, \quad \vartheta_{r2} > \vartheta_i\\
&\overline{A_1C_5} < \overline{A_2B_2} + \overline{B_2C_6}, \quad \vartheta_{r3} < \vartheta_i\\
&\overline{A_1C_1} = \overline{A_2B_2} + \overline{B_2C_2}, \quad \vartheta_{r1} = \vartheta_i
\end{aligned} \tag{1}$$

Hence, the waves on the wave fronts WF_2 and WF_3 will not be synchronous and thus will not add up or *interfere constructively* like the waves on the wave front WF_1. If waves are reflected from every point y on the boundary $z = 0$ we will expect that the reflection angle $\vartheta_r = \vartheta_{r1} = \vartheta_i$ will be distinguished.

The reflection of signals is different. According to Fig.2.1-1 a signal arriving at the boundary $z = 0$ first in the point $y = 0$ rather than $y \to -\infty$ will excite a cylinder wave first in the point $y = 0$ and then in the points $y > 0$. No cylinder wave should be excited at any point $y < 0$ since a signal is by definition zero before a certain, finite time and this implies that there must be a finite point on a boundary

[1] The more common phrase would be *in phase*. However, the concept of phase applies only to waves with periodic sinusoidal time variation while Snell's law applies to waves with much more general time variation. Snell's experiments were done with what we would call colored noise today. Sinusoidal electromagnetic waves became prominent with the development of radio transmission after 1900. It took another 60 years before the laser permitted the generation of periodic sinusoidal light waves.

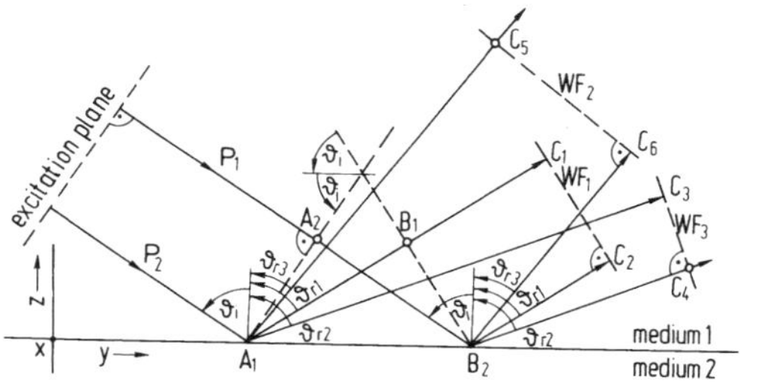

FIG.2.2-1. The sum of waves reflected with an angle $\vartheta_r = \vartheta_{r1} = \vartheta_i$ yields a larger peak amplitude than the sum of waves reflected with an angle $\vartheta_{r2} > \vartheta_i$ or $\vartheta_{r3} < \vartheta_i$.

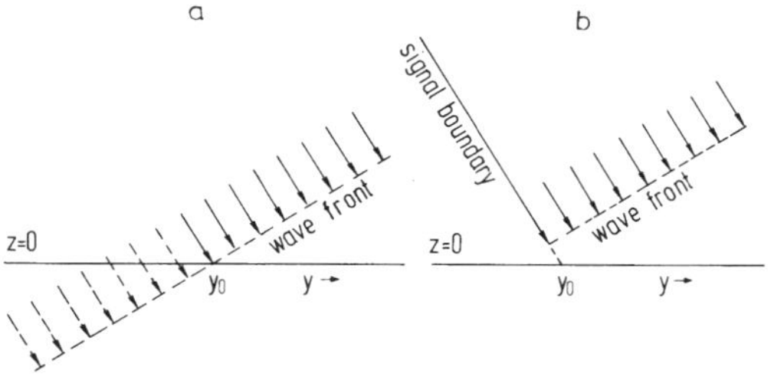

FIG.2.2-2. Planar wave front of a wave infinitely extended in time (a), and of a signal wave that is zero before a finite time and reaches the boundary plane $z = 0$ first in a finite point y_0.

plane that is reached first by a signal; Figure 2.2-2 makes this statement evident. We see in Fig.2.2-2a the usual planar wave front reaching a planar boundary $z = 0$. The boundary is reached first at $y \to -\infty$ at the time $t \to -\infty$. The introduction of a *signal boundary* in Fig.2.2-2b assures that a finite point y_0 is reached first at a finite time t_0. The signal boundary in Fig.2.2-2b is on the left end of the wave front, but for a signal incident from the right it would be on the right end. Those not happy with the introduction of a signal boundary may try to start with a cylinder or a spherical wave and see what the transition from an infinitely extended wave to a signal wave implies as the cylinder or spherical wave approaches a planar wave[2].

The signal boundary in Fig.2.2-2b is one reason why the reflection of signals is not completely described by Snell's law. When the wave front reaches the boundary plane $z = 0$ first in the point y_0, a cylinder wave will be excited and there will be no distinguished angle of reflection. As time increases the wave front in Fig.2.2-2b reaches points $y > y_0$ and for $t \to \infty$ the signal will excite cylinder waves in all points of the interval $y_0 \leq y < \infty$. This is no different from Fig.2.2-2a where cylinder waves are excited in all points $-\infty < y \leq y_0$ of the boundary $z = 0$. Hence, Snell's law of reflection can only apply to signals when the steady state $t \to \infty$ is approached.

Refer to Fig.2.2-3a that shows the moment the wave front reaches the boundary plane $z = 0$ first in the point $y = 0$. A cylinder wave is emitted from the point $y = 0$

[2]We note that the need for a signal boundary is due to the infinite extension of the boundary plane in Fig.2.2-2. There is no need for this signal boundary if we study the reception of a signal by a point-like sensor.

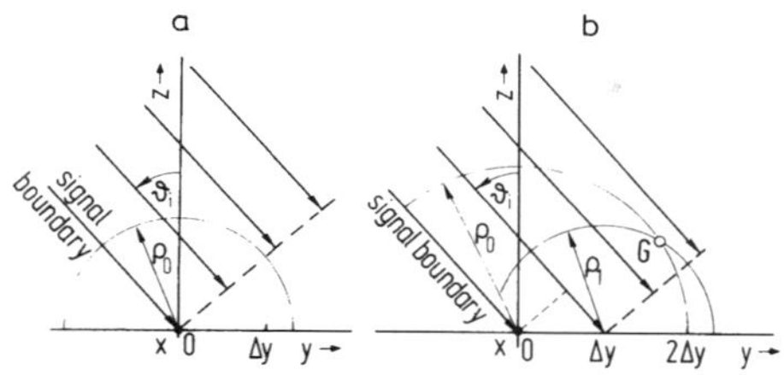

FIG.2.2-3. Arrival of a signal at the point $y = 0$ (a) and at the point $y = \Delta y$ (b).

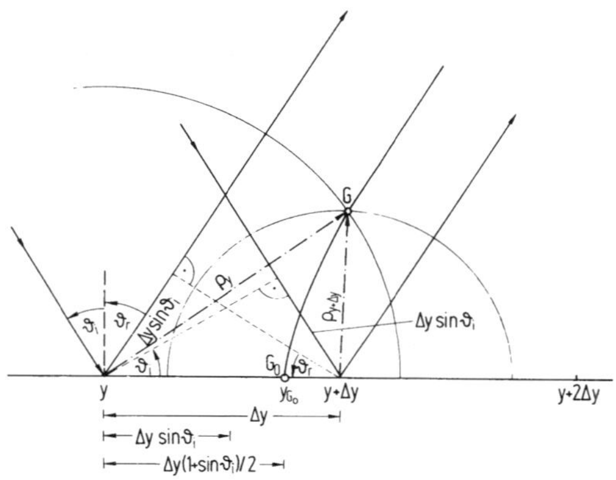

FIG.2.2-4. Locus G as function of time where the leading edges of the waves reflected at the points y and $y + \Delta y$ arrive at the same time.

and no distinguished reflection angle ϑ_r exists. A short time later the point $y = \Delta y$ is reached by the wave front and a second cylinder wave is emitted from this point as shown in Fig.2.2-3b. The leading edge of the waves emitted from $y = 0$ and $y = \Delta y$ will reach the point G at the same time if the condition

$$\rho_0 = \Delta y \sin \vartheta_i + \rho_1 \tag{2}$$

is satisfied.

The locus of G as function of time is of interest. According to Eq.(2) and its generalization the locus is a hyperbola

$$\rho_y - \rho_{y+\Delta y} = \Delta y \sin \vartheta_i \tag{3}$$

with center point G_0 located at

$$y_{G_0} = y + \Delta y(1 + \sin \vartheta_i)/2 \tag{4}$$

as shown in Fig.2.2-4. The direction of the asymptote of the hyperbola is seen to be

$$\sin \vartheta_r = \frac{\Delta y \sin \vartheta_i}{\Delta y} \tag{5}$$

which means that Snell's law is approached asymptotically at great distance from G_0. We note that Eq.(5) is independent of the arbitrarily chosen value of Δy.

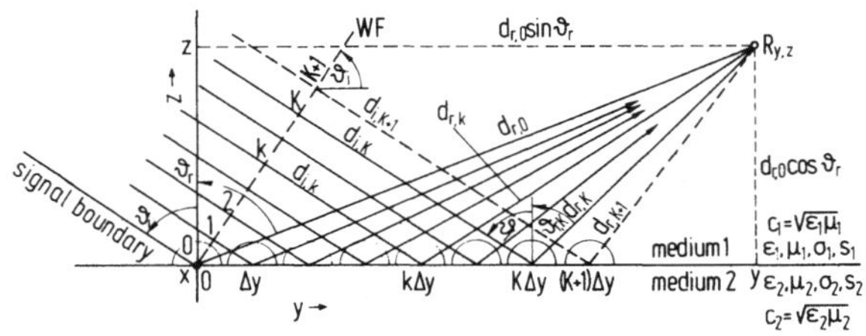

FIG.2.2-5. Reflection of a planar signal wave close to the reflecting plane $z = 0$.

The independence of Eq.(5) from Δy implies that the relation $\sin\vartheta_{\mathrm{r}} = \sin\vartheta_{\mathrm{i}}$ for the direction of the asymptote does not depend on the use of two reflecting points y and $y+\Delta y$ as in Fig.2.2-4. In Fig.2.2-5 we have $K+1$ reflection points. We may pair any two of them with distances Δy, $2\Delta y$, ... , $K\Delta y$ and get Eq.(5) with Δy replaced by $2\Delta y$, ... , $K\Delta y$ and the relation $\sin\vartheta_{\mathrm{r}} = \sin\vartheta_{\mathrm{i}}$ unchanged. Hence, we may expect to get Snell's law only at sufficiently great distance from the reflecting plane $z = 0$ in Figs.2.2-3 or 2.2-5.

Let us consider an incident wave with the time variation of a sinusoidal step function:

$$\begin{aligned} F(t) = S(t)\sin 2\pi t/T &= 0 && \text{for } t \leq 0 \\ &= \sin 2\pi t/T && \text{for } t > 0 \end{aligned} \tag{6}$$

According to Fig.2.2-3b one would then choose

$$\Delta y = cT/\sin\vartheta_{\mathrm{i}} \tag{7}$$

to make the time variation of the waves excited at the points $y = 0$, Δy, $2\Delta y$, ... equal, The distance from the reflecting plane where the hyperbola in Fig.2.2-4 approaches a straight line would then be of the order of the wave length cT. Other time variations of $F(t)$ will yield significantly different results.

We turn to the determination of the time variation of a reflected signal with general time variation $F(t)$ close to the reflecting boundary. Figure 2.2-5 shows a signal arriving from the left with angle of incidence ϑ_{i}. At every point of the wave front the time variation shall be $F(t)$. From the point k on the wave front WF the signal has to travel the distance

$$d_{\mathrm{i},k} = k\Delta y\sin\vartheta_{\mathrm{i}} \tag{8}$$

to reach the point $y = k\Delta y$ where it excites a cylinder wave. This cylinder wave then has to travel the distance

$$d_{\mathrm{r},k} = \left[(y - k\Delta y)^2 + z^2\right]^{1/2} \tag{9}$$

to reach the point $R_{y,z}$ for which we want to calculate the time variation of the reflected signal. If the time variation of the signal in point k of the wave front WF is $F(t)$, it will be $F(t - d_{\mathrm{i},k}/c_1 - d_{\mathrm{r},k}/c_1)$ after the signal has travelled the distance

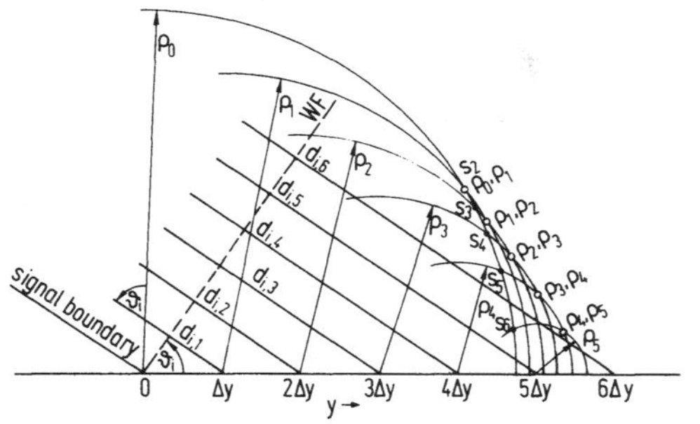

FIG.2.2-6. Relations between the radii ρ_k with $k = 0, 1, \ldots, K$ of the cylinder waves excited at the points $y = k\Delta y$ of a reflecting plane by an incident planar signal wave. The value $K = 5$ is used.

$d_{i,k} + d_{r,k}$ to the point $R_{y,z}$. Summation of the partial signals over $k = 0, 1, \ldots, K$ yields the time variation $F_{r,y,z}(t)$ in the point $R_{y,z}$:

$$F_{r,y,z}(t) = \frac{1}{K+1}\sum_{k=0}^{K} F\left(t - \frac{k\Delta y \sin\vartheta_i}{c_1} - \frac{[(y-k\Delta y)^2 + z^2]^{1/2}}{c_1}\right)$$
$$c_1 = 1/\sqrt{\epsilon_1\mu_1} \qquad (10)$$

We still have to derive some constraints on this formula. First, we can ignore the reflections at the points $y > (K+1)\Delta y$ only if the distance $d_{i,K+1} + d_{r,K+1}$ is larger than the distance $d_{i,K} + d_{r,K}$:

$$d_{i,K+1} + d_{r,K+1} > d_{i,K} + d_{r,K} \qquad (11)$$

Using Eqs.(8) and (9) we obtain for $\Delta y \ll y$ and $K \gg 1$:

$$\sin\vartheta_i > \sin\vartheta_{r,K} = \frac{y - y_K}{[(y-y_K)^2 + z^2]^{1/2}}, \quad y_K = K\Delta y$$
$$\vartheta_{r,K} < \vartheta_i \qquad (12)$$

The angle $\vartheta_{r,K}$ is shown in Fig.2.2-5 at $y_K = K\Delta y$. Equation (10) can be used only to the left of $d_{r,K}$ in Fig.2.2-5. If one wants $F_{r,y,z}(t)$ to the right of this line one must increase K until Eq.(12) is satisfied.

A further constraint applies to the time variable t in Eq.(10). Let us denote the time when the wave front WF in Fig.2.2-5 reaches the point $y = 0$ with $t = 0$. Equation (10) can then be used in the time interval $0 \le t < t_{r,K}$, where $t_{r,K}$ is defined by:

$$t_{r,K} = (d_{i,K+1} + d_{r,K+1})/c_1$$
$$\approx \frac{1}{c_1}\left\{y_K \sin\vartheta_i + \left[(y-y_K)^2 + z^2\right]^{1/2}\right\}, \quad y_K = K\Delta y \qquad (13)$$

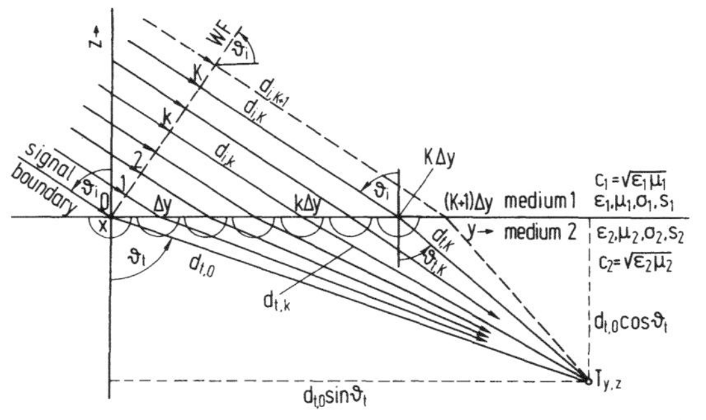

FIG.2.2-7. Transmission of a planar signal wave close to the boundary plane $z = 0$.

Figure 2.2-6 shows certain relations between the radii ρ_k of the cylinder waves excited at the points $y = k\Delta y$ of a reflecting plane by an incident planar signal wave. A radius

$$\rho_K \leq \Delta y \sin \vartheta_{\rm i} \tag{14}$$

is chosen sufficiently small to make sure the cylinder wave at $y = (K+1)\Delta y$ is not yet excited and thus can be ignored. The other radii ρ_k then have the values

$$\rho_k = \rho_K + (K-k)d_{{\rm i},1} = \rho_K + (K-k)\Delta y \sin \vartheta_{\rm i}, \quad k = 0,\ 1, \ldots,\ K \tag{15}$$

The points denoted ρ_k, ρ_{k+1} show where two adjacent cylinder waves intersect while the points denoted s_i with $i = 2,\ 3,\ 4,\ 5,\ 6$ show the limit where i cylinder waves are superimposed. This illustration requires some studying to see the finer details of the production of a reflected wave from the superposition of cylinder waves.

We turn to Fig.2.2-7 that shows the production of the transmitted wave at the point $T_{y,z}$ in medium 2. A signal arriving from the left shall have again the time variation $F(t)$ in every point of the wave plane WF. From the point k on the wave front WF the signal travels the distance $d_{{\rm i},k}$ defined by Eq.(8) to reach the point $y = k\Delta y$. The cylinder wave excited in that point travels the distance

$$d_{{\rm t},k} = \left[(y - k\Delta y)^2 + z^2\right]^{1/2} \tag{16}$$

to reach the point $T_{y,z}$. If the time variation of the signal in point k of the wave front WF is $F(t)$, it will be $F(t - d_{{\rm i},k}/c_1 - d_{{\rm t},k}/c_2)$ after the signal has travelled the distance $d_{{\rm i},k} + d_{{\rm t},k}$ to the point $T_{y,z}$. Summation of the partial signals over $k = 0$, $1, \ldots, K$ yield the time variation $F_{{\rm t},y,z}(t)$ in the point $T_{y,z}$:

$$F_{{\rm t},y,z}(t) = \frac{1}{K+1}\sum_{k=0}^{K} F\left(t - \frac{k\Delta y \sin\vartheta_{\rm i}}{c_1} - \frac{\left[(y-k\Delta y)^2 + z^2\right]^{1/2}}{c_2}\right)$$
$$c_1 = 1/\sqrt{\epsilon_1\mu_1}, \quad c_2 = 1/\sqrt{\epsilon_2\mu_2} \tag{17}$$

Again we have to derive constraints on this formula. The transmitted waves that are excited at the points $y \geq (K+1)\Delta y$ of Fig.2.2-7 can only be ignored if the time $d_{\mathrm{i},K+1}/c_1 + d_{\mathrm{t},K+1}/c_2$ is larger than the time $d_{\mathrm{i},K}/c_1 + d_{\mathrm{t},K}/c_2$:

$$d_{\mathrm{i},K+1}/c_1 + d_{\mathrm{t},K+1}/c_2 > d_{\mathrm{i},K}/c_1 + d_{\mathrm{t},K}/c_2 \tag{18}$$

Using Eqs.(8) and (16) we obtain for $\Delta y \ll y$, $K \gg 1$:

$$\begin{aligned} \sin\vartheta_\mathrm{i} &> \frac{c_1}{c_2}\sin\vartheta_{\mathrm{t},K} = \frac{c_1}{c_2}\frac{y - y_K}{[(y-y_K)^2 + z^2]^{1/2}}, \quad y_K = K\Delta y \\ \vartheta_{\mathrm{t},K} &< \arcsin\left(\frac{c_2}{c_1}\sin\vartheta_\mathrm{i}\right) \end{aligned} \tag{19}$$

We still need the constraint on the time variable t in Eq.(17). The time when the wave front WF in Fig.2.2-7 reaches the point $y = 0$ is again denoted $t = 0$. Equation (17) can then be used in the time interval $0 \leq t < t_{\mathrm{t},K}$, where $t_{\mathrm{t},K}$ is defined by

$$\begin{aligned} t_{\mathrm{t},K} &= d_{\mathrm{i},K+1}/c_1 + d_{\mathrm{t},K+1}/c_2 \\ &\approx \frac{1}{c_1}y_K\sin\vartheta_\mathrm{i} + \frac{1}{c_2}\left[(y-y_K)^2 + z^2\right]^{1/2} \\ y_K &= K\Delta y, \quad c_1 = 1/\sqrt{\epsilon_1\mu_1}, \quad c_2 = 1/\sqrt{\epsilon_2\mu_2} \end{aligned} \tag{20}$$

2.3 Reflection of a Step Wave

In Section 2.2 we have shown that there are deviations from Snell's law for reflected or transmitted signals close to the boundary plane. A careful reading of that section shows that we have only shown that the relation $\vartheta_\mathrm{i} = \vartheta_\mathrm{r}$ yields the direction in which waves reflected from various points of the boundary plane will be synchronous. This does *not* imply that significant energy could *not* be reflected in other directions. We will show that such reflections are indeed possible for a step wave. The derived principles are applicable to many other waveforms since two step functions $S(\theta)$ and $S(\theta - \Delta\theta)$ can be used to represent a rectangular pulse $S(\theta) - S(\theta - \Delta\theta)$.

Consider Fig.2.3-1. A planar wave with the time variation of a step function approaches from the left with angle of incidence $\vartheta_\mathrm{i} = 60°$ and is reflected at the reflection plane. This reflection plane may be infinitely extended but the signal boundary makes the incident wave reach it first in the point $x/\Delta L = -4$. Alternately, the reflection plane may have a finite width $L = 9\Delta L$, like the wing of an airplane.

The wavefront WF in Fig.2.3-1 reaches the point $k = x/\Delta L = -4$ first and a cylinder wave is excited in this point. The beginning of this cylinder wave at the time t_0 is represented by the semicircle with radius R_{-4}.

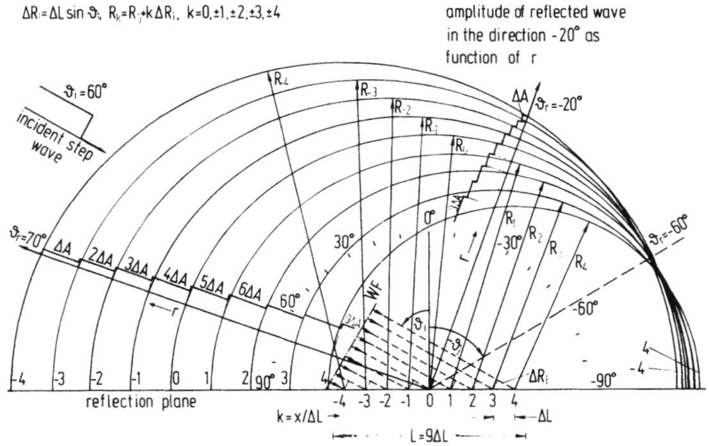

FIG.2.3-1. Reflection of an incident planar step wave with angle of incidence $\vartheta_i = 60°$ at the points $k = -4$, -3, ... , 4 of a reflecting plane into the directions $\vartheta_r = -20°$ and $\vartheta_r = +70°$. Note that all angles are measured in the mathematically positive sense which requires to write Snell's law of reflection as $\vartheta_r = -\vartheta_i$ or $\vartheta_r = 360° - \vartheta_i$. WF: wavefront.

With a delay $\Delta t = \Delta R_i/c = (\Delta L \sin\vartheta_i)/c$ the wavefront reaches the point $k = -3$. Again a cylinder wave is excited. Its beginning at the time t_0 is represented by the semicircle with radius $R_{-3} = R_{-4} - \Delta R_i$.

We may progress in this way to the points $k = -2, \ldots, 4$ on the reflection plane. Writing is simplified if we refer all radii R_k to the center radius R_0:

$$R_k = R_0 + k\Delta R_i, \qquad k = 0,\ \pm 1,\ \pm 2, \ldots \tag{1}$$

Let us construct the wave reflected in the direction $\vartheta = -20°$ at the time t_0 as function of the distance r from the point $k = 0$ on the reflection plane. First we draw a straight line from $k = 0$ in the direction $-20°$. At distances r beyond the semicircle R_{-4} the amplitude of the reflected wave is 0 since the beginning of the wave has not yet reached these distances. Between the semicircles R_{-4} and R_{-3} only the wave reflected from point $k = -4$ has arrived; we assign its amplitude the value ΔA. Between the semicircles R_{-3} and R_{-2} the amplitude is $2\Delta A$ since reflections from points $k = -4$ and $k = -3$ have arrived. We may proceed in this fashion until distances r inside the semicircle R_4 are reached. Reflected waves from all nine points $k = -4$ to $k = +4$ are now superimposed and the amplitude is $9\Delta A$. The plotted 'amplitude of reflected wave in the direction $-20°$ as function of r' in Fig.2.3-1 is obtained in this way.

Consider next the direction $\vartheta_r = 70°$ on the left of Fig.2.3-1. Again we have the amplitude zero for distances r beyond the semicircle defined by R_{-4}, the amplitude ΔA between the semicircles defined by R_{-4} and R_{-3}, etc. For distances r inside the semicircle defined by R_4 we get again the amplitude $9\Delta A$.

Some more studying of Fig.2.3-1 shows that the reflected wave will have eventually the value $9\Delta A$ for any reflection angle. What varies is how fast the amplitude increases from 0 to $9\Delta A$. This increase is very slow in the direction 70°—both as

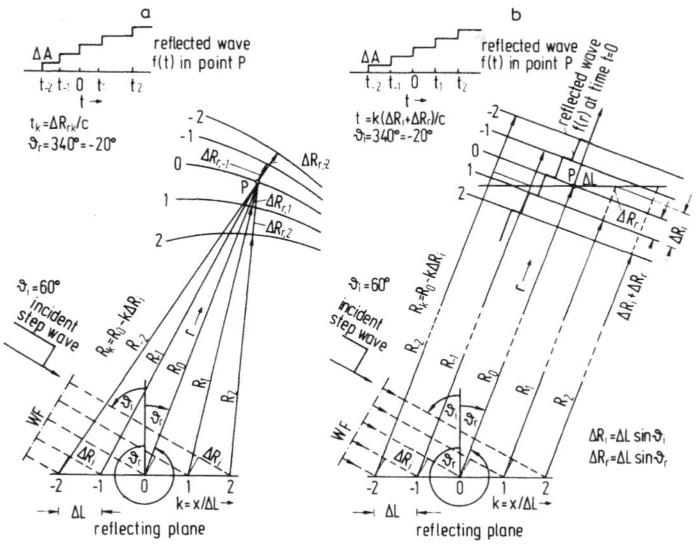

FIG.2.3-2. Near zone (a) and far zone (b) reflection of an incident planar step wave with angle of incidence $\vartheta_i = 60°$ at the points $k = 0, \pm1, \pm2$ of a reflecting plane in the direction $\vartheta_r = -20°$. Note that all angles are measured from the direction perpendicular to the reflection plane in the mathematically positive sense which yields $\vartheta_r = 340° = -20°$ rather than $\vartheta_r = 20°$. WF: wavefront.

function of the spatial variable r or a time variable t—but much faster in the direction $-20°$. The fastest increase is evidently close to[1] $\vartheta_r \approx -\vartheta_i = -60°$. The reason why this fastest rise is not exactly at $\vartheta_r = -\vartheta_i$ is that the hyperbola of Fig.2.2-4 cannot be replaced by its asymptote for the short distances used in Fig.2.3-1. We have to extend the principles of Fig.2.3-1 to the far zone, $r \to \infty$, to recognize more clearly the meaning of a reflection angle $\vartheta_r = -\vartheta_i$ for waves with general time variation.

Figure 2.3-2a shows with a larger scale than Fig.2.3-1 the reflection of a step wave in the direction $\vartheta_r = -20°$. There is a slight difference: Figure 2.3-2a yields the time variation $f(t)$ of the reflected wave in the point P at the fixed distance r. In Fig.2.3-1, on the other hand, we had the spatial variation $g(r)$ of the reflected wave at a certain time t. If the wave requires the time t_{i0} to propagate from the wavefront WF to the center point $k = 0$ on the reflection plane, it requires the time

$$t_{ik} = t_{i0} + k\Delta R_i/c, \quad \Delta R_i = \Delta L \sin\vartheta_i, \quad k = 0, \pm1, \pm2, \ldots \tag{2}$$

to reach a general point k. The wave reflected at the point $k = 0$ requires the additional time $t_{r0} = R_0/c$ to reach the point P. We choose the time scale so that $t_{i0} + t_{r0}$ yield the time $t = 0$:

$$t - (t_{i0} + t_{r0}) = 0 \tag{3}$$

[1] Snell's law in the form $\vartheta_r = \vartheta_i$ requires that ϑ_r and ϑ_i are measured differently. If both angles are measured equally in the mathematically positive sense one must write $\vartheta_r = -\vartheta_i$ or $\vartheta_r = 360° - \vartheta_i$. We use the notation $\vartheta_r = -\vartheta_i$ for Snell's law from here on.

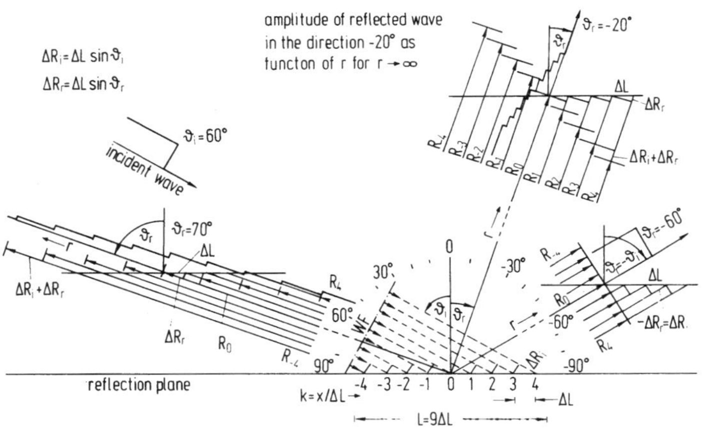

FIG.2.3-3. Extension of Fig.2.3-1 from the near zone to the far zone $r \to \infty$. WF: wavefront.

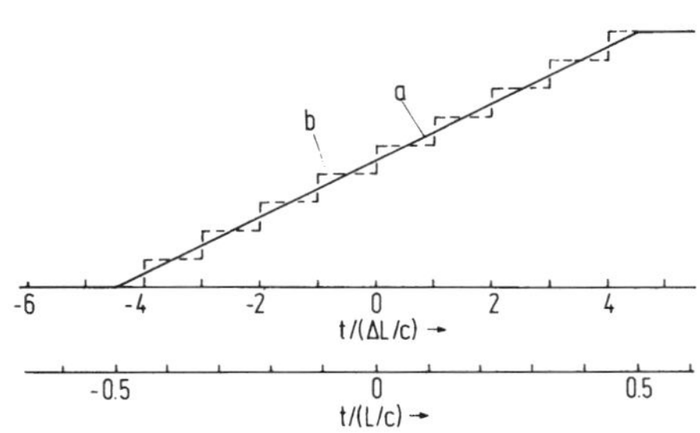

FIG.2.3-4. Ramp function (a) approximated by a step function (b). Note that the time scale $t/(L/c)$ applies to any value of ΔL including $\Delta L \to 0$.

At this time the wave reflected at the general point k will have propagated the distance R_k beyond the reflection point k

$$R_k = R_0 - k\Delta R_{\mathrm{i}}, \quad k = 0,\ \pm 1,\ \pm 2 \tag{4}$$

and still has to propagate the additional distance $\Delta R_{\mathrm{r}k}$ to reach point P in Fig.2.3-2a. A distance $\Delta R_{\mathrm{r}k}$ beyond P receives a negative sign. The time of arrival at point P becomes:

$$t_k = \Delta R_{\mathrm{r}k}/c \tag{5}$$

Rather than calculate the distances $\Delta R_{\mathrm{r}k}$ we take their values from the geometric construction of Fig.2.3-2a. If every wave reflected at one of the points k is a step function with amplitude ΔA we get the reflected wave $f(t)$ shown on top of Fig.2.3-2a.

In order to extend our results from the near zone to the far zone we make the radii R_k of Fig.2.3-2a parallel as shown in Fig.2.3-2b. The circular wavefronts $-2, -1, 0, 1, 2$ become the planar wavefronts $-2, -1, 0, 1, 2$. We are no longer interested in the distance from a point k on the reflecting plane to the reception point P but in the distance from point k to the planar wavefront k through the

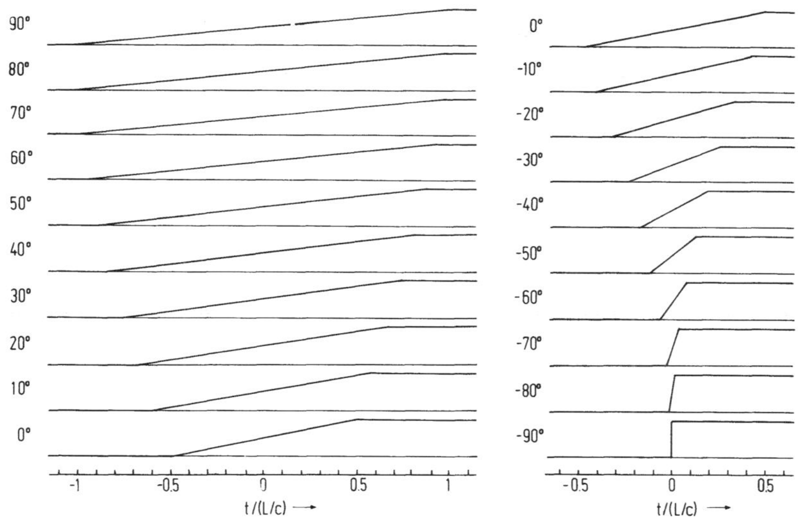

FIG.2.3-5. Waves with the time variation of ramp functions reflected in the directions $\vartheta_r = 90°, 80°, \ldots, -90°$ as defined in Fig.2.3-3 due to an incident wave with the time variation of a step function and angle of incidence $\vartheta_i = 90°$. Note the step function obtained for $\vartheta_r = -90°$. This illustration is a limiting case, but angles of incidence close to 90°—or elevation angles close to 0°—are common in radar.

reception point P. When the wavefront has propagated the distance R_k of Eq.(4) it still has to propagate the additional distance

$$\Delta R_{rk} = k(\Delta R_i + \Delta R_r) = k\Delta L(\sin\vartheta_i + \sin\vartheta_r), \quad \Delta R_r = \Delta L \sin\vartheta_r \tag{6}$$

to reach the planar wavefront 0 through P. The time of arrival at this wavefront becomes

$$t_k = k(\Delta R_i + \Delta R_r)/c = k\Delta L(\sin\vartheta_i + \sin\vartheta_r)/c \tag{7}$$

The step function on top of Fig.2.3-2b can now be plotted. We note that all steps have the same duration $(\Delta R_i + \Delta R_r)/c$ which is not so in Fig.2.3-2a.

Let us point out that the reflection angle $\vartheta_r = -20°$ in Fig.2.3-2b is negative. Hence, we get

$$\Delta R_i = \Delta L \sin\vartheta_i = \Delta L \sin 60°, \quad \Delta R_r = \Delta L \sin\vartheta_i = -\Delta L \sin 20°$$

and we see that ΔR_r becomes negative for reflection angles $0 > \vartheta_r \geq -90°$.

Using the results of Fig.2.3-2 we may redraw Fig.2.3-1 for the far zone as shown in Fig.2.3-3. The waves reflected in the directions $\vartheta_r = -20°$ and $\vartheta_r = 70°$ are similar to those in Fig.2.3-1. The important new result is the wave reflected in the direction $\vartheta_r = -\vartheta_i = -60°$. We get now an undisturbed step wave propagating in the direction demanded by Snell's law due to the relation $\Delta R_i = -\Delta R_r$.

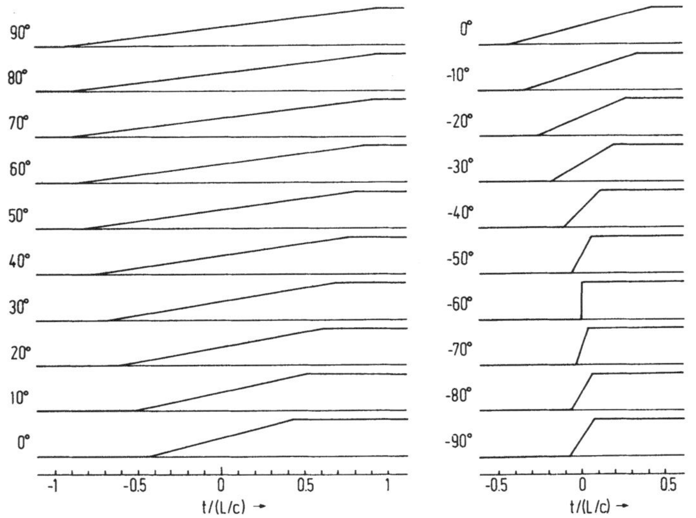

FIG.2.3-6. Waves with the time variation of ramp functions reflected in the directions $\vartheta_r = 90°, 80°, \dots, -90°$ as defined in Fig.2.3-3 due to an incident wave with the time variation of a step function and angle of incidence $\vartheta_i = 60°$. Note the step function obtained for $\vartheta_r = -60°$.

When the distance ΔL is reduced to zero in Figs.2.3-1 to 2.3-3 the reflected waves become continuous. The analytical functions representing these continuous waves are difficult to derive for the near zone but the equality of the width of the steps in the far zone according to Fig.2.3-2b means that we get in the limit $\Delta L \to 0$ the ramp function shown in Fig.2.3-4.

According to Fig.2.3-3 the beginning of the ramp of the function of Fig.2.3-4 comes from the location $x/\Delta L = -4.5$ or $-L/2$, the end of the ramp function from the location $x/\Delta L = +4.5$ or $+L/2$. This suggests the time scale $t/(L/c)$ shown in Fig.2.3-4, which holds for any value of ΔL including $\Delta L \to 0$. From Eq.(7) we get the following times t_b and t_e for the beginning and end of the ramp:

$$\begin{aligned} t_b &= -\frac{4.5\Delta L}{c}\Big(\sin\vartheta_i + \sin\vartheta_r\Big) = -\frac{L}{2c}\Big(\sin\vartheta_i + \sin\vartheta_r\Big) \\ \frac{t_b}{L/c} &= -\frac{1}{2}\Big(\sin\vartheta_i + \sin\vartheta_r\Big) \\ \frac{t_e}{L/c} &= +\frac{1}{2}\Big(\sin\vartheta_i + \sin\vartheta_r\Big) \end{aligned} \qquad (8)$$

Figure 2.3-5 shows the time variation of reflected waves for reflection angles $\vartheta_t = 90°, 80°, \dots, -90°$ due to an incident wave with the time variation of a step function and angle of incidence $\vartheta_i = 90°$. The amplitude of the reflected wave rises to the same value regardless of the angle of reflection. But the time $2L/c$ is required for $\vartheta_r = 90°$, while the rise time is zero for $\vartheta_r = -90° = -\vartheta_i$.

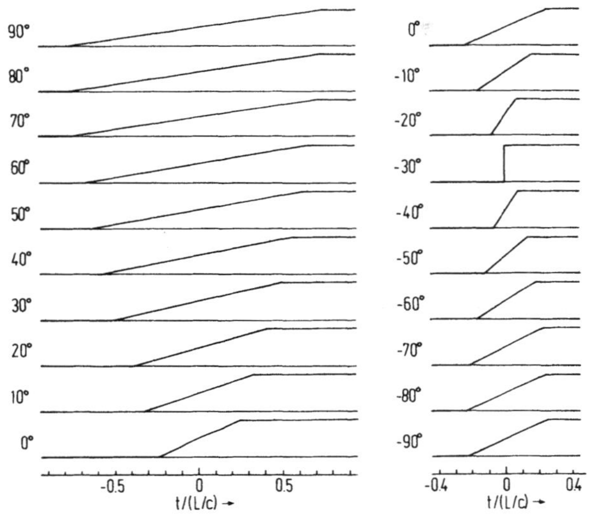

FIG.2.3-7. Waves with the time variation of ramp functions reflected in the directions $\vartheta_r = 90°, 80°, \ldots, -90°$ as defined in Fig.2.3-3 due to an incident wave with the time variation of a step function and angle of incidence $\vartheta_i = 30°$. Note the step function obtained for $\vartheta_r = -30°$.

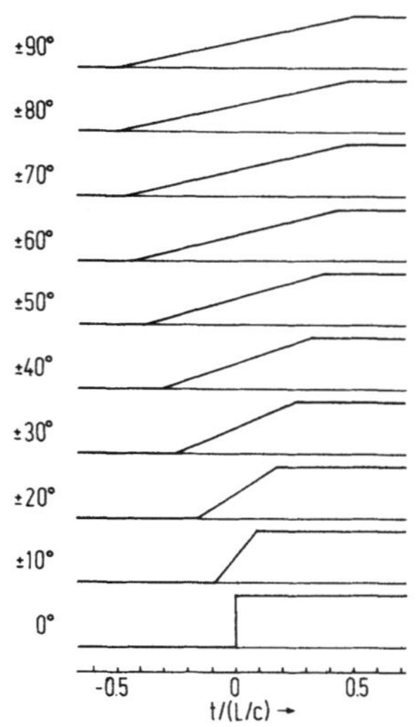

FIG.2.3-8. Waves with the time variation of ramp functions reflected in the directions $\vartheta_r = 90°, 80°, \ldots, -90°$ as defined in Fig.2.3-3 due to an incident wave with the time variation of a step function and angle of incidence $\vartheta_i = 0°$. Note the step function obtained for $\vartheta_r = 0°$.

Consider a rectangular pulse $R(t)$ represented by two step functions $S(t)$ with delay ΔT:

$$R(t) = S(t) - S(t - \Delta T) \tag{9}$$

For $\Delta T \geq 2L/c$ the reflected pulse will reach the same peak amplitude for any value of the reflection angle ϑ_r. For a pulse of duration $\Delta T = L/c$ the angle of reflection will have to be in the approximate sector $10° \geq \vartheta_\mathrm{r} \geq -90°$ to let the reflected pulse reach the full peak amplitude. For the limit $\Delta T \to 0$ a reflection will only occur in the direction $\vartheta_\mathrm{r} = -\vartheta_\mathrm{i}$. Hence, Snell's law of reflection will be satisfied by very short pulses like the Dirac delta function.

Figure 2.3-6 is a repetition of Fig.2.3-5 but the angle of incidence is reduced to $\vartheta_\mathrm{i} = 60°$. The main difference compared with Fig.2.3-5 is that the ramps have become shorter. As expected, a step function occurs now for $\vartheta_\mathrm{r} = -60°$.

Figure 2.3-7 shows corresponding plots for an angle of incidence $\vartheta_\mathrm{i} = 30°$. The longest ramp for $\vartheta_\mathrm{r} = 90°$ is now only about half as long as for $\vartheta_\mathrm{r} = 90°$ in Fig.2.3-5.

Figure 2.3-8 shows more plots, these ones holding for an angle of incidence $\vartheta_\mathrm{i} = 0$. The ramps have become still shorter than in Fig.2.3-7.

Snell's law of reflection plays an important role in the design of *stealth* airplanes. The surfaces of such airplanes are supposed to reflect any incident radar pulses primarily in the direction $\vartheta_\mathrm{r} = -\vartheta_\mathrm{i}$ and not in the direction $\vartheta_\mathrm{r} = \vartheta_\mathrm{i}$ back to the radar that sent the pulses. This works for radar pulses consisting of many cycles of sinusoidal waves as used by the classical radar. We have seen that it would also work for radars using pulses similar to the Dirac delta function. Without proof we mention that signals resembling thermal noise would also be reflected according to Snell's law. But there are evidently plenty of perfectly practical signals that defeat the design of stealth airplanes according to Snell's law of reflection.

We shall investigate in the following section the reflection of rectangular pulses. The energy of such pulses is finite. This makes it possible to compare the energy reflected in different directions. The infinite energy of step waves according to Figs.2.3-5 to 2.3-8 does not permit such a comparison.

2.4 Reflection of Rectangular Pulses

In this section we shall apply the results of Section 2.3 to the reflection of TEM waves having electric field strengths with the time variation of rectangular pulses. The goal is to derive some numerical results for the design of anti-stealth radars.

First we develop another derivation of the ramp function of Fig.2.3-4. Figure 2.4-1a shows a reflector with length L in the direction of the x-axis and in principle infinite extension in the direction of the y-axis. A planar wavefront of a TEM wave with field strengths $\mathbf{E}(x, z, t)$ and $\mathbf{H}(x, z, t)$, having perpendicular polarization, angle of incidence ϑ_i, and the time variation of a step function shall arrive at the center $x/L = 0$ of the reflector at the time $t = t_\mathrm{i}$. Consider the wave reflected at $x/L = 0$ with the angle of reflection ϑ_r at a large distance where the reflected wave has practically a planar wavefront. The time to reach this distant wavefront from $x/L = 0$ is denoted t_r so that the total time becomes $t = t_\mathrm{i} + t_\mathrm{r}$. A wave reflected at $x/L = -1/2$ will have to propagate a distance shorter by $(L/2)\sin\vartheta_\mathrm{i}$ to reach the reflector and a distance shorter by $(L/2)\sin\vartheta_\mathrm{r}$ to reach the reflected wavefront. Hence, it will reach the wavefront at the time

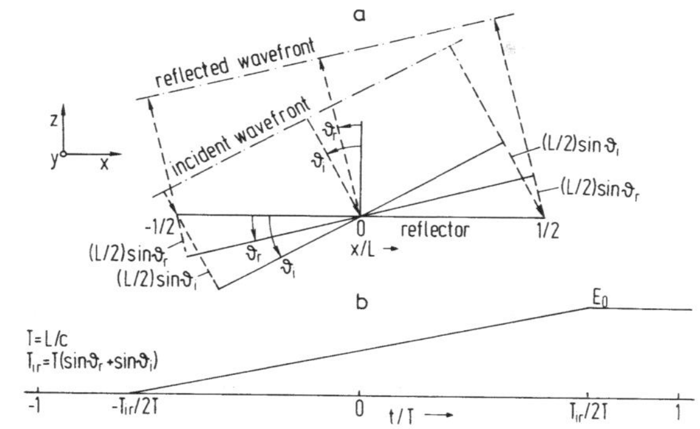

FIG.2.4-1. General derivation of the ramp function of Fig.2.3-4 produced by the reflection of a step function at a reflector of length L with angles of incidence ϑ_i and reflection ϑ_r. Note that the angles are measured from $+90°$ to $-90°$ as shown in Fig.2.3-3.

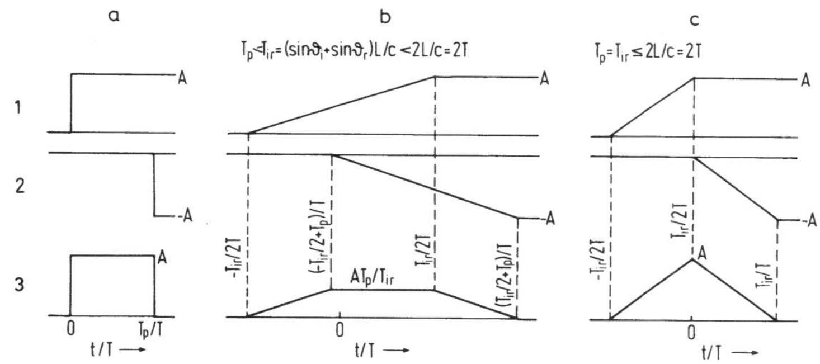

FIG.2.4-2. Transition from a step wave to a rectangular pulse wave. (a) incident wave; (b) reflected wave for $T_p < T_{ir}$; (c) reflected wave for $T_p = T_{ir}$.

$$t = t_i + t_r - T_{ir}/2$$
$$T_{ir} = T(\sin\vartheta_i + \sin\vartheta_r), \quad T = L/c \tag{1}$$

On the other hand, a wave reflected at $x/L = +1/2$ will have to propagate the additional distances $(L/2)\sin\vartheta_i$ and $(L/2)\sin\vartheta_r$ to reach the reflected wavefront at the time

$$t = t_i + t_r + T_{ir}/2 \tag{2}$$

If we normalize our time scale so that $t_i + t_r = 0$ we get the increase of the electric field strength as function of time at the reflected wavefront shown in Fig.2.4-1b. The field strength begins to rise at $t = -T_{ir}/2$ and rises linearly until $t = +T_{ir}/2$; from then on it has a constant value denoted E_0. We observe that the ramp function becomes equal to that of Fig.2.3-4 for $T = T_{ir}$ or $\sin\vartheta_i + \sin\vartheta_r = 1$.

The transition from a step wave to a rectangular pulse wave is made in Fig.2.4-2a. The sum of two step functions $AS(t)$ and $-AS(t - T_p)$ produces a rectangular pulse of duration T_p having the amplitude A.

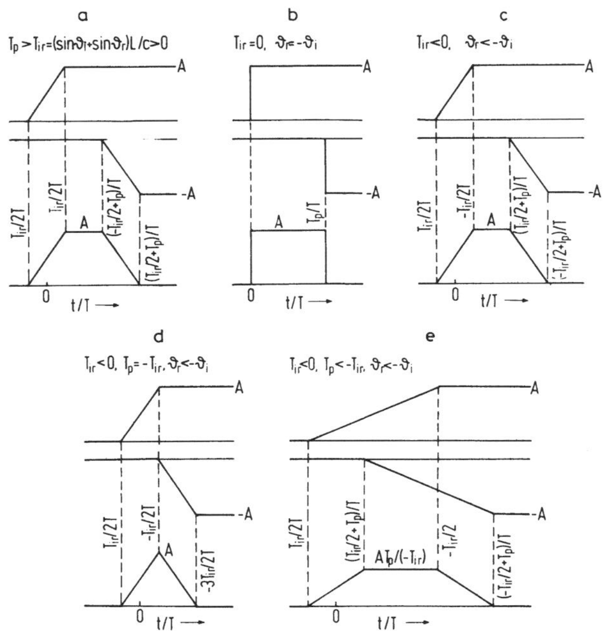

FIG.2.4-3. Time variation of a reflected rectangular pulse. (a) $T_p > T_{ir} > 0$; (b) $T_{ir} = 0$; (c) $T_{ir} < 0$, $T_p > -T_{ir}$; (d) $T_{ir} < 0$, $T_p = -T_{ir}$; (e) $T_{ir} < 0$, $T_p < -T_{ir}$.

Let the duration of the rectangular pulse be shorter than T_{ir}. This is possible only for sufficiently short pulses, $T_p < 2T$, since $\sin\vartheta_i$ and $\sin\vartheta_r$ cannot be larger than 1. The step functions of Fig.2.4-2a become the ramp functions of Fig.2.4-2b if we use the result of Fig.2.4-1b that the ramp function rises in the time interval $-T_{ir}/2 \leq t \leq T_{ir}/2$. The rectangular pulse of Fig.2.4-2a becomes the trapezoidal pulse with amplitude AT_p/T_{ir} of Fig.2.4-2b.

For a fixed angle of incidence ϑ_i but ϑ_r varying from $+90°$ to $-90°$ the time T_{ir} will drop until the condition

$$
\begin{aligned}
-T_{ir}/2 + T_p &= T_{ir}/2 \\
T_p &= T_{ir} = T(\sin\vartheta_i + \sin\vartheta_r) \\
\sin\vartheta_r &= T_p/T - \sin\vartheta_i, \quad T = L/c
\end{aligned}
\tag{3}
$$

is satisfied. This special case is shown in Fig.2.4-2c. The rectangular pulse with amplitude A has become a triangular pulse with amplitude A.

When ϑ_r keeps dropping toward $-90°$ we reach the general case $T_p > T_{ir}$ shown in Fig.2.4-3a, which applies to rectangular pulses of any duration T_p. We get again a trapezoidal pulse but it has now the amplitude A rather than AT_p/T_{ir} as in Fig.2.4-2b.

As ϑ_r decreases further toward $-90°$ it reaches the value $\vartheta_r = -\vartheta_i$ for which Fig.2.4-3b and $T_{ir} = 0$ applies. The electric field strength of the reflected wave has now the same time variation as that of the incident wave in Fig.2.4-2a.

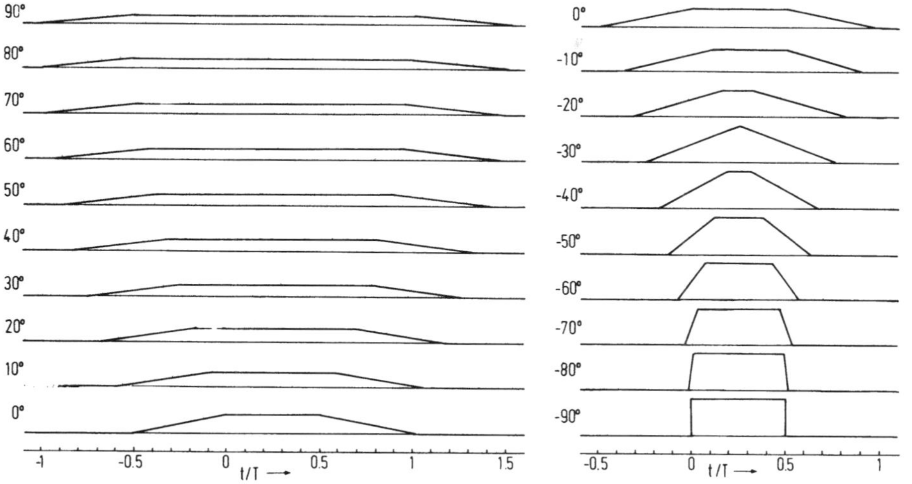

FIG.2.4-4. Pulses reflected into the directions $\vartheta_r = 90°, 80°, \ldots, -90°$ due to a rectangular pulse with angle of incidence $\vartheta_i = 90°$ and a duration $T_p = T/2 = L/2c$. See Fig.2.3-3 for the definition of ϑ_i and ϑ_r.

For $\vartheta_r < -\vartheta_i$ we get again a trapezoidal pulse as shown in Fig.2.4-3c. Compared with the trapezoidal pulse of Fig.2.4-3a we see an exchange of $-T_{ir}$ and $+T_{ir}$. However, T_{ir} is now negative so that $T_{ir}/2T$ is smaller than $-T_{ir}/T$!

For still smaller values of ϑ_r we reach the case $T_p = -T_{ir}$ shown in Fig.2.4-3d. We note once more that T_{ir} is negative and $-T_{ir}$ positive. Using the definition of T_{ir} in Eq.(1) and $T_p = -T_{ir}$ we obtain the value of ϑ_r for which Fig.2.4-3d applies:

$$\sin\vartheta_r = -T_p/T - \sin\vartheta_i \tag{4}$$

As ϑ_r decreases further we get the general case of Fig.2.4-3e that corresponds to the one of Fig.2.4-2b. The cases of Figs.2.4-3a and c occur for any values of ϑ_i and T_p. All the other cases require sufficiently small values of T_p.

Using the results of Figs.2.4-2 and 2.4-3 we can plot the time variation of reflected rectangular pulses. Figure 2.4-4 holds for an incident rectangular pulse with duration $T_p = T/2 = L/2c$ and angle of incidence $\vartheta_i = 90°$ as defined in Fig.2.3-3. The reflected pulses are shown for $\vartheta_r = 90°, 80°, \ldots, -90°$. The angle of incidence $\vartheta_i = 90°$ is a limit case but angles close to $\vartheta_i = 90°$ are important since they correspond to elevation angles close to 0° in terms of radar and search radars typically operate at low elevation angles.

The trapezoidal pulses for $90° \le \vartheta \le -20°$ in Fig.2.4-4 correspond to the trapezoidal pulse in Fig.2.4-2b. From Eq.(3) we get for $T_p = T/2$ and $\sin\vartheta_i = 1$

$$\begin{aligned}\sin\vartheta_r &= -0.5\\ \vartheta_r &= -30°\end{aligned} \tag{5}$$

The triangular pulse corresponding to the one in Fig.2.4-2c is shown for $\vartheta_r = -30°$ in Fig.2.4-4. For $\vartheta_r = -40°, -50°, \ldots, -80°$ the pulses in Fig.2.4-4 correspond to

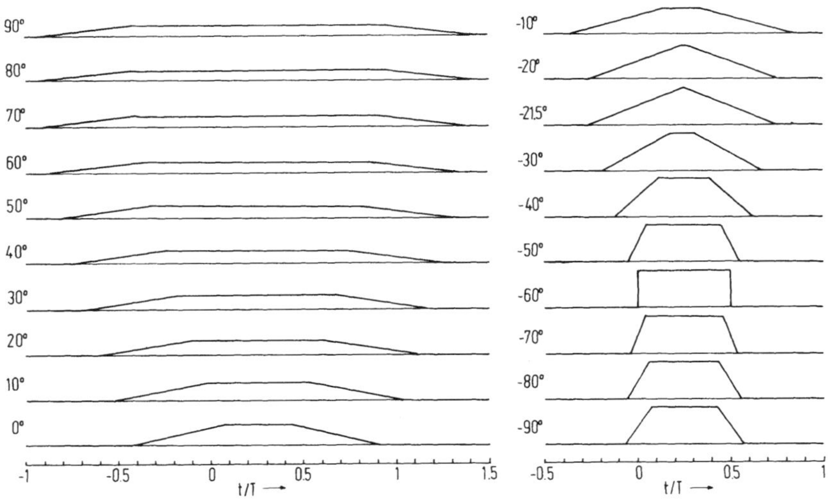

FIG.2.4-5. Pulses reflected into the directions $\vartheta_r = 90°, 80°, \ldots, -90°$ due to a rectangular pulse with angle of incidence $\vartheta_i = 60°$ and a duration $T_p = T/2 = L/2c$. See Fig.2.3-3 for the definition of ϑ_i and ϑ_r.

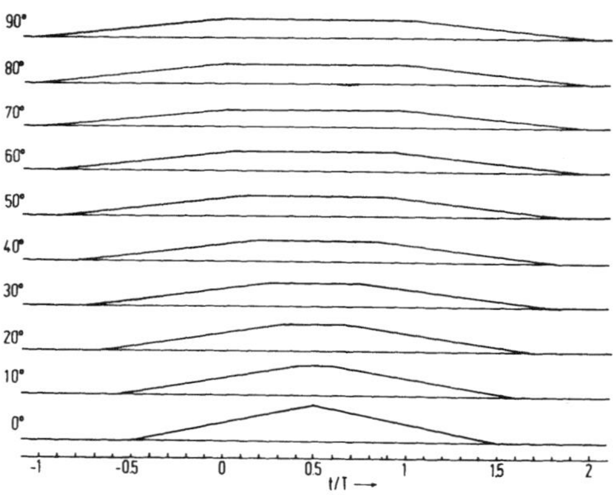

FIG.2.4-6. Pulses reflected into the directions $\vartheta_r = 90°, 80°, \ldots, 0°$ due to a rectangular pulse with angle of incidence $\vartheta_i = 90°$ and a duration $T_p = T = L/c$. See Fig.2.3-3 for the definition of ϑ_i and ϑ_r.

the trapezoidal pulse in Fig.2.4-3a. Finally, for $\vartheta_r = -90°$ we get the rectangular pulse of Fig.2.4-3b. The case $\vartheta_r < -\vartheta_i$ of Fig.2.4-3c is not possible for $\vartheta_i = 90°$.

Figure 2.4-5 is a repetition of Fig.2.4-4 for an angle of incidence $\vartheta_i = 60°$. For the reflection direction of the triangular pulse we get from Eq.(3)

$$\begin{aligned} \sin\vartheta_r &= 0.5 - \sin 60° = -0.3660 \\ \vartheta_r &= -21.47° \end{aligned} \tag{6}$$

The undistorted pulse is reflected in the direction $\vartheta = -60°$. The pulses for $\vartheta_r = -70°$, $-80°$, and $-90°$ correspond to the trapezoidal pulse of Fig.2.4-3c.

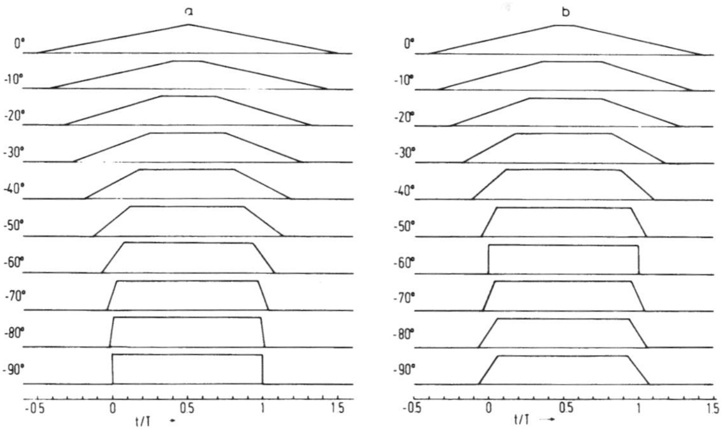

FIG.2.4-7. Pulses reflected into the directions $\vartheta_r = 90°, 80°, \ldots, 7.7°$ due to a rectangular pulse with angle of incidence $\vartheta_i = 60°$ and a duration $T_p = T = L/c$. See Fig.2.3-3 for the definition of ϑ_i and ϑ_r.

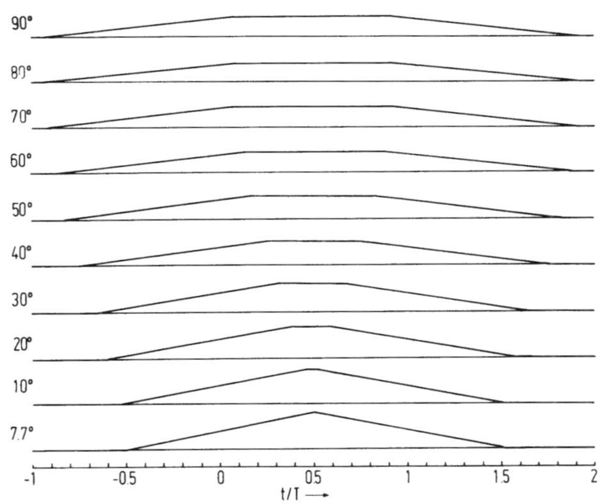

FIG.2.4-8. Pulses reflected into the direction $\vartheta_r = 0°, -10°, \ldots, -90°$ due to a rectangular pulse with angle of incidence $\vartheta_i = 90°$ (a) or $\vartheta_i = 60°$ (b) and a duration $T_p = T = L/c$.

In order to see how the duration T_p of the incident pulse affects the reflected pulses we repeat Figs.2.4-4 and 2.4-5 in Figs.2.4-6 to 2.4-8 for $T_p = T$ and in Figs.2.4-9 to 2.4-12 for $T_p = 2T$. The scale in Figs.2.4-4 to 2.4-12 is the same in order to make the comparison easier, but this results in such wide illustrations that the reflection angles $\vartheta_r = 90°, 80°, \ldots, 0°$ have to be shown separated from the reflection angles $\vartheta_r = 0°, -10°, \ldots, -90°$.

Using once more Eq.(3) we find that the triangular pulse in Fig.2.4-6 occurs for

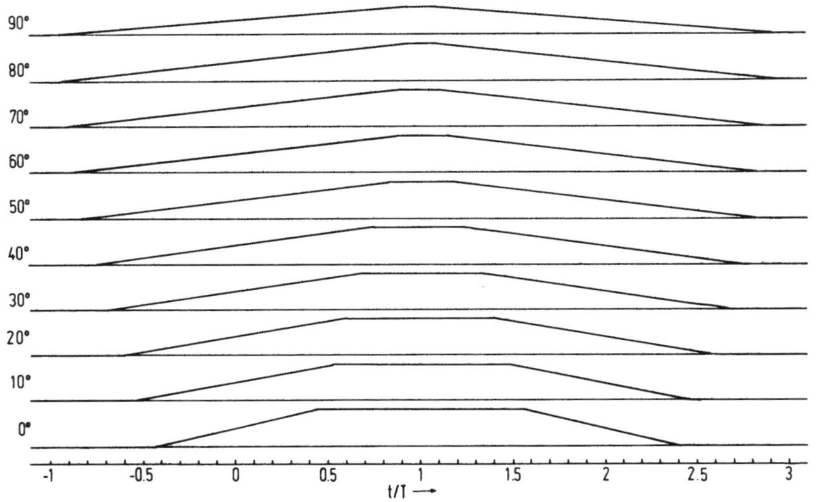

FIG.2.4-9. Pulses reflected into the direction $\vartheta_r = 90°, 80°, \ldots, 0°$ due to a rectangular pulse with angle of incidence $\vartheta_i = 90°$ and a duration $T_p = 2T = 2L/c$.

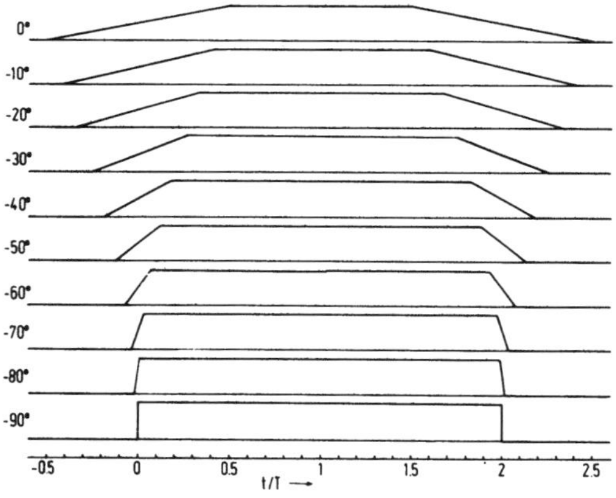

FIG.2.4-10. Pulses reflected into the direction $\vartheta_r = 0°, -10°, \ldots, -90°$ due to a rectangular pulse with angle of incidence $\vartheta_i = 90°$ and a duration $T_p = 2T = 2L/c$.

$$\begin{aligned} \sin\vartheta_r &= 1 - \sin 90° = 0 \\ \vartheta_r &= 0° \end{aligned} \tag{7}$$

while in Fig.2.4-7 it occurs for

$$\begin{aligned} \sin\vartheta_r &= 1 - \sin 60° = 0.1340 \\ \vartheta_r &= 7.70° \end{aligned} \tag{8}$$

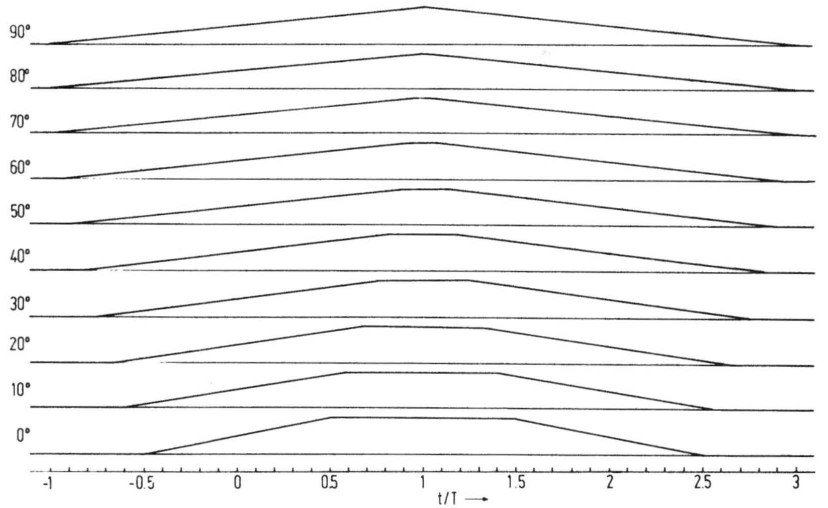

FIG.2.4-11. Pulses reflected into the direction $\vartheta_r = 90°, 80°, \ldots, 0°$ due to a rectangular pulse with angle of incidence $\vartheta_i = 60°$ and a duration $T_p = 2T = 2L/c$.

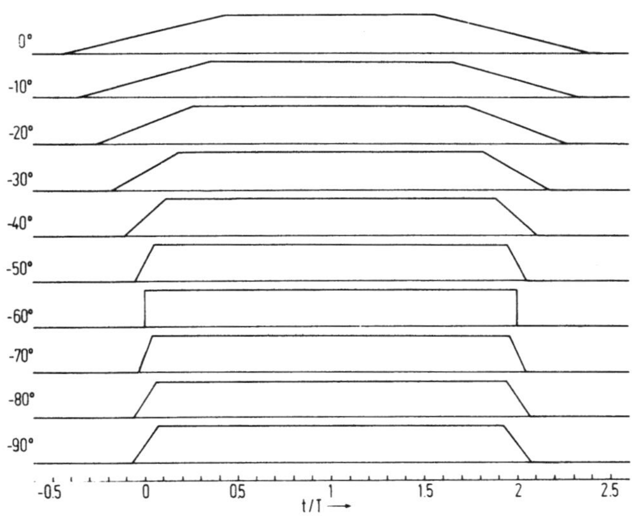

FIG.2.4-12. Pulses reflected into the direction $\vartheta_r = 0°, -10°, \ldots, -90°$ due to a rectangular pulse with angle of incidence $\vartheta_i = 60°$ and a duration $T_p = 2T = 2L/c$.

Nothing new occurs in Fig.2.4-8, which holds for an angle of reflection $0° \geq \vartheta_r \geq -90°$ and angles of incidence $\vartheta_i = 90°$ or $\vartheta_i = 60°$.

A further extension to incident pulses of duration $T_p = 2T = 2L/c$ is shown in Figs.2.4-9 to 2.4-12. The triangular pulse in Fig.2.4-9 occurs now for

$$\begin{aligned} \sin\vartheta_r &= 2 - \sin 90° = 1 \\ \vartheta_r &= 90° \end{aligned} \tag{9}$$

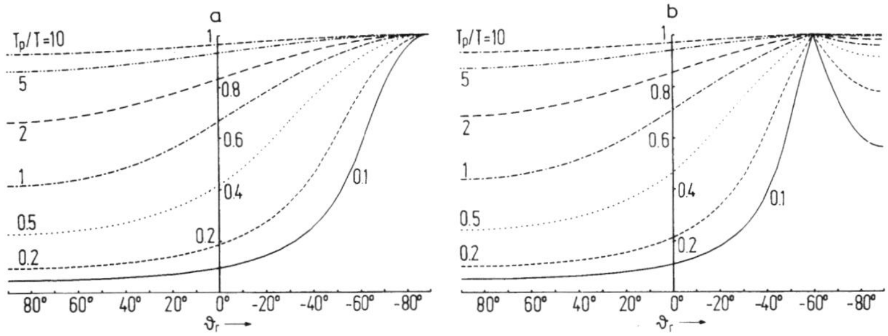

FIG.2.4-13. Plots of the reflected relative energy $w(T_p/T, \vartheta_i, \vartheta_r)$ according to Eq.(16) as function of the reflection angle ϑ_r for incident angles $\vartheta_i = 90°$ (a) and 60° (b) for rectangular pulses with duration $T_p/T = 0.1$, 0.2, 0.5, 1, 2, 5, 10.

while the condition $2 - \sin\vartheta_i \leq 1$ cannot be satisfied for $\vartheta_i = 60°$; the cases of Fig.2.4-2b and c do not occur for $T_p = 2T$ and $\vartheta_i = 60°$.

We turn to the calculation of the energy of the reflected pulses relative to the undistorted pulse reflected according to Snell's law $\vartheta_r = -\vartheta_i$. According to Fig.2.4-3b we define the nominal reference energy

$$W_0 = A^2 T_p \tag{10}$$

The nominal energy of a trapezoidal pulse according to Fig.2.4-2b becomes

$$\begin{aligned} W_1 &= 2\int_0^{T_p} \left(\frac{AT_p}{T_{ir}}\right)^2 \frac{t}{T_p}\,dt + \left(\frac{AT_p}{T_{ir}}\right)^2 (T_{ir} - T_p) \\ &= A^2 T_p \frac{T_p}{T_{ir}}\left(1 - \frac{1}{3}\frac{T_p}{T_{ir}}\right) \quad \text{for } T_p \leq T_{ir} = T(\sin\vartheta_i + \sin\vartheta_r) \end{aligned} \tag{11}$$

and we obtain the relative energy w_1:

$$\begin{aligned} w_1 = \frac{W_1}{W_0} = \frac{T_p}{T_{ir}}\left(1 - \frac{1}{3}\frac{T_p}{T_{ir}}\right) = \frac{T_p}{T(\sin\vartheta_i + \sin\vartheta_r)}\left(1 - \frac{1}{3}\frac{T_p}{T(\sin\vartheta_i + \sin\vartheta_r)}\right) \\ \text{for } 1 \geq \sin\vartheta_r \geq \frac{T_p}{T} - \sin\vartheta_i, \ \vartheta_r > -\vartheta_i \end{aligned} \tag{12}$$

For the relative energy of the trapezoidal pulses in Figs.2.4-3a, c, and e we get in analogy:

$$\begin{aligned} w_2 = 1 - \frac{1}{3}\frac{T}{T_p}(\sin\vartheta_i + \sin\vartheta_r) \\ \text{for } \frac{T_p}{T} - \sin\vartheta_i \geq \sin\vartheta_r \geq -\sin\vartheta_i, \ \vartheta_r > -\vartheta_i \end{aligned} \tag{13}$$

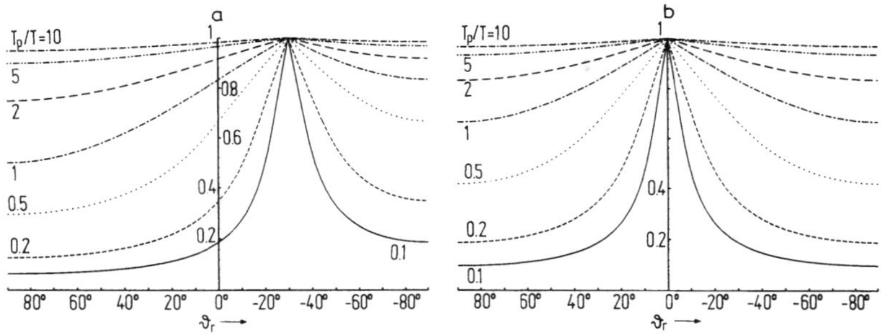

FIG.2.4-14. Plots of the reflected relative energy $w(T_p/T, \vartheta_i, \vartheta_r)$ according to Eq.(16) as function of the reflection angle ϑ_r for incident angles $\vartheta_i = 30°$ (a) and $0°$ (b) for rectangular pulses with duration $T_p/T = 0.1, 0.2, 0.5, 1, 2, 5, 10$.

$$w_3 = 1 + \frac{1}{3}\frac{T}{T_p}(\sin\vartheta_i + \sin\vartheta_r)$$

$$\text{for } -\sin\vartheta_i \geq \sin\vartheta_r \geq -\frac{T_p}{T} - \sin\vartheta_i,\ \vartheta_r < -\vartheta_i \quad (14)$$

$$w_4 = \frac{T_p}{-T(\sin\vartheta_i + \sin\vartheta_r)}\left(1 - \frac{1}{3}\frac{T_p}{-T(\sin\vartheta_i + \sin\vartheta_r)}\right)$$

$$\text{for } -\frac{T_p}{T} - \sin\vartheta_i \geq \sin\vartheta_r \geq -1,\ \vartheta_r < -\vartheta_i \quad (15)$$

We denote with $w(T_p/T, \vartheta_i, \vartheta_r)$ the relative energy for the whole range $1 \geq \sin\vartheta_r \geq -1$:

$$\begin{aligned}
w(T_p/T, \vartheta_i, \vartheta_r) = & \\
= w_1 \quad \text{for} \quad & 1 \geq \sin\vartheta_r \geq \frac{T_p}{T} - \sin\vartheta_i, \quad \vartheta_r > -\vartheta_i \\
= w_2 \quad \text{for} \quad & \frac{T_p}{T} - \sin\vartheta_i \geq \sin\vartheta_r \geq -\sin\vartheta_i, \quad \vartheta_r > -\vartheta_i \\
= w_3 \quad \text{for} \quad & -\sin\vartheta_i \geq \sin\vartheta_r \geq -\frac{T_p}{T} - \sin\vartheta_i, \quad \vartheta_r < -\vartheta_i \\
= w_4 \quad \text{for} \quad & -\frac{T_p}{T} - \sin\vartheta_i \geq \sin\vartheta_r \geq -1, \quad \vartheta_r < -\vartheta_i \qquad (16)
\end{aligned}$$

Plots of $w(T_p/T, \vartheta_i, \vartheta_r)$ are shown for $\vartheta_i = 90°$ and $60°$ for various values of T_p/T in Fig.2.4-13 while Fig.2.4-14 shows corresponding plots for $\vartheta_i = 30°$ and $0°$. We note that for $T_p/T = 10$ the reflected energy is essentially independent of the angle of reflection ϑ_r. The smallest relative energy occurs at $\vartheta_r = 90°$ for all values $\vartheta_i = 90°$, $60°$, $30°$, and $0°$. For decreasing values of T_p/T we see that the reflected energy becomes more and more concentrated in the direction $\vartheta_r = -\vartheta_i$ as demanded by Snell's law.

Table 2.4-1 lists the relative energy $w(T_p/T, \vartheta_i, \vartheta_r = \vartheta_i)$ reflected back to a radar that produces an incident wave with angle of incidence $\vartheta_i = 90°$ or $60°$ and various values of T_p/T. The value of w is also shown in decibel, since this is the

TABLE 2.4-1

RELATIVE ENERGY $w(T_p/T, \sin\vartheta_i, \vartheta_r = \vartheta_i)$ REFLECTED BACK IN THE DIRECTION OF THE INCIDENT WAVE ACCORDING TO EQ.(16) FOR $\vartheta_i = 90°$ OR $60°$ AND VARIOUS VALUES OF T_p/T.

	$\vartheta_i = 90°$		$\vartheta_i = 60°$	
T_p/T	$w(T_p/T, 90°, 90°)$	w [dB]	$w(T_p/T, 60°, 60°)$	w [dB]
10	0.933	−0.30	0.942	−0.26
5	0.867	−0.62	0.885	−0.53
2	0.667	−1.76	0.711	−1.48
1	0.417	−3.80	0.466	−3.31
0.5	0229	−6.40	0.261	−5.84
0.2	0.0967	−10.15	0.111	−9.55
0.1	0.0492	−13.08	0.0566	−12.47

preferred measure of the radar engineer. It is evident that the reduction of reflected energy in the direction $\vartheta_r = \vartheta_i$ compared with the direction $\vartheta_r = -\vartheta_i$ is insignificant for $T_p/T = 10$, 5, and 2. Since T_p represents in essence a radar range cell and T the size of the target, we may conclude that the use of reflection to decrease the energy returned by a target toward the radar does not work as counter measure against a carrier-free radar if a range cell about twice as large as the target is acceptable. Range cells of this or larger size are generally used for search radars. A tracking radar may need shorter range cells, if details of the target are to be resolved, but this exception applies only to radars that use pulses with a duration of 10 ns or less, in which case the absorption losses due to rain or fog above 10 GHz limit the usefulness of the classical radar using a sinusoidal carrier with upwards of 100 cycles per pulse[1].

It is worth looking back how the stealth technology came to use the reflection principle for the reduction of the *radar cross section* or the detectability of targets. First, Snell's law of reflection was assumed to hold generally when in reality it applies only to waves with certain time variation. It is not clear why this assumption was ever made since any radar engineer knows that a target becomes a point scatterer rather than a reflector for sinusoidal waves if the frequency is sufficiently low or the wavelength sufficiently long. A second difficult to understand assumption was that electromagnetic waves *must* have the sinusoidal time variation used by the classical radar. There is nothing in Maxwell's equations that supports such an assumption. Those who are more experimental than theoretical minded may look at a lightning discharge and ponder the question why such a discharge should produce electromagnetic waves with sinusoidal time variation.

Sinusoidal waves became distinguished because the technology available at the beginning of radio transmission around 1900—based on coils, capacitors, and res-

[1]Considerable additional effort is required before the principles discussed here can be applied to the design of a radar. For instance, the large noise temperature below 500 MHz favors the use of pulses with a duration of less than 10 ns. One must use coded sequences of such short pulses to obtain plots of the relative reflected energy comparable to the ones for $T_p/T = 2$, 5, 10 in Figs.2.4-13 and 2.4-14. Such sequences of pulses rather than one pulse are also required to permit the radar receiver to discriminate between returned signals that were radiated by its radiator rather than some other radiator.

onating dipoles—favored them. But the technology based on semiconductors favors step functions and rectangular pulses. This point has not sunk in yet after half a century of semiconductor technology. The flat Earth, the geocentric system, and the circular orbits of planets had once a similar grip on our thinking.

The uselessness of absorbing materials, designed to defeat the classical radar using sinusoidal waves, against radars using more sophisticated waveforms has been known publicly since 1981 and more versatile absorbing materials have been developed[2]. In the case of reflection the laws of nature were temporarily suspended by more persuasive facts of life[3].

2.5 Partial Differential Equations for Parallel Polarization

Figure 2.5-1 shows a planar signal wave excited at the plane of excitation, beginning at the time $t = t_0$, and having a signal boundary. The primary difference with Fig.2.1-1 is that $\mathbf{E}_\mathrm{i}$ is now in the plane of incidence and $\mathbf{H}_\mathrm{i}$ points out of this plane in the direction of the viewer. The calculations of Section 2.1 apply again until Eq.(2.1-31) is reached. This equation is now replaced with the help of Fig.2.5-1 by the following relations:

$$\begin{gathered}\mathbf{E}_\mathrm{i}(y_\mathrm{i},t) = E_{\mathrm{i}y}\mathbf{e}_y + E_{\mathrm{i}z}\mathbf{e}_z\\ \mathbf{H}_\mathrm{i}(y_\mathrm{i},t) = H_{\mathrm{i}x}\mathbf{e}_x\\ E_{\mathrm{i}x} = 0,\ E_{\mathrm{i}y} = E_\mathrm{i}(y,z,t)\cos\vartheta_\mathrm{i},\ E_{\mathrm{i}z} = E_\mathrm{i}(y,z,t)\sin\vartheta_\mathrm{i}\\ H_{\mathrm{i}x} = H_\mathrm{i}(y,z,t),\ H_{\mathrm{i}y} = 0,\ H_{\mathrm{i}z} = 0 \end{gathered}\tag{1}$$

Let this wave reach the boundary plane $x = 0$ in Fig.2.5-1. As in the case of perpendicular polarization we assume again that both media 1 and 2 are homogeneous, isotropic, and time-invariant. As in the case of perpendicular polarization, a cylinder wave with axis parallel to the x-axis will be excited in every point y of

[2] Natio and Takahashi 1989

[3] Historic note: The first papers challenging the dominant role of periodic sinusoidal waves in communication and radar were published in 1960 (Cook 1960, Harmuth 1960). The development of the carrier-free ground-probing radar by C.E.Cook, R.M.Morey (1974), J.Chapman, and others started at that time. Geophysical Survey Systems Inc. was founded in 1970 for the commercial production of such ground-probing radars. From 1981 on the problems caused by the carrier-free radar for the stealth technology were publicly discussed (Harmuth 1981, footnote p. 46). This created enormous opposition to the further development of the carrier-free technology. Like all attempts to prevent advancement of science and knowledge it achieved a delay of progress at great cost, particularly since the time honored method of burning on the stake no longer enjoyed popular support. In 1995 it became publicly known that an *Accelerated Initiative on Ultrawideband Electromagnetics and Signals* had been initiated at the Radar Division of Naval Research Laboratory. Work was also started at the Army Research Laboratory and Eglin Air Force Base Wright Laboratory (Ground probing radar triggering deep penetrating bombs for the efficient clearing of air raid shelters; Brinson, Min, and Willis Jr., 1996). Results of this work were presented at a meeting at Naval Research Laboratory in Washington, DC on 5 December 1995 by C.Baum, K.Gerlach, J.Hansen, M.Kragaloff, J.McCorkle, K.Min, E.Mokole, M.Steiner, and T.Tice. This was 35 years after the first publications in 1960. 'Ultrawideband' is one of the synonyms for 'nonsinusoidal', 'carrier-free', or 'large relative bandwidth'. A book by Shvartsburg (1996) provides information about developments in the Soviet Union.

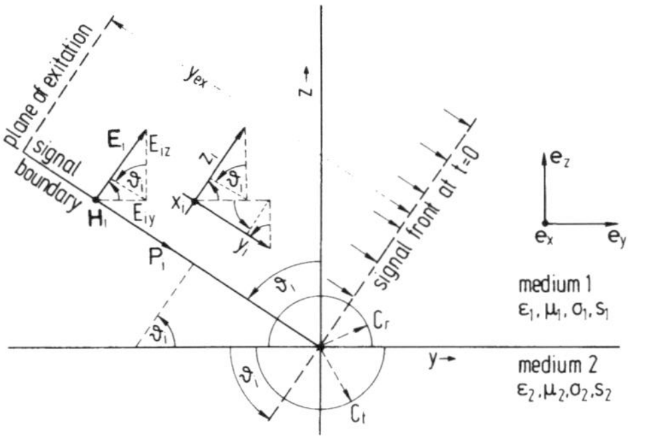

FIG.2.5-1. Reflection and transmission of a planar TEM signal wave with parallel, linear polarization at a boundary plane between two media 1 and 2.

the boundary plane. No electric field strength in the direction of x and no magnetic field strengths in the directions of y and z will be excited due to Eq.(1):

$$E_x = E_x(y,0,t) = 0,\; H_y = H_y(y,0,t) = 0,\; H_z = H_z(y,0,t) = 0 \tag{2}$$

The following relations are obtained with these three conditions from Eqs.(2.1-1)–(2.1-8). The relations hold both in medium 1 and 2; no indices 1 and 2 are written for this reason:

$$0 = 0 \tag{3}$$

$$\frac{\partial H_x}{\partial z} = \epsilon\frac{\partial E_y}{\partial t} + \sigma E_y \tag{4}$$

$$-\frac{\partial H_x}{\partial y} = \epsilon\frac{\partial E_z}{\partial t} + \sigma E_z \tag{5}$$

$$-\frac{\partial E_z}{\partial y} + \frac{\partial E_y}{\partial z} = \mu\frac{\partial H_x}{\partial t} + sH_x \tag{6}$$

$$\frac{\partial E_z}{\partial x} = 0 \tag{7}$$

$$-\frac{\partial E_y}{\partial x} = 0 \tag{8}$$

$$\frac{\partial E_y}{\partial y} + \frac{\partial E_z}{\partial z} = 0 \tag{9}$$

$$\frac{\partial H_x}{\partial x} = 0 \tag{10}$$

From Eqs.(2), (7), (8), and (10) we obtain relations that are always satisfied by cylinder or planar waves:

$$\partial E_x/\partial x = \partial E_y/\partial x = \partial E_z/\partial x = 0 \tag{11}$$

$$\partial H_x/\partial x = \partial H_y/\partial x = \partial H_z/\partial x = 0 \tag{12}$$

Equations (4) and (5) are equal for TEM waves. One may be left out. To show this we use again Eq.(2.1-43)

$$\frac{dz}{dy} = \operatorname{tg} \vartheta_{\mathrm{r}} \tag{13}$$

derived from Fig.2.1-2a. Instead of Fig.2.1-2b we use now Fig.2.5-2 which shows the vectors $\mathbf{E}_{\mathrm{r}}$ and $\mathbf{H}_{\mathrm{r}}$ of a parallel polarized TEM wave reflected in the direction of the angle ϑ_{r}. The relation

$$\frac{E_{\mathrm{r}z}}{E_{\mathrm{r}y}} = \frac{E_z}{E_y} = -\operatorname{tg} \vartheta_{\mathrm{r}} \tag{14}$$

is readily recognizable. Substitution of Eqs.(13) and (14) into Eq.(5) yields

$$-\frac{\partial H_{\mathrm{x}}}{\partial y} = -\frac{\partial H_{\mathrm{x}}}{\partial z}\frac{\partial z}{\partial y} = -\frac{\partial H_x}{\partial y} \operatorname{tg} \vartheta_{\mathrm{r}} = -\left(\epsilon \frac{\partial E_{\mathrm{y}}}{\partial t} + \sigma E_y\right) \operatorname{tg} \vartheta_{\mathrm{r}} \tag{15}$$

which is Eq.(4). Furthermore, Eq.(9) is automatically satisfied for a TEM wave as can be seen by substituting Eqs.(13) and (14):

$$\frac{\partial E_y}{\partial y} + \frac{\partial E_z}{\partial z} = \frac{\partial E_y}{\partial z}\frac{\partial z}{\partial y} + \frac{\partial E_z}{\partial z} = \frac{\partial E_{\mathrm{y}}}{\partial z} \operatorname{tg} \vartheta_{\mathrm{r}} + \frac{\partial E_y}{\partial z} \operatorname{tg} \vartheta_{\mathrm{r}} = 0 \tag{16}$$

Only Eqs.(4) and (6) remain for a cylinder TEM wave. Using Eqs.(13) and (14) we can eliminate E_z in Eq.(6). We add the index r to indicate a reflected wave and write ϵ_1, μ_1, σ_1, s_1 to indicate that the wave propagates in medium 1 in Fig.2.5-1:

$$\frac{1}{\cos^2 \vartheta_{\mathrm{r}}} \frac{\partial E_{\mathrm{r}y}}{\partial z} = \mu_1 \frac{\partial H_{\mathrm{r}x}}{\partial t} + s_1 H_{\mathrm{r}x} \tag{17}$$

$$\frac{\partial H_{\mathrm{r}x}}{\partial z} = \epsilon_1 \frac{\partial E_{\mathrm{r}y}}{\partial x} + \sigma_1 E_{\mathrm{r}y} \tag{18}$$

Figure 2.5-2 shows the relation

$$\frac{E_{\mathrm{t}z}}{E_{\mathrm{t}y}} = \operatorname{tg} \vartheta_{\mathrm{t}} \tag{19}$$

Using furthermore Eq.(2.1-50) we may write the transmitted wave with the index t and the material constants ϵ_2, μ_2, σ_2, s_2:

$$\frac{1}{\cos^2 \vartheta_{\mathrm{t}}} \frac{\partial E_{\mathrm{t}y}}{\partial z} = \mu_2 \frac{\partial H_{\mathrm{t}x}}{\partial t} + s_2 H_{\mathrm{t}x} \tag{20}$$

$$\frac{\partial H_{\mathrm{t}x}}{\partial z} = \epsilon_2 \frac{\partial E_{\mathrm{t}y}}{\partial t} + \sigma_2 E_{\mathrm{t}y} \tag{21}$$

We have the four unknowns $E_{\mathrm{r}y}$, $H_{\mathrm{r}x}$, $E_{\mathrm{t}y}$, and $H_{\mathrm{t}x}$ in the four Eqs.(17), (18), (20), and (21) as well as the two angles ϑ_{r} and ϑ_{t}. Two additional equations are

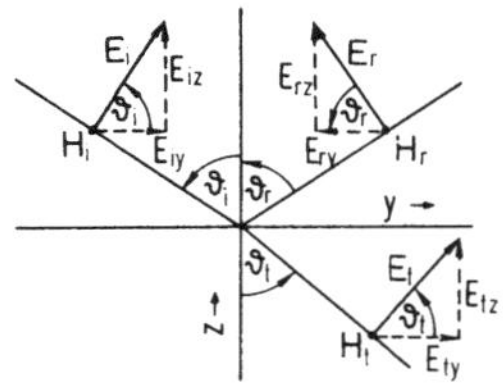

FIG.2.5-2. The relations $E_{rz} = -E_{ry}\,\mathrm{tg}\,\vartheta_r$ and $E_{tz} = E_{ty}\,\mathrm{tg}\,\vartheta_t$ hold for cylinder TEM waves with parallel polarization for any value of ϑ_r and ϑ_t.

obtained from boundary conditions that must be met at the boundary plane $z = 0$ in Fig.2.5-1:

(a) The tangential components ($\mathbf{e}_y$) of the electric field strengths in both media must be equal at $z = 0$:

$$E_i(y, z=0, t)\cos\vartheta_i = E_{iy}(y, z=0, t) = -E_{ry}(y, z=0, t) + E_{ty}(y, z=0, t) \quad (22)$$

(b) The tangential components ($\mathbf{e}_x$) of the magnetic field strengths in both media must be equal at $z = 0$:

$$H_i(y, z=0, t) = H_{ix}(y, z=0, t) = H_{rx}(y, z=0, t) + H_{tx}(y, z=0, t) \quad (23)$$

If we make the substitutions

$$\begin{array}{lll} -H_{iy} \to E_{iy} & & E_{ix} \to H_{ix} \\ -H_{ry} \to E_{ry}, & -H_{rz} \to E_{rz}, & E_{rx} \to H_{rx} \\ -H_{ty} \to E_{ty}, & -H_{tz} \to E_{tz}, & E_{tx} \to H_{tx} \end{array}$$

$$\begin{array}{llll} \epsilon_1 \to \mu_1, & \sigma_1 \to s_1, & \mu_1 \to \epsilon_1, & s_1 \to \sigma_1 \\ \epsilon_2 \to \mu_2, & \sigma_2 \to s_2, & \mu_2 \to \epsilon_2, & s_2 \to \sigma_2 \end{array} \quad (24)$$

in Eqs.(2.1-44), (2.1-47)–(2.1-49), and (2.1-51)–(2.1-54) we obtain the following transformed equations:

$$\frac{E_{rz}}{E_{ry}} = -\,\mathrm{tg}\,\vartheta_r \quad (25)$$

$$\frac{1}{\cos^2\vartheta_r}\frac{\partial E_{ry}}{\partial z} = \mu_1\frac{\partial H_{rx}}{\partial t} + \sigma_1 H_{rx} \quad (26)$$

$$\frac{\partial H_{rx}}{\partial z} = \epsilon_1\frac{\partial E_{ry}}{\partial t} + \sigma_1 E_{ry} \quad (27)$$

$$\frac{E_{tz}}{E_{ty}} = \mathrm{tg}\,\vartheta_t \quad (28)$$

$$\frac{1}{\cos^2\vartheta_t}\frac{\partial E_{ty}}{\partial z} = \mu_2\frac{\partial H_{tx}}{\partial t} + s_2 H_{tx} \quad (29)$$

$$\frac{\partial H_{\mathrm{t}x}}{\partial z} = \epsilon_2 \frac{\partial E_{\mathrm{t}y}}{\partial t} + \sigma_2 E_{\mathrm{t}y} \tag{30}$$

$$H_{\mathrm{i}x}(y, z = 0, t) = H_{\mathrm{r}x}(y, z = 0, t) + H_{\mathrm{t}x}(y, z = 0, t) \tag{31}$$

$$E_{\mathrm{i}y}(y, z = 0, t) = -E_{\mathrm{r}y}(y, z = 0, t) + E_{\mathrm{t}y}(y.z = 0, t) \tag{32}$$

A comparison of Eqs.(14), (17)–(21), (23), and (22) with Eqs.(25)–(32) shows that the equations for perpendicular polarization are transformed into the equations for parallel polarization by the substitutions of Eq.(24). Hence, we may restrict ourselves to the analysis of waves with perpendicular polarization and use Eq.(24) to produce the corresponding results for waves with parallel polarization.

2.6 General Polarization of TEM Waves

In Section 2.1, Eqs.(16)–(28) we have touched briefly on the topic of polarization but postponed a more detailed discussion. We return now to this topic.

Consider the electric field strength $\mathbf{E} = \mathbf{E}(x, z, t)$ of a TEM wave propagating in the direction y. We may write it in component form

$$\mathbf{E} = E_x \mathbf{e}_x + E_z \mathbf{e}_z$$
$$E_x = E \cos\chi, \ E_z = E \sin\chi \tag{1}$$

where $\mathbf{e}_x$ and $\mathbf{e}_z$ are unit vectors in the direction of the x- and z-axes. Let us choose $|\mathbf{E}| = E =$ constant and $\chi = \pi/2$ to obtain:

$$\mathbf{E} = E\mathbf{e}_z \tag{2}$$

This field strength is plotted in Fig.2.6-1a. It represents a TEM wave linearly polarized in the direction z. The choice $\chi = 0$ produces the electric field strength

$$\mathbf{E} = E\mathbf{e}_x \tag{3}$$

of a TEM wave linearly polarized in the direction x. An electric field strength according to Eq.(3) is shown in Fig.2.6-1b.

Instead of the plots of Fig.2.6-1 we usually see plots according to Fig.2.6-2 where the magnitude of $|\mathbf{E}|$ has a sinusoidal time variation:

$$|\mathbf{E}| = E = E_0 \sin\omega t \tag{4}$$

We will concentrate here on the angle of polarization of $\mathbf{E}$, which is much easier to study with a constant magnitude $|\mathbf{E}|$ according to Fig.2.6-1 than with a time variable magnitude $|\mathbf{E}|$ according to Fig.2.6-2.

If we choose the polarization angle proportionate to time, $\chi = \chi(t) = \omega t$, and substitute in Eq.(1)

$$E_x = E \sin\omega t, \ E_z = E \cos\omega t \tag{5}$$

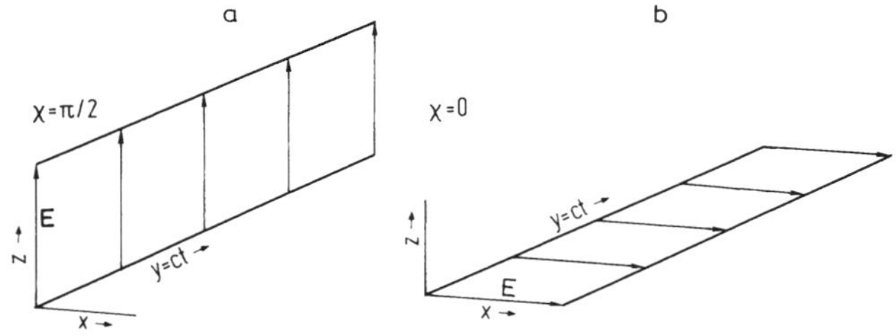

FIG.2.6-1. Linearly polarized TEM waves with the electric field strength $\mathbf{E} = E\mathbf{e}_z$ in the direction z (a) and $\mathbf{E} = E\mathbf{e}_x$ in the direction x (b). The magnitude $|\mathbf{E}| = E$ of the electric field strength is constant.

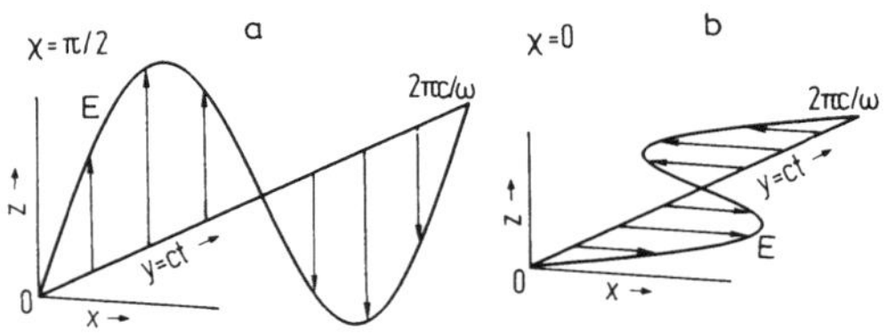

FIG.2.6-2. Linearly polarized TEM waves with the electric field strength $\mathbf{E} = E\mathbf{e}_z$ in the direction z (a) and $\mathbf{E} = E\mathbf{e}_x$ in the direction x (b). The magnitude $|\mathbf{E}| = E = E_0 \sin \omega t$ of the electric field strength has a sinusoidal time variation.

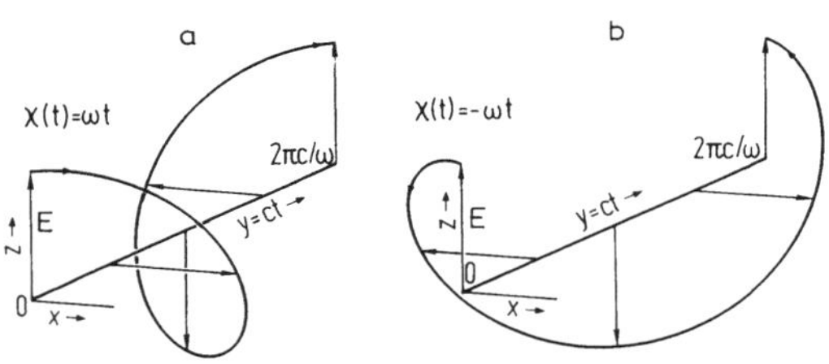

FIG.2.6-3. Electric field strength of a circularly polarized TEM wave having right polarization (a) according to Eq.(5) and left polarization (b) according to Eq.(6), with one cycle in the interval $0 \leq y < 2\pi c/\omega$. The magnitude $|\mathbf{E}|$ of the electric field strength is constant.

we get the plot of Fig.2.6-3a, which represents the electric field strength of a TEM wave with right circular polarization, while the choice $\chi = -\omega t$ produces the plot of Fig.2.6-3b which represents the electric field strength of a TEM wave with left circular polarization:

$$E_x = -E \sin \omega t, \ E_z = E \cos \omega t \tag{6}$$

Figure 2.6-4 shows the electric field strength of a right and a left circularly polarized wave as in Fig.2.6-3 but the interval $0 \leq y < 4\pi c/\omega$ is twice as large as in Fig.2.6-3. The similarity between right and left polarization is more evident than in Fig.2.6-3.

In order to see the effect of a variable magnitude $|\mathbf{E}|$ of the electric field strength of circularly polarized waves we replace E in Eqs.(5) and (6) by $E_0 \cos \omega t$:

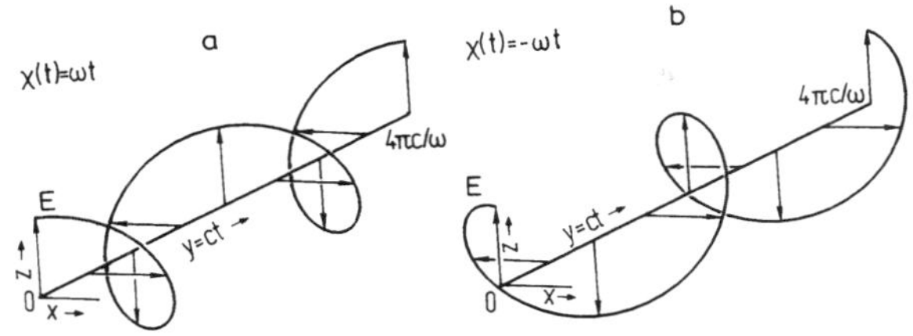

FIG.2.6-4. Electric field strength of a circularly polarized TEM wave having right polarization (a) and left polarization (b) as in Fig.2.6-3 but in a twice-as-long interval $0 \leq y < 4\pi c/\omega$. The magnitude $|\mathbf{E}|$ of the electric field strength is constant.

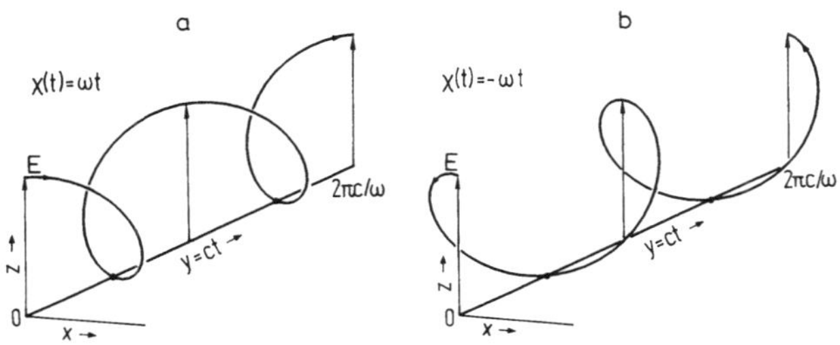

FIG.2.6-5. Electric field strength of a circularly polarized TEM wave with right (a) and left (b) polarization having one cycle in the interval $0 \leq y < 2\pi c/\omega$ as in Fig.2.6-3 but the magnitude $|\mathbf{E}|$ of the electric field strength has a sinusoidal time variation $E_0 \cos\omega t$ according to Eqs.(8) and (9).

$$E = E_0 \cos\omega t \tag{7}$$

$$E_x = +E_0 \cos\omega t \sin\omega t, \quad E_z = +E_0 \cos\omega t \cos\omega t \tag{8}$$

$$E_x = -E_0 \cos\omega t \sin\omega t, \quad E_z = +E_0 \cos\omega t \cos\omega t \tag{9}$$

The electric field strength of a right circularly polarized wave according to Eq.(8) is shown in Fig.2.6-5a, while Fig.2.6-5b shows the electric field strength of the corresponding left circularly polarized wave according to Eq.(9). We note that the electric field strength is zero for $y = \pi c/2\omega$ and $y = 3\pi c/2\omega$.

The polarization angle can be modulated to transmit information very much like the amplitude of an electric field strength. Let the polarization angle $\chi = \pi/2$ in Fig.2.6-1 stand for a digit 1 and $\chi = 0$ for a digit 0. By changing from $\chi = \pi/2$ to $\chi = 0$ as shown in Fig.2.6-6a we may transmit the sequence 10 by means of *linear polarization modulation*:

$$\begin{aligned} \mathbf{E} &= E\mathbf{e}_z, & 0 \leq y < y_0/2 \quad &\text{for digit 1} \\ \mathbf{E} &= E\mathbf{e}_x, & y_0/2 \leq y < y_0 \quad &\text{for digit 0} \end{aligned} \tag{10}$$

A variation of this concept is the choice

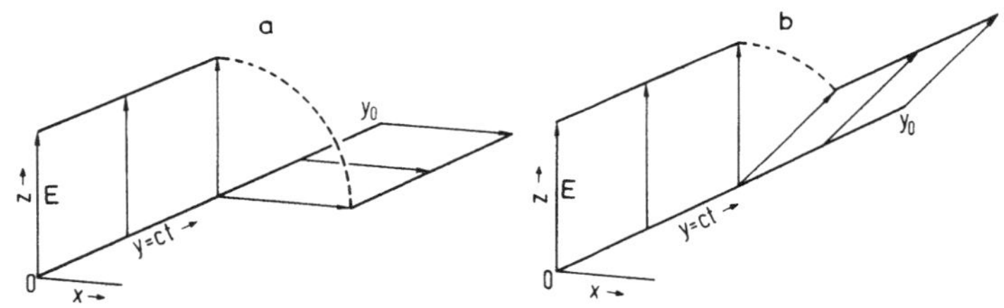

FIG.2.6-6. Modulation of the polarization angle of linearly polarized TEM waves. (a) $\mathbf{E} = E\mathbf{e}_z$ for $0 \leq y < y_0/2$ and $\mathbf{E} = E\mathbf{e}_x$ for $y_0/2 \leq y < y_0$; (b) $\mathbf{E} = E\mathbf{e}_z$ for $0 \leq y < y_0/2$ and $\mathbf{E} = (\sqrt{2}/2)E(\mathbf{e}_x + \mathbf{e}_z)$ for $y_0/2 \leq y < y_0$. The magnitude $|\mathbf{E}|$ of the electric field strength is constant.

$$\begin{aligned}
\mathbf{E} &= E\mathbf{e}_z, && 0 \leq y < y_0/2 && \text{for digit 1} \\
\mathbf{E} &= \frac{1}{2}\sqrt{2}E(\mathbf{e}_x + \mathbf{e}_z), && y_0/2 \leq y < y_0 && \text{for digit 0}
\end{aligned} \tag{11}$$

as shown In Fig.2.6-6b. This illustration implies that the modulation of the polarization angle is not restricted to binary signals.

The extension of linear polarization modulation to *circular polarization modulation* is shown in Fig.2.6-7a. We recognize the following relations for the electric field strength:

$$\begin{aligned}
\mathbf{E} &= E(\mathbf{e}_x \sin\omega t + \mathbf{e}_z \cos\omega t), && 0 \leq y < \pi c/\omega && \text{for digit 1} \\
\mathbf{E} &= E(-\mathbf{e}_x \sin\omega t + \mathbf{e}_z \cos\omega t), && \pi c/\omega \leq y < 2\pi c/\omega && \text{for digit 0}
\end{aligned} \tag{12}$$

If we want to carry over the variation of linear polarization modulation according to Eq.(11) and Fig.2.6-6b to circular polarization we must rewrite Eq.(12) as follows:

$$\begin{aligned}
\mathbf{E} &= E(\mathbf{e}_x \sin\omega t + \mathbf{e}_z \cos\omega t), && 0 \leq y < \pi c/\omega && \text{for digit 1} \\
\mathbf{E} &= \frac{1}{2}E[(\mathbf{e}_x \sin\omega t + \mathbf{e}_z \cos\omega t) + (-\mathbf{e}_x \sin\omega t + \mathbf{e}_z \cos\omega t)] \\
&= \frac{1}{2}E\mathbf{e}_z \cos\omega t, && \pi c/\omega \leq y < 2\pi c/\omega && \text{for digit 0}
\end{aligned} \tag{13}$$

The result is shown in Fig.2.6-7b. In the interval $\pi c/\omega \leq y < 2\pi c/\omega$ we see what looks like the electric field strength of a linearly polarized TEM wave with time-varying amplitude. One must keep in mind that linear and circular polarization are two representations of the same thing. The electric field strengths of circularly polarized waves with constant magnitude $|\mathbf{E}|$ can be seen as the sum of two linearly polarized waves with time varying amplitudes E_x and E_z according to Eqs.(5) and (6), but a linearly polarized wave with time-varying amplitude can also be seen as the sum of two circularly polarized waves with constant magnitude $|\mathbf{E}|$ of the amplitude.

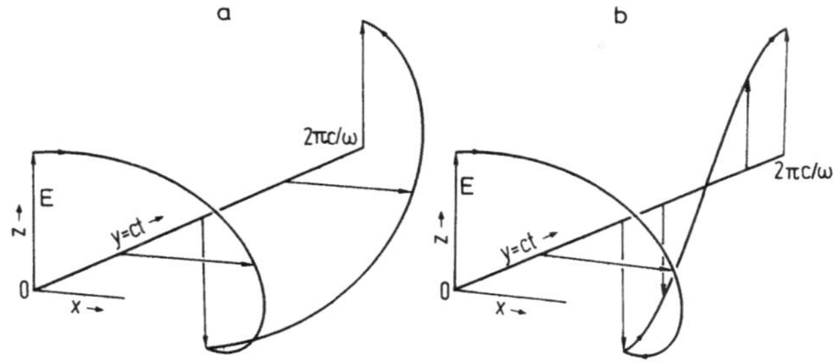

FIG.2.6-7. Modulation of the polarization angle of circularly polarized TEM waves. (a) $\mathbf{E} = E\mathbf{e}_r$ for $0 \leq y < \pi c/\omega$ and $\mathbf{E} = E\mathbf{e}_l$ for $\pi c/\omega \leq y < 2\pi c/\omega$; (b) $\mathbf{E} = E\mathbf{e}_r$ for $0 \leq y < \pi c/\omega$ and $\mathbf{E} = (1/2)E(\mathbf{e}_r + \mathbf{e}_l)$ for $\pi c/\omega \leq y < 2\pi c/\omega$. The magnitude $|\mathbf{E}|$ of the electric field strength is constant if rotational unit vectors $\mathbf{e}_r$ and $\mathbf{e}_l$ are used rather than $\mathbf{e}_x$ and $\mathbf{e}_z$.

Instead of using unit vectors $\mathbf{e}_x$ and $\mathbf{e}_z$ we can use right and left rotating unit vectors $\mathbf{e}_r$ and $\mathbf{e}_l$

$$\mathbf{e}_r = +\mathbf{e}_x \sin\omega t + \mathbf{e}_z \cos\omega t \tag{14}$$
$$\mathbf{e}_l = -\mathbf{e}_x \sin\omega t + \mathbf{e}_z \cos\omega t \tag{15}$$
$$\mathbf{e}_z = (\mathbf{e}_r + \mathbf{e}_l)/2\cos\omega t \tag{16}$$
$$\mathbf{e}_x = (\mathbf{e}_r - \mathbf{e}_l)/2\cos\omega t \tag{17}$$

to rewrite Eq.(13)

$$\begin{aligned} \mathbf{E} &= E\mathbf{e}_r, & 0 \leq y < \pi c/\omega \quad &\text{for digit 1} \\ \mathbf{E} &= \frac{1}{2}E(\mathbf{e}_r + \mathbf{e}_l), & \pi c/\omega \leq y < 2\pi c/\omega \quad &\text{for digit 0} \end{aligned} \tag{18}$$

or Eq.(11):

$$\begin{aligned} \mathbf{E} &= \frac{1}{2\cos\omega t}E(\mathbf{e}_r + \mathbf{e}_l) & 0 \leq y < \pi c/\omega \quad &\text{for digit 1} \\ \mathbf{E} &= \frac{\sqrt{2}}{2\cos\omega t}E\mathbf{e}_r & \pi c/\omega \leq y < 2\pi c/\omega \quad &\text{for digit 0} \end{aligned} \tag{19}$$

There is now no amplitude variation in Eq.(18) but the amplitudes in Eq.(19) are functions of time.

As an example of a *general polarization* we choose $\chi(t) = \pi \sin\omega t$ and substitute in Eq.(1):

$$E_x = E\sin(\pi\sin\omega t), \quad E_z = E\cos(\pi\sin\omega t)$$
$$\mathbf{E} = E_x\mathbf{e}_x + E_z\mathbf{e}_z \tag{20}$$

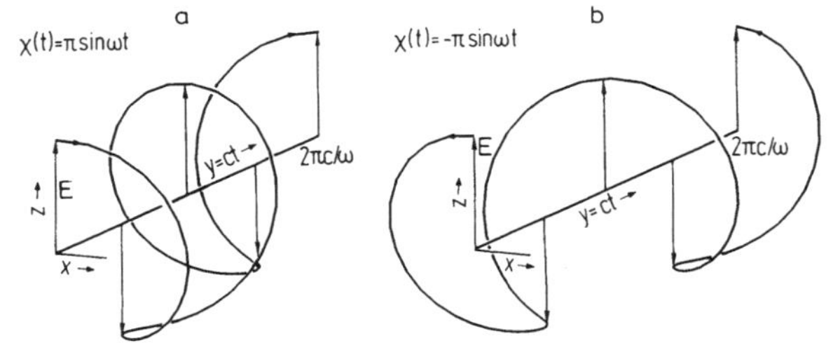

FIG.2.6-8. Examples of general polarization with $\chi(t) = \pi \sin \omega t$ (a) and $\chi(t) = -\pi \sin \omega t$ (b) but constant magnitude $|\mathbf{E}|$ of the electric field strength.

The resulting electric field strength $\mathbf{E}$ of a TEM wave with *general right polarization* is shown in Fig.2.6-8. The corresponding field strength $\mathbf{E}$ with *general left polarization* is obtained by the substitution

$$E_x = -E \sin(\pi \sin \omega t), \quad E_z = E \cos(\pi \sin \omega t) \tag{21}$$

in Eq.(20), which yields Fig.2.6-8b.

Using the right and left rotational unit vectors $\mathbf{e}_r$ and $\mathbf{e}_l$ instead of $\mathbf{e}_x$ and $\mathbf{e}_y$ we may rewrite Eq.(20) in the form

$$\mathbf{E} = E\left(\frac{\sin(\pi \sin \omega t) + \cos(\pi \sin \omega t)}{2 \cos \omega t}\mathbf{e}_r - \frac{\sin(\pi \sin \omega t) - \cos(\pi \sin \omega t)}{2 \cos \omega t}\mathbf{e}_l\right) \tag{22}$$

and Eq.(21) in the form

$$\mathbf{E} = E\left(-\frac{\sin(\pi \sin \omega t) - \cos(\pi \sin \omega t)}{2 \cos \omega t}\mathbf{e}_r + \frac{\sin(\pi \sin \omega t) + \cos(\pi \sin \omega t)}{2 \cos \omega t}\mathbf{e}_l\right) \tag{23}$$

which shows that the factors of $\mathbf{e}_r$ and $\mathbf{e}_l$ have been exchanged. We note that the factors of $\mathbf{e}_x$ and $\mathbf{e}_z$ in Eqs.(20) and (21) as well as those of $\mathbf{e}_r$ and $\mathbf{e}_l$ in Eqs.(22) and (23) vary with time but the magnitude $|\mathbf{E}|$ of the electric field strength in Figs.2.6-8a and b is constant.

3 Analytic Solution for Cylinder Waves

3.1 Cylinder Waves Excited at a Boundary

The two field strengths $E_{\mathrm{i}x}$ and $H_{\mathrm{i}y}$ of Eq.(2.1-31) in a point $z = 0$, y excite cylinder waves—or rather half-cylinder waves—in medium 1 and 2 according to Fig.2.2-5 and 2.2-7. The coordinates of cylinder waves are usually denoted r, φ, z but the cylinder axes in Fig.2.2-5 are in the direction of the x-axis rather than the z-axis, which calls for the coordinates r, φ, x. A further notational problem is caused by the letter r that was used for 'reflected', as in $E_{\mathrm{r}x}$. Following common practice the r denoting reflected is set in roman while the r denoting the radial variable is set in italic. Figure 3.1-1 shows r, φ, x for reflected and transmitted waves. We note that the x-axis points out of the paper plane toward the viewer.

Instead of Eqs.(2.1-1)–(2.1-8) we obtain in cylinder coordinates the following equations:

$$\frac{1}{r}\frac{\partial H_x}{\partial \varphi} - \frac{\partial H_\varphi}{\partial x} = \epsilon\frac{\partial E_r}{\partial t} + \sigma E_r \tag{1}$$

$$\frac{\partial H_r}{\partial x} - \frac{\partial H_x}{\partial r} = \epsilon\frac{\partial E_\varphi}{\partial t} + \sigma E_\varphi \tag{2}$$

$$\frac{1}{r}\left(\frac{\partial (rH_\varphi)}{\partial r} - \frac{\partial H_r}{\partial \varphi}\right) = \epsilon\frac{\partial E_x}{\partial t} + \sigma E_x \tag{3}$$

$$-\frac{1}{r}\frac{\partial E_x}{\partial \varphi} + \frac{\partial E_\varphi}{\partial x} = \mu\frac{\partial H_r}{\partial t} + sH_r \tag{4}$$

$$-\frac{\partial E_r}{\partial x} + \frac{\partial E_x}{\partial r} = \mu\frac{\partial H_\varphi}{\partial t} + sH_\varphi \tag{5}$$

$$-\frac{1}{r}\left(\frac{\partial (rE_\varphi)}{\partial r} - \frac{\partial E_r}{\partial \varphi}\right) = \mu\frac{\partial H_x}{\partial t} + sH_x \tag{6}$$

$$\frac{1}{r}\frac{\partial (rE_r)}{\partial r} + \frac{1}{r}\frac{\partial E_\varphi}{\partial \varphi} + \frac{\partial E_x}{\partial x} = 0 \tag{7}$$

$$\frac{1}{r}\frac{\partial (rH_r)}{\partial r} + \frac{1}{r}\frac{\partial H_\varphi}{\partial \varphi} + \frac{\partial H_x}{\partial x} = 0 \tag{8}$$

We look for a TEM wave solution. This implies that neither E nor H have a component in the direction of r:

$$E_r = 0,\ H_r = 0 \tag{9}$$

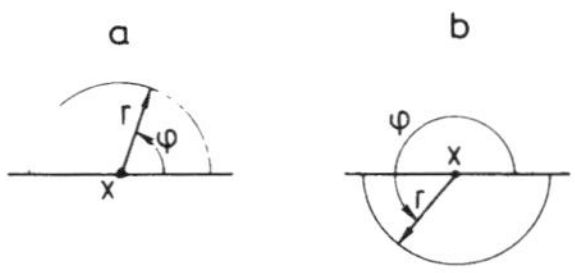

FIG.3.1-1. The parameters r, φ, x of a reflected cylinder wave (a) and of a transmitted cylinder wave (b).

A cylinder wave implies that the following derivatives are zero:

$$\frac{\partial E_\varphi}{\partial \varphi} = \frac{\partial E_\varphi}{\partial x} = \frac{\partial E_x}{\partial \varphi} = \frac{\partial E_x}{\partial x} = 0 \tag{10}$$

$$\frac{\partial H_\varphi}{\partial \varphi} = \frac{\partial H_\varphi}{\partial x} = \frac{\partial H_x}{\partial \varphi} = \frac{\partial H_x}{\partial x} = 0 \tag{11}$$

Finally, we specialize to a perpendicularly polarized wave which implies the relations

$$E_\varphi = 0, \; E_r = 0, \; H_x = 0 \tag{12}$$

Substitution of Eqs.(9)–(12) into Eqs.(1)–(8) leaves only Eqs.(3) and (5) unequal zero:

$$\frac{1}{r}\frac{\partial (rH_\varphi)}{\partial r} = \frac{1}{r}H_\varphi + \frac{\partial H_\varphi}{\partial r} = \epsilon\frac{\partial E_x}{\partial t} + \sigma E_x \tag{13}$$

$$\frac{\partial E_x}{\partial r} = \mu\frac{\partial H_\varphi}{\partial t} + sH_\varphi \tag{14}$$

We add the index r to indicate a reflected wave and write ϵ_1, μ_1, σ_1, s_1 to indicate that the wave propagates in medium 1 in Fig.2.2-5:

$$\frac{1}{r}\frac{\partial (rH_{\mathrm{r}\varphi})}{\partial r} = \frac{1}{r}H_{\mathrm{r}\varphi} + \frac{\partial H_{\mathrm{r}\varphi}}{\partial r} = \epsilon_1\frac{\partial E_{\mathrm{r}x}}{\partial t} + \sigma_1 E_{\mathrm{r}x} \tag{15}$$

$$\frac{\partial E_{\mathrm{r}x}}{\partial r} = \mu_1\frac{\partial H_{\mathrm{r}\varphi}}{\partial t} + s_1 H_{\mathrm{r}\varphi} \tag{16}$$

For the transmitted wave in medium 2 in Fig.2.2-7 we write the index t and ϵ_2, μ_2, σ_2, s_2:

$$\frac{1}{r}\frac{\partial (rH_{\mathrm{t}\varphi})}{\partial r} = \frac{1}{r}H_{\mathrm{t}\varphi} + \frac{\partial H_{\mathrm{t}\varphi}}{\partial r} = \epsilon_2\frac{\partial E_{\mathrm{t}x}}{\partial t} + \sigma_2 E_{\mathrm{t}x} \tag{17}$$

$$\frac{\partial E_{\mathrm{t}x}}{\partial r} = \mu_2\frac{\partial H_{\mathrm{t}\varphi}}{\partial t} + s_2 H_{\mathrm{t}\varphi} \tag{18}$$

Two additional relations are obtained from the boundary conditions at $z = 0$ of Eqs.(2.1-53) and (2.1-54):

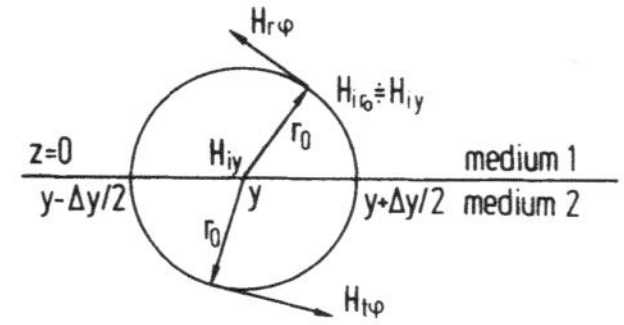

FIG.3.1-2. Replacement of a straight line $y - \Delta y/2 \leq y \leq y + \Delta y/2$ in either Fig.2.2-5 or 2.2-7 by a circle with radius $r_0 = \Delta y/2$. For $r_0 \to 0$ the field strength $H_{\mathrm{i}r_0}$ on the circle approaches the field strength $H_{\mathrm{i}y}$ in the point y.

(a) The tangential components of the electric field strengths in both media must be equal at $z = 0$ or $r = 0$:

$$E_{\mathrm{i}x}(y, z = 0, t) = E_{\mathrm{i}}(y, z = 0, t) = E_{\mathrm{r}x}(r = 0, t) + E_{\mathrm{t}x}(r = 0, t) \tag{19}$$

(b) The tangential components of the magnetic field strengths in both media be equal at $z = 0$ or $r = 0$.

Here we run into a difficulty. The incident wave produces a magnetic field strength $H_{\mathrm{i}y}(y, z = 0, t) = -H_{\mathrm{i}}(y, z = 0, t)\cos\vartheta_{\mathrm{i}}$ in the point y of the plane $z = 0$ but the boundary values of a cylinder wave need to be specified on a cylinder not in a plane. Refer to Fig.3.1-2 which shows an enlargement of the neighborhood of a point y in both Figs.2.2-5 and 2.2-7. The field strength $H_{\mathrm{i}y}$ is defined for $z = 0$ in the point $y = k\Delta y$. We need it on a cylinder with radius $r_0 = \Delta y/2$. Let us choose the field strength $H_{\mathrm{i}r_0}$ on a cylinder with radius r_0

$$H_{\mathrm{i}r_0} = H_{\mathrm{i}y} \tag{20}$$

For $r_0 \to 0$ the circle with radius r_0 in Fig.3.1-2 approaches the point y and the relation $H_{\mathrm{i}r_0} = H_{\mathrm{i}y}$ will be exactly satisfied. Hence, we write the boundary condition in the form

$$\begin{aligned} &H_{\mathrm{i}y}(y, z = 0, t) \\ &\quad = -H_{\mathrm{i}}(y, z = 0, t)\cos\vartheta_{\mathrm{i}} = H_{\mathrm{i}r_0} = H_{\mathrm{r}\varphi}(r = r_0, t) + H_{\mathrm{t}\varphi}(r = r_0, t) \end{aligned} \tag{21}$$

Equations (15) and (16) have the two unknowns $H_{\mathrm{r}\varphi}$ and $E_{\mathrm{r}x}$ while Eqs.(17) and (18) have the unknowns $H_{\mathrm{t}\varphi}$ and $E_{\mathrm{t}x}$. Hence, we can solve each pair of equations but the solutions must be general enough that we can satisfy Eqs.(19) and (21).

We return to Eqs.(13), (14) and eliminate E_x by differentiation of Eq.(13) with respect to r and Eq.(14) with respect to t. This yields a partial differential equation for H_φ. The associated electric field strength E_x follows then by integration of Eq.(14):

$$\frac{\partial^2 H_\varphi}{\partial r^2} - \epsilon\mu\frac{\partial^2 H_\varphi}{\partial t^2} - (\epsilon s + \mu\sigma)\frac{\partial H_\varphi}{\partial t} + \frac{1}{r}\frac{\partial H_\varphi}{\partial r} - (\frac{1}{r^2} + \sigma s)H_\varphi = 0 \tag{22}$$

$$E_x = E_x(r, t) = \int \left(\mu\frac{\partial H_\varphi}{\partial t} + sH_\varphi\right) dr + E_{x1}(t) \tag{23}$$

Instead of obtaining E_x from Eq.(14) we may also obtain it from Eq.(13). This is more difficult but just as valid and the two methods for the determination of E_x must yield the same result. This requirement helps determine integration constants. We first solve the homogeneous part of Eq.(13)

$$\frac{\partial E_x}{\partial t} + \frac{\sigma}{\epsilon} E_x = 0 \tag{24}$$

and obtain:

$$E_x = C e^{-\sigma t/\epsilon} \tag{25}$$

The inhomogeneous equation is solved by the method of variation of the constant. We assume C is a function of t:

$$E_x(t) = C(t) e^{-\sigma t/\epsilon}, \quad dE_x/dt = e^{-\sigma t/\epsilon}\left[dC(t)/dt - \sigma\epsilon^{-1}C(t)\right] \tag{26}$$

Substitution into Eq.(13) yields

$$C(t) = \frac{1}{\epsilon r}\int \frac{\partial(rH_\varphi)}{\partial r} e^{\sigma t/\epsilon} dt + E_{x2}(r) \tag{27}$$

and E_x follows from Eq.(25):

$$E_x(r,t) = e^{-\sigma t/\epsilon}\left[\frac{1}{\epsilon r}\int \frac{\partial(rH_\varphi)}{\partial r} e^{\sigma t/\epsilon} dt + E_{x2}(r)\right] \tag{28}$$

We still must add the general solution of the homogeneous equation according to Eq.(25) but this changes the integration constant E_{x2} only.

The solution defined by Eqs.(22), (23), and (28) holds for an excitation by a magnetic field strength such as H_{iy} of Eq.(2.1-31). The solution for excitation by an electric field strength such as E_{ix} of Eq.(2.1-31) requires that one eliminates first H_φ from Eq.(14) and then calculates the associated magnetic field strength by integration of Eq.(13) and (14). For the elimination of H_φ from Eq.(14) one may first differentiate Eq.(13) with respect to t; Eq.(14) is multiplied with r and then differentiated with respect to r. One obtains:

$$\frac{\partial^2 E_x}{\partial r^2} - \epsilon\mu\frac{\partial^2 E_x}{\partial t^2} - (\epsilon s + \mu\sigma)\frac{\partial E_x}{\partial t} + \frac{1}{r}\frac{\partial E_x}{\partial r} - \sigma s E_x = 0 \tag{29}$$

The associated magnetic field strength H_φ follows from Eq.(13) by direct integration

$$H_\varphi = H_\varphi(r,t) = \frac{1}{r}\int r\left(\epsilon\frac{\partial E_x}{\partial t} + \sigma E_x\right) dr + H_{\varphi 1}(t) \tag{30}$$

or from Eq.(14) by variation of the constant in analogy to Eqs.(24)–(28):

$$H_\varphi(r,t) = e^{-st/\mu}\left[\frac{1}{\mu}\int \frac{\partial E_x}{\partial r} e^{st/\mu}\, dt + H_{\varphi 2}(t)\right] \tag{31}$$

For a numerical solution we rewrite Eqs.(13), (14), (22), (23), and (28)–(31) in normalized notation. In medium 1 we replace ϵ, μ, σ, s according to Fig.2.1-1 by ϵ_1, μ_1, σ_1, s_1 and define ρ_r and θ for medium 1, where the subscript r denotes 'reflected':

$$\theta = \sigma_1 t/2\epsilon_1, \ \rho_\mathrm{r} = \sigma_1 r/2\epsilon_1 c_1, \ c_1 = 1/\sqrt{\mu_1\epsilon_1} \tag{32}$$

Equations (13) and (14) assume the following form for $H_\varphi \to H_{\mathrm{r}\varphi}$ and $E_x \to E_{\mathrm{r}x}$:

$$\frac{1}{\rho_\mathrm{r}}\frac{\partial(\rho_\mathrm{r} Z_1 H_{\mathrm{r}\varphi})}{\partial\rho_\mathrm{r}} = \frac{1}{\rho_\mathrm{r}} Z_1 H_{\mathrm{r}\varphi} + \frac{\partial(Z_1 H_{\mathrm{r}\varphi})}{\partial\rho_\mathrm{r}} = \frac{\partial E_{\mathrm{r}x}}{\partial\theta} + 2E_{\mathrm{r}x} \tag{33}$$

$$\frac{\partial E_{\mathrm{r}x}}{\partial\rho_\mathrm{r}} = \frac{\partial(Z_1 H_{\mathrm{r}\varphi})}{\partial\theta} + 2\iota(Z_1 H_{\mathrm{r}\varphi}) \tag{34}$$

$$Z_1 = \sqrt{\mu_1/\epsilon_1}, \ \iota = \epsilon_1 s_1/\mu_1\sigma_1 \tag{35}$$

Furthermore, Eqs.(22), (23), and (28)–(31) become:

$$\frac{\partial^2(Z_1 H_{\mathrm{r}\varphi})}{\partial\rho_\mathrm{r}^2} - \frac{\partial^2(Z_1 H_{\mathrm{r}\varphi})}{\partial\theta^2} - 2(\iota+1)\frac{\partial(Z_1 H_{\mathrm{r}\varphi})}{\partial\theta} + \frac{1}{\rho_\mathrm{r}}\frac{\partial(Z_1 H_{\mathrm{r}\varphi})}{\partial\rho_\mathrm{r}} - \left(\frac{1}{\rho_\mathrm{r}^2} + 4\iota\right)(Z_1 H_{\mathrm{r}\varphi}) = 0 \tag{36}$$

$$E_{\mathrm{r}x} = \int\left(\frac{\partial(Z_1 H_{\mathrm{r}\varphi})}{\partial\theta} + 2\iota(Z_1 H_{\mathrm{r}\varphi})\right) d\rho_\mathrm{r} + E_{\mathrm{r}1}(\theta) \tag{37}$$

$$E_{\mathrm{r}x} = e^{-2\theta}\left(\frac{1}{\rho_\mathrm{r}}\int\frac{\partial(\rho_\mathrm{r} Z_1 H_{\mathrm{r}\varphi})}{\partial\rho_\mathrm{r}} e^{2\theta}\, d\theta + E_{\mathrm{r}2}(\rho_\mathrm{r})\right) \tag{38}$$

$$\frac{\partial^2 E_{\mathrm{r}x}}{\partial\rho_\mathrm{r}^2} - \frac{\partial^2 E_{\mathrm{r}x}}{\partial\theta^2} - 2(\iota+1)\frac{\partial E_{\mathrm{r}x}}{\partial\theta} + \frac{1}{\rho_\mathrm{r}}\frac{\partial E_{\mathrm{r}x}}{\partial\rho_\mathrm{r}} - 4\iota E_{\mathrm{r}x} = 0 \tag{39}$$

$$Z_1 H_{\mathrm{r}\varphi} = \frac{1}{\rho_\mathrm{r}}\int\rho_\mathrm{r}\left(\frac{\partial E_{\mathrm{r}x}}{\partial\theta} + E_{\mathrm{r}x}\right) d\rho_\mathrm{r} + Z_1 H_{\mathrm{r}1}(\theta) \tag{40}$$

$$Z_1 H_{\mathrm{r}\varphi} = e^{-2\iota\theta}\left(\int\frac{\partial E_{\mathrm{r}x}}{\partial\rho_\mathrm{r}} e^{2\iota\theta}\, d\theta + Z_1 H_{\mathrm{r}2}(\rho_\mathrm{r})\right) \tag{41}$$

For the transmitted signal in medium 2 we replace ϵ, μ, σ, s in Eqs.(13), (14), (22), (23) and (28)–(31) according to Fig.2.1-1 by ϵ_2, μ_2, σ_2, s_2. The normalized time variable θ of Eq.(32) is retained but a 'transmitted' variable ρ_t is used instead of ρ_r:

$$\theta = \sigma_1 t/2\epsilon_1, \ \rho_\mathrm{t} = \sigma_2 r/2\epsilon_2 c_2, \ c_2 = 1/\sqrt{\mu_2\epsilon_2} \tag{42}$$

Equations (13) and (14) assume the following form for $H_\varphi \to H_{\mathrm{t}\varphi}$ and $E_x \to E_{\mathrm{t}x}$:

$$\frac{1}{\rho_t}\frac{\partial(\rho_t Z_2 H_{t\varphi})}{\partial\rho_t} = \frac{1}{\rho_t}(Z_2 H_{t\varphi}) + \frac{\partial(Z_2 H_{t\varphi})}{\partial\rho_t} = \frac{\gamma_\sigma}{\gamma_\epsilon}\frac{\partial E_{tx}}{\partial\theta} + 2E_{tx} \tag{43}$$

$$\frac{\partial E_{tx}}{\partial\rho_t} = \frac{\gamma_\sigma}{\gamma_\epsilon}\frac{\partial(Z_2 H_{t\varphi})}{\partial\theta} + 2\iota\frac{\gamma_\sigma\gamma_\mu}{\gamma_s\gamma_\epsilon}(Z_2 H_{t\varphi}) \tag{44}$$

$$Z_2 = \sqrt{\frac{\mu_2}{\epsilon_2}},\ \iota = \frac{\epsilon_1 s_1}{\mu_1\sigma_1},\ \gamma_s = \frac{s_1}{s_2},\ \gamma_\sigma = \frac{\sigma_1}{\sigma_2},\ \gamma_\epsilon = \frac{\epsilon_1}{\epsilon_2},\ \gamma_\mu = \frac{\mu_1}{\mu_2} \tag{45}$$

Equations (22), (23), and (28)–(31) are rewritten for a transmitted wave in medium 2 of Fig.2.1-1:

$$\frac{\partial^2(Z_2 H_{t\varphi})}{\partial\rho_t^2} - \frac{\gamma_\sigma^2}{\gamma_\epsilon^2}\frac{\partial^2(Z_2 H_{t\varphi})}{\partial\theta^2} - 2\frac{\gamma_\sigma}{\gamma_\epsilon}\left(\iota\frac{\gamma_\mu\gamma_\sigma}{\gamma_\epsilon\gamma_s} + 1\right)\frac{\partial(Z_2 H_{t\varphi})}{\partial\theta} + \frac{1}{\rho_t}\frac{\partial(Z_2 H_{t\varphi})}{\partial\rho_t} - \left(\frac{1}{\rho_t^2} + 4\iota\frac{\gamma_\mu\gamma_\sigma}{\gamma_\epsilon\gamma_s}\right)(Z_2 H_{t\varphi}) = 0 \tag{46}$$

$$E_{tx} = \frac{\gamma_\sigma}{\gamma_\epsilon}\int\left(\frac{\partial(Z_2 H_{t\varphi})}{\partial\theta} + 2\iota\frac{\gamma_\mu}{\gamma_s}(Z_2 H_{t\varphi})\right)d\rho_t + E_{t1}(\theta) \tag{47}$$

$$E_{tx} = e^{-2\theta\gamma_\epsilon/\gamma_\sigma}\left(\frac{\gamma_\epsilon}{\gamma_\sigma}\frac{1}{\rho_t}\int\frac{\partial(\rho_t Z_2 H_{t\varphi})}{\partial\rho_t}e^{2\theta\gamma_\epsilon/\gamma_\sigma}\,d\theta + E_{t2}(\rho_r)\right) \tag{48}$$

$$\frac{\partial^2 E_{tx}}{\partial\rho_t^2} - \frac{\gamma_\sigma^2}{\gamma_\epsilon^2}\frac{\partial^2 E_{tx}}{\partial\theta^2} - 2\frac{\gamma_\sigma}{\gamma_\epsilon}\left(\iota\frac{\gamma_\mu\gamma_\sigma}{\gamma_\epsilon\gamma_s} + 1\right)\frac{\partial E_{tx}}{\partial\theta} + \frac{1}{\rho_t}\frac{\partial E_{tx}}{\partial\rho_t} - 4\iota\frac{\gamma_\mu\gamma_\sigma}{\gamma_\epsilon\gamma_s}E_{tx} = 0 \tag{49}$$

$$Z_2 H_{t\varphi} = \frac{1}{\rho_t}\int\rho_t\left(\frac{\gamma_\sigma}{\gamma_\epsilon}\frac{\partial E_{tx}}{\partial\theta} + 2E_{tx}\right)d\rho_t + Z_2 H_{t1}(\theta) \tag{50}$$

$$Z_2 H_{t\varphi} = e^{-2\iota\theta\gamma_\mu/\gamma_s}\left(\frac{\gamma_\epsilon}{\gamma_\sigma}\int\frac{\partial E_{tx}}{\partial\rho_t}e^{2\iota\theta\gamma_\mu/\gamma_s}\,d\theta + Z_2 H_{t2}(\theta)\right) \tag{51}$$

3.2 Magnetic Excitation of Signal Solutions

We turn to the analytic determination of signal solutions for perpendicularly polarized cylinder waves as defined by Fig.2.1-1 and Eqs.(3.1-36)–(3.1-38) as well as (3.1-39)–(3.1-41). Starting with Eq.(3.1-36) we assume an excitation at $\rho_r = 0$ by a magnetic step function:

$$\begin{aligned} H_{r\varphi}(0,\theta) = H_{iy} = H_0 S(\theta) &= 0 \quad \text{for } \theta < 0 \\ &= H_0 \quad \text{for } \theta \geq 0 \end{aligned} \tag{1}$$

At great distance $\rho_r \to \infty$ we have the further boundary condition

$$H_{r\varphi}(\rho_r \to \infty, \theta) = \text{finite} \tag{2}$$

Let $H_{r\varphi}$ and E_{rx} be zero for $\rho_r > 0$ at the time $\theta = 0$. This yields the initial conditions[1]

$$H_{r\varphi}(\rho_r, 0) = E_{rx}(\rho_r, 0) = 0 \tag{3}$$

If $H_{r\varphi}(\rho_r, 0)$ and $E_{rx}(\rho_r, 0)$ are zero for $\rho_r > 0$, their derivatives with respect to ρ_r must be zero too:

$$\partial H_{r\varphi}(\rho_r, 0)/\partial\rho_r = \partial E_{rx}(\rho_r, 0)/\partial\rho_r = 0 \tag{4}$$

Equations (3) and (4) imply the further initial conditions

$$\partial H_{r\varphi}(\rho_r, \theta)/\partial\theta = \partial E_{rx}(\rho_r, \theta)/\partial\theta = 0 \tag{5}$$

for $\rho_r > 0$ and $\theta = 0$ due to Eqs.(3.1-33) and (3.1-34).

Let us assume the solution of Eq.(3.1-36) can be written as the sum of the steady state solution $F(\rho_r)$ plus the deviation $w(\rho_r, \theta)$ from the steady state solution:

$$H_{r\varphi}(\rho_r, \theta) = H_{H,r\varphi}(\rho_r, \theta) = H_H(\rho, \theta) = H_0[w(\rho, \theta) + F(\rho)] \tag{6}$$

We have introduced the subscript H to indicate the magnetic field strength is due to a magnetic force function according to Eq.(1), since an additional associated magnetic field strength is produced by an electric excitation force according to Eq.(3.1-39). The subscripts r and φ are left out until the end of the calculation to simplify writing.

Since Eq.(3.1-36) is a linear equation it must satisfy the two components of Eq.(6) individually. Substitution of $F(\rho)$ into Eq.(3.1-36) and multiplication with ρ^2 yields the ordinary differential equation

$$\rho^2 \frac{d^2F}{d\rho^2} + \rho \frac{dF}{d\rho} - (4\iota\rho^2 + 1)F = 0 \tag{7}$$

Using the substitutions

$$\chi = 2\sqrt{\iota}\rho, \quad \frac{d\chi}{d\rho} = 2\sqrt{\iota}, \quad \frac{dF}{d\rho} = 2\sqrt{\iota}\frac{dF}{d\chi}, \quad \frac{d^2F}{d\rho^2} = 4\iota\frac{d^2F}{d\chi^2} \tag{8}$$

we obtain the differential equation of the modified Bessel functions of order 1

$$\chi^2 \frac{d^2F}{d\chi^2} + \chi \frac{dF}{d\chi} - (\chi^2 + 1)F = 0 \tag{9}$$

with the general solution:

[1]The initial condition holds for $\theta \leq 0$ too since the excitation force $H_0S(\theta)$ is zero for $\theta < 0$ according to Eq.(1).

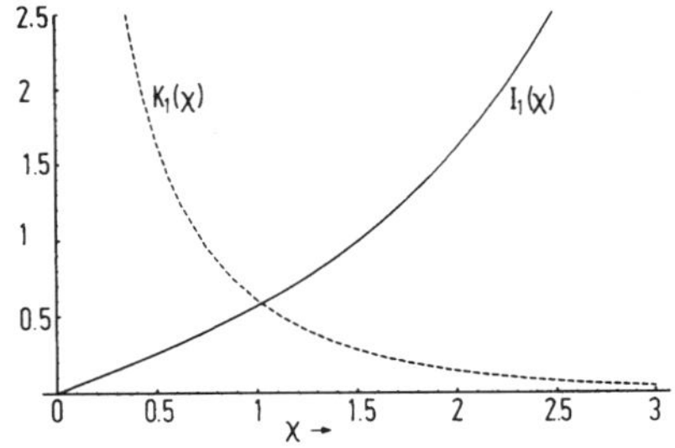

FIG.3.2-1. The modified Bessel functions $K_1(\chi)$ and $I_1(\chi)$.

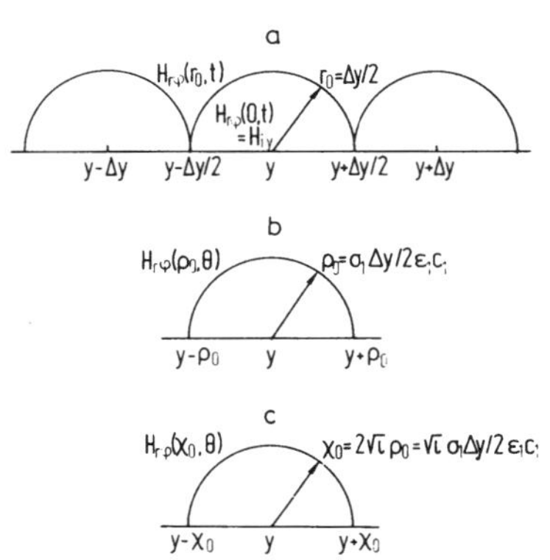

FIG.3.2-2. Enlarged section of Fig.2.2-5 in the neighborhood of the point $y = k\Delta y$ showing the spatial variable r_0 (a), and the transition to the normalized variables ρ_0 (b) and $\chi_0(c)$.

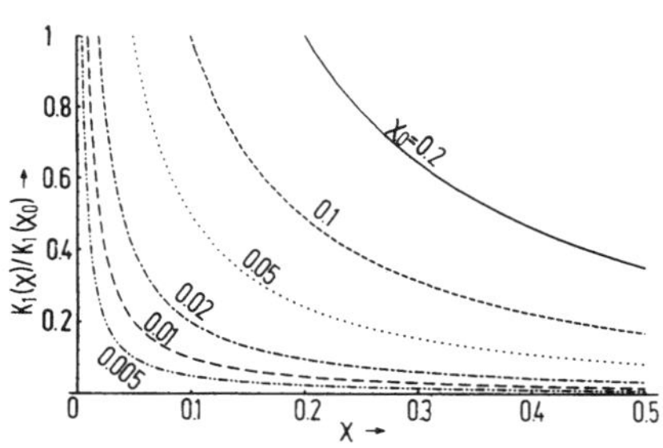

FIG.3.2-3. Plots of $F(\chi) = K_1(\chi)/K_1(\chi_0)$ according to Eq (14) for $\chi_0 = 0.2$, 0.1, 0.05, 0.02, 0.01, 0.005. All plots have the maximum value 1 and all are defined in the interval $\chi_0 \leq \chi < \infty$.

$$F(\chi) = A_{00}K_1(\chi) + A_{01}I_1(\chi) \tag{10}$$

The functions $K_1(\chi)$ and $I_1(\chi)$ are plotted in Fig.3.2-1. The function $I_1(\chi)$ must be eliminated since it is not finite at $\chi = 2\sqrt{\iota\rho} \to \infty$ as demanded by Eq.(2). The other function $K_1(\chi)$ seems useless too due to its pole at $\chi = 0$, but a closer investigation shows that this is not so.

To see why $K_1(\chi)$ is acceptable refer to Fig.2.2-5. An incident wave produces

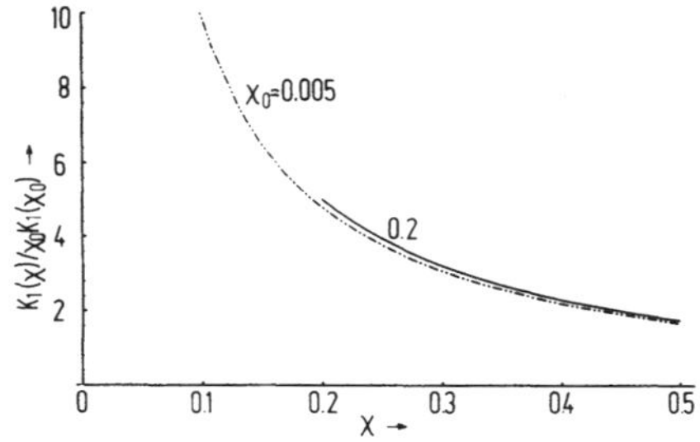

FIG.3.2-4. The function $F(\chi)/\chi_0 = K_1(\chi)/\chi_0 K_1(\chi_0)$ for $\chi_0 = 0.2$ and $\chi_0 = 0.005$. The functions are defined for $\chi \geq \chi_0$.

the excitation force $H_{r\varphi}(0,\theta) = H_{iy}$ in the point $y = k\Delta y$ that represents the line interval $y - \Delta y/2 \leq y \leq y + \Delta y/2$. This interval is shown enlarged in Fig.3.2-2a. Instead of the line interval we want the half circle with radius $r_0 = \Delta y/2$ for cylinder waves with the field strength $H_{r\varphi}(r_0,t)$ on it. For $r_0 \to 0$ the field strength $H_{r\varphi}(r_0,t)$ approaches $H_{r\varphi}(0,t)$. For a numerical solution we are only interested in sufficiently small but finite values of r_0 just as the values of Δy in Fig.2.2-5 need to be only sufficiently small but finite.

For the particular value $\rho = \rho_0$ or $\chi = \chi_0 = 2\sqrt{\iota\rho_0}$ we obtain from Eqs.(6) and (9) the following result:

$$H_{\mathrm{H}}(\chi_0 \to 0,\theta) = H_0[w(\chi_0,\theta) + A_{00}K_1(\chi_0)] = H_0 \quad \text{for } \theta \geq 0,\ \chi \geq \chi_0 \tag{11}$$

$$A_{00} = 1/K_1(\chi_0) \tag{12}$$

$$w(\chi_0,\theta) = 0 \tag{13}$$

$$F(\chi) = K_1(\chi)/K_1(\chi_0) \tag{14}$$

As χ_0 and thus r_0 in Fig.3.2-2a become smaller and smaller the half circle around y approximates the point y closer and closer but $F(\chi)$ will never be larger than 1 in its range of definition $\chi_0 \leq \chi < \infty$. Figure 3.2-3 shows $F(\chi)$ for various values of χ_0. Each plot holds in the interval $\chi_0 \leq \chi < \infty$. The choice of χ_0 determines the accuracy of the calculation and also how close to $\chi = 0$ the result applies.

In order to estimate the accuracy of the calculation we observe that the number of reflecting points in Fig.2.2-5 increases proportionate to $1/\Delta y$ as the value of Δy decreases. Since χ_0 is proportionate to Δy according to Fig.3.2-2c the value

$$F(\chi)/\chi_0 = K_1(\chi)/\chi_0 K_1(\chi_0) \tag{15}$$

is a good measure of accuracy. Figure 3.2-4 shows plots of Eq.(15) for $\chi_0 = 0.2$ and $\chi_0 = 0.005$. The plots are very close in the interval $0.2 \leq \chi < 0.5$ where both functions are defined, but the function $F(\chi)/\chi_0 = F(\chi)/0.005$ is defined for $\chi_0 \geq 0.005$.

The boundary condition of Eq.(13) requires $w(\chi_0,\theta) \to 0$ for $\chi_0 \to 0$. For large values of χ we get from Eq.(2):

$$w(\infty, \theta) = \text{finite} \tag{16}$$

The initial conditions of Eqs.(3) and (4) yield:

$$w(\rho, 0) + F(\rho) = 0$$

$$w(\rho, 0) = -K_1(2\sqrt{\iota\rho})/K_1(2\sqrt{\iota\rho_0}) \quad \text{or} \quad w(\chi, 0) = -K_1(\chi)/K_1(\chi_0) \tag{17}$$

$$\partial w(\rho, \theta)/\partial\theta = \partial w(\chi, \theta)/\partial\theta = 0 \quad \text{for } \theta = 0, \ \rho > 0, \ \chi > 0 \tag{18}$$

Substitution of Eq.(6) into Eq.(3.1-36) yields for $w(\rho, \theta)$ the same equation as for $Z_1 H_{r\varphi}$:

$$\frac{\partial^2 w}{\partial\rho^2} - \frac{\partial^2 w}{\partial\theta^2} - 2(\iota + 1)\frac{\partial w}{\partial\theta} + \frac{1}{\rho}\frac{\partial w}{\partial\rho} - \left(\frac{1}{\rho^2} + 4\iota\right) w = 0 \tag{19}$$

Particular solutions $w_\omega(\rho, \theta)$ are obtained by the separation of variables using Bernoulli's product method,

$$w_\omega(\rho, \theta) = u(\rho)v(\theta) \tag{20}$$

$$\frac{1}{u}\frac{\partial^2 u}{\partial\rho^2} + \frac{1}{\rho}\frac{1}{u}\frac{\partial u}{\partial\rho} - \frac{1}{\rho^2} = \frac{1}{v}\frac{\partial^2 v}{\partial\theta^2} + 2(\iota + 1)\frac{1}{v}\frac{\partial v}{\partial\theta} + 4\iota = -\omega^2 \tag{21}$$

which yields two ordinary differential equations

$$\frac{d^2 u}{d\rho^2} + \frac{1}{\rho}\frac{du}{d\rho} + (\omega^2 - \frac{1}{\rho^2})u = 0 \tag{22}$$

and

$$\frac{d^2 v}{d\theta^2} + 2(\iota + 1)\frac{dv}{d\theta} + (\omega^2 + 4\iota)v = 0 \tag{23}$$

Equation (23) has the solution

$$v(\theta) = A_{10}e^{\gamma_1\theta} + A_{11}e^{\gamma_2\theta} \tag{24}$$

where the coefficients γ_1 and γ_2 are the roots of the equation

$$\gamma^2 + 2(\iota + 1)\gamma + \omega^2 + 4\iota = 0 \tag{25}$$

which we write in the following form:

$$\begin{aligned}
\gamma_1 &= -a + (a^2 - b^2)^{1/2} \qquad \text{for } a^2 > b^2 \\
\gamma_2 &= -a - (a^2 - b^2)^{1/2} \\
\gamma_1 &= -a + j(b^2 - a^2)^{1/2} \qquad \text{for } a^2 < b^2 \\
\gamma_2 &= -a - j(b^2 - a^2)^{1/2} \\
a &= \iota + 1, \ b^2 = \omega^2 + 4\iota
\end{aligned} \tag{26}$$

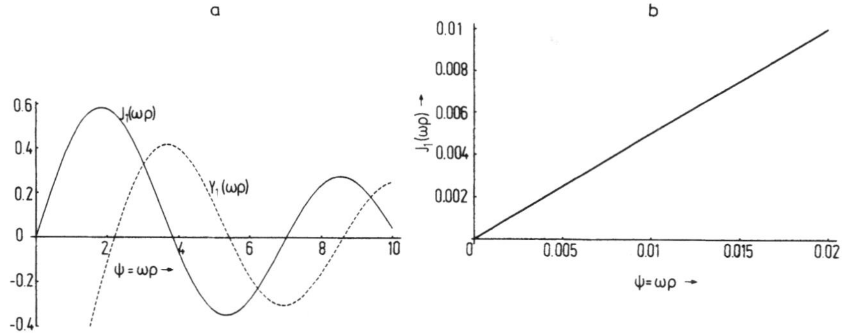

FIG.3.2-5. The Bessel functions of first kind $J_1(\omega\rho)$ and of second kind $Y_1(\omega\rho)$ in the interval $0 \leq \omega\rho \leq 10$ (a) as well as $J_1(\omega\rho)$ in the expanded interval $0 \leq \omega\rho \leq 0.0175$ (b).

The substitution

$$\psi = \omega\rho \tag{27}$$

is used to rewrite Eq.(22):

$$\psi^2 \frac{d^2u}{d\psi^2} + \psi \frac{du}{d\psi} + (\psi^2 - 1)u = 0 \tag{28}$$

The solutions of Eq.(28) are the Bessel functions of first kind $J_1(\psi)$ and of second kind $Y_1(\psi)$:

$$u(\psi) = A_{20}J_1(\psi) + A_{21}Y_1(\psi) \tag{29}$$

Only $J_1(\psi)$ approaches zero for $\psi \to 0$ and can thus approximate the condition of Eq.(13):

$$u(\psi) = u(\omega\rho) = A_{20}J_1(\omega\rho) \tag{30}$$

$$\begin{aligned} J_1(0) = 0,\ J_1(\omega\rho \to \infty) &\approx \sqrt{\frac{2}{\pi\omega\rho}} \sin\left(\omega\rho - \frac{\pi}{4}\right) \\ &\approx \frac{\sin\omega\rho - \cos\omega\rho}{\sqrt{\pi\omega\rho}} \end{aligned} \tag{31}$$

Plots of $J_1(\omega\rho)$ close to $\omega\rho = 0$ are shown in Figs.3.2-5a and b. Choosing for $\omega\rho$ the values of χ_0 in Fig.3.2-3 we see in Fig.3.2-5b or compute the values $J_1(0.02) = 0.0099995$, $J_1(0.001) = 0.00499994$, and $J_1(0.005) = 0.00249999$. The maximum value of $J_1(\omega\rho)$ in Fig.3.2-5a is about 0.582 at $\omega\rho = 1.8$. Hence, the error $J_1(\chi_0)/0.582$ relative to the maximum value of $J_1(\omega\rho)$ made by choosing $J_1(\chi_0) = 0$ is 1.7% for $\chi_0 = 0.02$, 0.86% for $\chi_0 = 0.01$, and 0.4% for $\chi_0 = 0.005$.

Leaving out the solution $A_{21}Y_1(\psi)$ in Eq.(29) the particular solution $w_\omega(\rho, \theta)$ of Eq.(20) becomes:

$$w_\omega(\rho, \theta) = \left(A_1 e^{\gamma_1\theta} + A_2 e^{\gamma_2\theta}\right) J_1(\omega\rho) \tag{32}$$

A general solution $w(\rho, \omega)$ is constructed by making A_1 and A_2 functions of ω and then integrating over all possible values of ω:

$$w(\rho, \theta) = \int_0^\infty \left[A_1(\omega)e^{\gamma_1\theta} + A_2(\omega)e^{\gamma_2\theta}\right] J_1(\omega\rho)\, d\omega \tag{33}$$

The time derivative $\partial w/\partial\theta$ equals:

$$\frac{\partial w}{\partial\theta} = \int_0^\infty \left[A_1(\omega)\gamma_1 e^{\gamma_1\theta} + A_2(\omega)\gamma_2 e^{\gamma_2\theta}\right] J_1(\omega\rho)\, d\omega \tag{34}$$

The initial conditions for $\theta = 0$ of Eqs.(17) and (18) demand:

$$\int_0^\infty [A_1(\omega) + A_2(\omega)]\, J_1(\omega\rho)\, d\omega = -\frac{K_1(2\sqrt{\iota\rho})}{K_1(2\sqrt{\iota\rho_0})} \tag{35}$$

$$\int_0^\infty [A_1(\omega)\gamma_1 + A_2(\omega)\gamma_2]\, J_1(\omega\rho)\, d\omega = 0 \tag{36}$$

These two equations must be solved for the two unknowns $A_1(\omega)$ and $A_2(\omega)$. If $J_1(\omega\rho)$ were replaced by $\sin\omega\rho$ we would immediately see that each equation was one half of a Fourier sine transform pair and there would be no difficulty eliminating the integrals on the left sides, which would yield two explicit equations for $A_1(\omega)$ and $A_2(\omega)$.

Instead of Fourier transform pairs we must consider the less well known Fourier-Bessel transform pairs (Nielsen 1904, pp. 369–370; Watson 1966, XIV, p. 453; Whittacker and Watson 1952, XVII, Example 44, p. 385):

$$g(\rho) = \int_0^\infty f(\omega) J_\mathrm{n}(\omega\rho)\omega\, d\omega \qquad n = 0,\ 1,\ 2, \ldots \tag{37}$$

$$f(\omega) = \int_0^\infty g(\rho) J_\mathrm{n}(\omega\rho)\rho\, d\rho \tag{38}$$

Bringing Eqs.(35) and (36) into the form of Eq.(37)

$$-\frac{K_1(2\sqrt{\iota\rho})}{K_1(2\sqrt{\iota\rho_0})} = \int_0^\infty \frac{1}{\omega}[A_1(\omega) + A_2(\omega)] J_1(\omega\rho)\omega\, d\omega \tag{39}$$

$$0 = \int_0^\infty \frac{1}{\omega}[A_1(\omega)\gamma_1 + A_2(\omega)\gamma_2] J_1(\omega\rho)\omega\, d\omega \tag{40}$$

yields with the help of Eq.(38) two explicit equations for $A_1(\omega)$ and $A_2(\omega)$

$$\frac{1}{\omega}[A_1(\omega)+A_2(\omega)] = -\frac{1}{K_1(2\sqrt{\iota\rho_0})}\int_0^\infty K_1(2\sqrt{\iota\rho})J_1(\omega\rho)\rho\,d\rho \tag{41}$$

$$\frac{1}{\omega}[A_1(\omega)\gamma_1+A_2(\omega)\gamma_2] = 0 \tag{42}$$

with the solutions:

$$A_1(\omega) = \frac{\omega\gamma_2 G(\omega)}{\gamma_1-\gamma_2},\quad A_2(\omega) = \frac{\omega\gamma_1 G(\omega)}{\gamma_2-\gamma_1}$$
$$G(\omega) = \frac{1}{K_1(2\sqrt{\iota\rho_0})}\int_0^\infty K_1(2\sqrt{\iota\rho})J_1(\omega\rho)\rho\,d\rho \tag{43}$$

The integral in Eq.(43) is tabulated (Gradshteyn and Ryzhik 1980; p. 672, 6.521/2):

$$\int_0^\infty K_1(2\sqrt{\iota\rho})J_1(\omega\rho)\rho\,d\rho = \frac{\omega}{2\sqrt{\iota}(\omega^2+4\iota)} \tag{44}$$

The functions $A_1(\omega)$ and $A_2(\omega)$ are rewritten:

$$(\iota+1)^2 > \omega^2+4\iota$$

$$A_1(\omega) = -\frac{1}{4\sqrt{\iota}K_1(2\sqrt{\iota\rho_0})}\frac{\omega^2}{\omega^2+4\iota}\left(1+\frac{\iota+1}{[(\iota+1)^2-\omega^2-4\iota]^{1/2}}\right) \tag{45}$$

$$A_2(\omega) = -\frac{1}{4\sqrt{\iota}K_1(2\sqrt{\iota\rho_0})}\frac{\omega^2}{\omega^2+4\iota}\left(1-\frac{\iota+1}{[(\iota+1)^2-\omega^2-4\iota]^{1/2}}\right) \tag{46}$$

$$(\iota+1)^2 < \omega^2+4\iota$$

$$A_1(\omega) = -\frac{1}{4\sqrt{\iota}K_1(2\sqrt{\iota\rho_0})}\frac{\omega^2}{\omega^2+4\iota}\left(1-j\frac{\iota+1}{[\omega^2+4\iota-(\iota+1)^2]^{1/2}}\right) \tag{47}$$

$$A_2(\omega) = -\frac{1}{4\sqrt{\iota}K_1(2\sqrt{\iota\rho_0})}\frac{\omega^2}{\omega^2+4\iota}\left(1+j\frac{\iota+1}{[\omega^2+4\iota-(\iota+1)^2]^{1/2}}\right) \tag{48}$$

For the limit $s_1 \to 0$, $\iota \to 0$ we get with

$$K_1(2\sqrt{\iota\rho_0}) \to 1/2\sqrt{\iota\rho_0}$$

the relations:

$$
\begin{aligned}
&1 > \omega \\
A_1(\omega) &= -\frac{\rho_0}{2}\left(1 + \frac{1}{(1-\omega^2)^{1/2}}\right) \\
A_2(\omega) &= -\frac{\rho_0}{2}\left(1 - \frac{1}{(1+\omega^2)^{1/2}}\right)
\end{aligned}
\tag{49}
$$

$$
\begin{aligned}
&1 < \omega \\
A_1(\omega) &= -\frac{\rho_0}{2}\left(1 - \frac{j}{(\omega^2-1)^{1/2}}\right) \\
A_2(\omega) &= -\frac{\rho_0}{2}\left(1 + \frac{j}{(\omega^2-1)^{1/2}}\right)
\end{aligned}
\tag{50}
$$

We substitute Eqs.(45) and (46) for

$$(\iota+1)^2 > \omega^2 + 4\iota \quad \text{or} \quad \iota^2 - 2\iota + 1 = (\iota-1)^2 = \omega_0^2 > \omega^2 \tag{51}$$

into Eq.(33) while Eqs.(47) and (48) are substituted for $\omega_0 < \omega$:

$$
\begin{aligned}
w(\rho,\theta) = &-\frac{e^{-a\theta}}{2\sqrt{\iota}K_1(2\sqrt{\iota}\rho_0)}\left(\int_0^{\omega_0}\left\{\operatorname{ch}\left[(a^2-b^2)^{1/2}\theta\right]\right.\right. \\
&\left. + \frac{a\operatorname{sh}\left[(a^2-b^2)^{1/2}\theta\right]}{(a^2-b^2)^{1/2}}\right\}\frac{\omega^2}{\omega^2+4\iota}J_1(\omega\rho)\,d\omega \\
&\left. + \int_{\omega_0}^{\infty}\left\{\cos\left[(b^2-a^2)^{1/2}\theta\right] + \frac{a\sin\left[(b^2-a^2)^{1/2}\theta\right]}{(b^2-a^2)^{1/2}}\right\}\frac{\omega^2}{\omega^2+4\iota}J_1(\omega\rho)\,d\omega\right) \\
&a = \iota+1,\ b^2 = \omega^2+4\iota,\ \omega_0 = |1-\iota| \\
&\frac{1}{2\sqrt{\iota}K_1(2\sqrt{\iota}\rho_0)} \approx \rho_0 \text{ for } 2\sqrt{\iota}\rho_0 \ll 1
\end{aligned}
\tag{52}
$$

The magnetic field strength $H_{\rm H}(\rho,\theta)$ of Eq.(6) assumes the following form with the help of Eqs.(14) and (52):

$$
\begin{aligned}
H_{\rm H}(\rho_{\rm r},\theta) = H_{{\rm H,r}\varphi}(\rho_{\rm r},\theta) = H_0\left(\frac{K_1(2\sqrt{\iota}\rho_{\rm r})}{K_1(2\sqrt{\iota}\rho_0)} + w(\rho_{\rm r},\theta)\right) \\
\rho_{\rm r} = \rho \geq \rho_0 > 0
\end{aligned}
\tag{53}
$$

In order to recognize the physical content of Eqs.(52) and (53) we investigate the limit

$$s_1 \to 0,\ \iota = 0,\ \omega_0 = 1,\ a = 1,\ b = \omega, \tag{54}$$

return to nonnormalized notation

$$\theta = \sigma_1 t/2\epsilon_1,\ \rho_{\mathrm{r}} = \sigma_1 r_{\mathrm{r}}/2\epsilon_1 c_1,\ \rho_0 = \sigma_1 r_0/2\epsilon_1 c_1, \tag{55}$$

and use some definitions that facilitate comparison with previous results[2]

$$\begin{gathered}\sigma_1/2\epsilon_1 = Z_1 c_1 \sigma_1/2 = \alpha,\ \alpha\omega/c = \beta,\ (1-\omega^2)^{1/2}\sigma_1/2\epsilon_1 = (\alpha^2 - \beta^2 c^2)^{1/2} \\ Z_1 = Z,\ c_1 = c,\ \sigma_1 = \sigma\end{gathered} \tag{56}$$

to write Eq.(52) in the following form:

$$\begin{aligned} w(r_{\mathrm{r}}, t) = & -r_0 e^{-\alpha t} \\ & \times \left\{ \int_0^{Z\sigma/2} \left[\operatorname{ch}(\alpha^2 - \beta^2 c^2)^{1/2} t + \frac{\alpha \operatorname{sh}(\alpha^2 - \beta^2 c^2)^{1/2} t}{(\alpha^2 - \beta^2 c^2)^{1/2}} \right] J_1(\beta r_{\mathrm{r}})\, d\beta \right. \\ & \left. + \int_{Z\sigma/2}^{\infty} \left[\cos(\beta^2 c^2 - \alpha^2)^{1/2} t + \frac{\alpha \sin(\beta^2 c^2 - \alpha^2)^{1/2} t}{(\beta^2 c^2 - \alpha^2)^{1/2}} \right] J_1(\beta r_{\mathrm{r}})\, d\beta \right\} \end{aligned} \tag{57}$$

The further limit

$$\sigma \to 0,\ \alpha = 0 \tag{58}$$

implies that there are no losses either due to magnetic (dipole or multipole) currents or electric (monopole, dipole, or multipole) currents. The first integral of Eq.(57) vanishes:

$$w(r_{\mathrm{r}}, t) = -r_0 \int_0^{\infty} \cos(\beta c t) J_1(\beta r_{\mathrm{r}})\, d\beta \tag{59}$$

With the limit

$$\lim_{\iota \to 0} \frac{K_1(2\sqrt{\iota}\rho_{\mathrm{r}})}{K_1(2\sqrt{\iota}\rho_0)} = \frac{2\sqrt{\iota}\rho_0}{2\sqrt{\iota}\rho_{\mathrm{r}}} = \frac{\rho_0}{\rho_{\mathrm{r}}} = \frac{r_0}{r_{\mathrm{r}}} \tag{60}$$

we can write Eq.(53) as follows:

$$\begin{gathered} H_{\mathrm{H},r\varphi}(r_{\mathrm{r}}, t) = \frac{H_0}{r_{\mathrm{r}}/r_0} \left(1 - r_{\mathrm{r}} \int_0^{\infty} \cos(\beta c t) J_1(\beta r_{\mathrm{r}})\, d\beta \right) \\ r_{\mathrm{r}} \geq r_0 > 0 \end{gathered} \tag{61}$$

[2] Harmuth 1986, p. 82

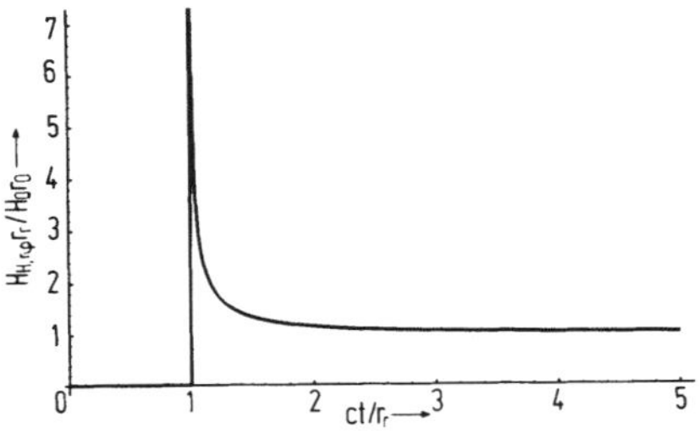

FIG.3.2-6. Plot of $H_{\mathrm{H},\mathrm{r}\varphi}(r_\mathrm{r},t)/(H_0 r_0/r_\mathrm{r})$ according to Eq.(64) as function of ct/r_r in the interval $0 \leq ct/r_\mathrm{r} \leq 5$.

In a table of integrals we find[3]:

$$
\begin{aligned}
r_\mathrm{r}\int_0^\infty \cos(ct\beta)J_1(r_\mathrm{r}\beta)\,d\beta &= \frac{\cos[\arcsin(ct/r_\mathrm{r})]}{(1-c^2t^2/r_\mathrm{r}^2)^{1/2}} && ct < r_\mathrm{r}\\
&= \infty \text{ or } 0 && ct = r_\mathrm{r}\\
&= -\frac{1}{(x^2-1)^{1/2}\left[x+(x^2-1)^{1/2}\right]} && ct > r_\mathrm{r}\\
&\quad x = ct/r_\mathrm{r}
\end{aligned}
\tag{62}
$$

With the substitution $ct/r_\mathrm{r} = \sin y$ we get

$$
\frac{\cos[\arcsin(ct/r_\mathrm{r})]}{(1-c^2t^2/r_\mathrm{r}^2)^{1/2}} = \frac{\cos[\arcsin(\sin y)]}{\left(1-\sin^2 y\right)^{1/2}} = \frac{\cos y}{\cos y} = 1 \tag{63}
$$

and Eq.(61) becomes:

$$
\begin{aligned}
\frac{H_{\mathrm{H},r\varphi}(r_\mathrm{r},t)}{H_0 r_0/r_\mathrm{r}} &= 0 && ct < r_\mathrm{r}\\
&= \text{undefined} && ct = r_\mathrm{r}\\
&= 1 + \frac{1}{\left(x^2-1\right)^{1/2}\left[x+\left(x^2-1\right)^{1/2}\right]} && ct > r_\mathrm{r}\\
&\quad x = ct/r_\mathrm{r},\ r_\mathrm{r} \geq r_0
\end{aligned}
\tag{64}
$$

Figure 3.2-6 shows $H_{\mathrm{H},r\varphi}/(H_0 r_0/r)$ in the interval $0 \leq ct/r_\mathrm{r} \leq 5$. There is no magnetic field strength for $ct/r_\mathrm{r} < 1$ in accordance with the requirements of special relativity. For large values of ct/r_r the function approaches 1, which means $H_{\mathrm{H},r\varphi}(r_\mathrm{r},t)$ varies like $1/r_\mathrm{r}$.

The associated electric field strength $E_{\mathrm{H},\mathrm{r}x}$ is defined by Eq.(3.1-37). In the loss-free case $s_1 = 0$, $\sigma_1 = 0$ it assumes the following form with $H_{\mathrm{r}\varphi} = H_{\mathrm{H},\mathrm{r}\varphi}$:

[3] Gradshteyn and Ryzhik 1980, p. 730, 6.671/2

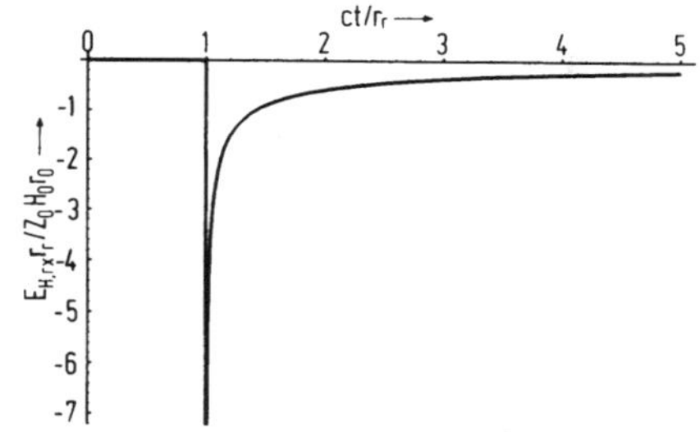

FIG.3.2-7. The electric field strength $E_{\mathrm{H,r}x}(r_{\mathrm{r}},t)$ according to Eq.(70) associated with the magnetic field strength $H_{\mathrm{H,r}\varphi}(r_{\mathrm{r}},t)$ of Fig.3.2-6.

$$\begin{aligned} E_{\mathrm{H,r}x} &= \int \frac{\partial(Z_1 H_{\mathrm{H,r}\varphi})}{\partial\theta} d\rho_{\mathrm{r}} + E_{\mathrm{cr}}(\theta) \\ &= \frac{Z_1}{c}\int \frac{\partial H_{\mathrm{H,r}\varphi}}{\partial t} dr_{\mathrm{r}} + E_{\mathrm{cr}}(t) \end{aligned} \tag{65}$$

Differentiation of Eq.(61) with respect to t yields

$$\frac{\partial H_{\mathrm{H,r}\varphi}}{\partial t} = H_0 c r_0 \int_0^\infty \sin(ct\beta) J_1(r_{\mathrm{r}}\beta)\beta\, d\beta \tag{66}$$

and $E_{\mathrm{H,r}x}$ becomes:

$$E_{\mathrm{H.r}x} = Z_1 H_0 r_0 \int_0^\infty \sin(ct\beta)\left(\int J_1(r_{\mathrm{r}}\beta)\, dr_{\mathrm{r}}\right)\beta\, d\beta + E_{\mathrm{cr}}(t) \tag{66}$$

The indefinite integral of $J_1(x)$ is known[4]

$$\int J_1(r_{\mathrm{r}}\beta) dr_{\mathrm{r}} = -\frac{J_0(r_{\mathrm{r}}\beta)}{\beta} \tag{67}$$

and the electric field strength $E_{\mathrm{H,r}x}$ becomes:

$$E_{\mathrm{H,r}x}(r_{\mathrm{r}},t) = -Z_1 H_0 r_0 \int_0^\infty \sin(ct\beta) J_0(r_{\mathrm{r}}\beta)\, d\beta + E_{\mathrm{cr}}(t) \tag{68}$$

With the tabulated integral[5]

$$\begin{aligned} \int_0^\infty J_0(r_{\mathrm{r}}\beta)\sin(ct\beta)\, d\beta &= 0 && ct < r_{\mathrm{r}} \\ &= \frac{1}{(c^2t^2 - r_{\mathrm{r}}^2)^{1/2}} && ct > r_{\mathrm{r}} \end{aligned} \tag{69}$$

[4] Gradshteyn and Ryzhik 1980, p. 634, 5.56/1
[5] Gradshteyn and Ryzhik 1980, p. 731, 6.67/7

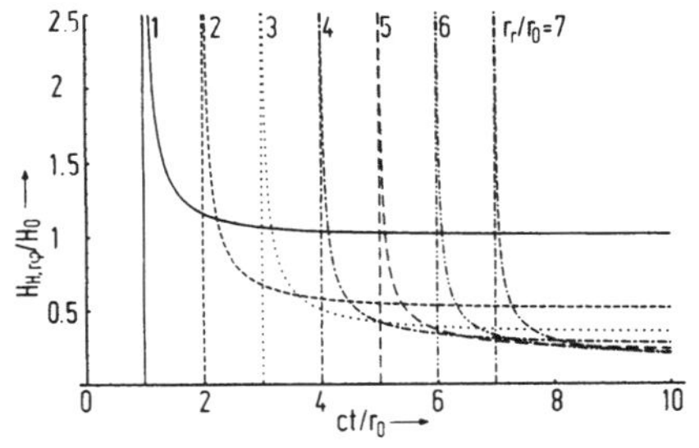

FIG.3.2-8. Plots of $H_{H,r\varphi}/H_0$ according to Eq.(61) for $r_r/r_0 = 1, 2, \dots, 7$ in the interval $0 \leq ct/r_0 \leq 10$.

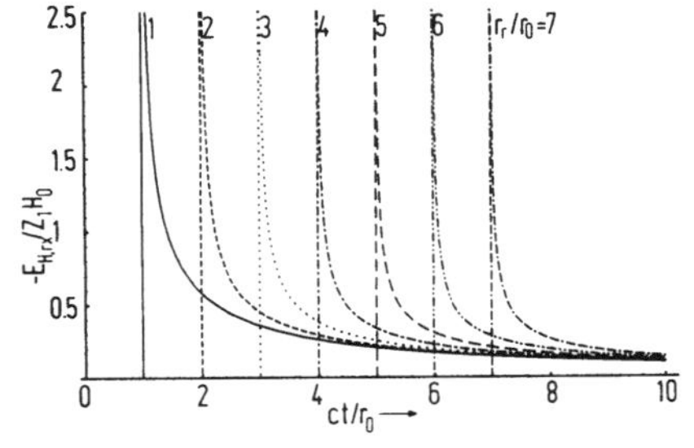

FIG.3.2-9. Plots of $-E_{H,rx}/Z_1H_0$ according to Eq.(70) for $r_r/r_0 = 1, 2, \dots, 7$ in the interval $0 \leq ct/r_0 \leq 10$.

we get:

$$
\begin{aligned}
E_{H,rx}(r_r, t) &= 0 && ct < r_r \\
&= -Z_1 H_0 \frac{r_0}{r_r} \frac{1}{(c^2t^2/r^2 - 1)^{1/2}} && ct > r_r
\end{aligned}
\tag{70}
$$

A plot of $E_{H,rx}(r_r, t)$ is shown in Fig.3.2-7. As one would expect there is no field strength for $ct/r_r < 1$ and the field strength drops to zero for large values of ct/r_r since the magnetic field strength in Fig.3.2-6 becomes time-independent.

The effect of the term r_r/r_0 in Eqs.(61) and (70) is made clear by Figs.3.2-8 and 3.2-9. In Fig.3.2-8 we see plots of $H_{H,r\varphi}/H_0$ according to Eq.(61) for $r_r/r_0 = 1, 2, \dots, 7$ in the interval $0 \leq ct/r_0 \leq 10$, while Fig.3.2-9 shows corresponding plots for $-E_{H,rx}/Z_1H_0$ according to Eq.(70).

Figure 3.2-10 shows plots of the magnetic field strength according to Eq.(53). The magnetic conductivity s_1 is not assumed to be zero and the ratio $\iota = \epsilon_1 s_1/\mu_1\sigma_1$ is chosen equal to $\iota = 0.01$. These plots should be compared with those of Fig.3.2-8 that hold for $\iota = 0$. The parameter r_r/r_0 in Fig.3.2-8 is replaced by $r_r/r_0 = \rho_r/\rho_0 = 10\rho_r$ for $\rho_0 = 0.1$. The infinite peaks and the smooth curves of Fig.3.2-8 are hard to duplicate for the complicated integral of Eq.(52) that defines $w(\rho_r, \theta)$ in Eq.(53). But we recognize that the monotonous *drop* to a constant value for $\theta \to \infty$ in Fig.3.2-8 is replaced by a drop followed by a monotonous *rise* to a constant value for $\theta \to \infty$.

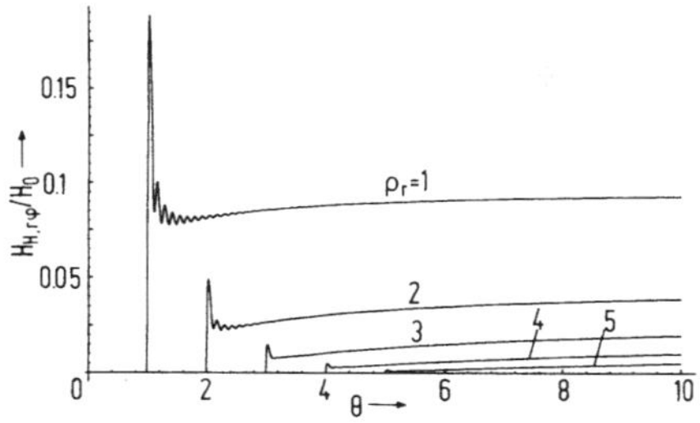

FIG.3.2-10. Plots of $H_{\mathrm{H,r}\varphi}/H_0$ according to Eq.(53) for $\rho_0 = 0.1$, $\iota = 0.01$, and $\rho_\mathrm{r} = 1, 2, 3, 4, 5$.

3.3 ASSOCIATED ELECTRIC FIELD STRENGTH

In Section 3.2 we derived the magnetic field strength $H_{\mathrm{H,r}\varphi}$ excited by a magnetic step function at the surface of a cylinder with radius ρ_0. This field strength is given by Eq.(3.2-53). To simplify writing we drop temporarily the subscripts r and φ:

$$Z_1 H_\mathrm{H}(\rho,\theta) = Z_1 H_0 \left(\frac{K_1(2\sqrt{\iota\rho})}{K_1(2\sqrt{\iota\rho_0})} + w(\rho,\theta) \right) \tag{1}$$

The associated electric field strength $E_{\mathrm{r}x}$ is determined by Eqs.(3.1-37) and (3.1-38). We write $H_{\mathrm{r}\varphi} = H_{\mathrm{H,r}\varphi} = H_\mathrm{H}$ and $E_{\mathrm{r}x} = E_{\mathrm{H,r}x} = E_\mathrm{H}$ to emphasize that the field strengths are due to magnetic excitation:

$$E_\mathrm{H} = \int \left(\frac{\partial (Z_1 H_\mathrm{H})}{\partial \theta} + 2\iota (Z_1 H_\mathrm{H}) \right) d\rho + E_{\mathrm{r}1}(\theta) \tag{2}$$

$$E_\mathrm{H} = e^{-2\theta} \left(\frac{1}{\rho} \int \frac{\partial (\rho Z_1 H_\mathrm{H})}{\partial \rho} e^{2\theta} d\theta + E_{\mathrm{r}2}(\rho) \right) \tag{3}$$

Consider first Eq.(3). The term $\partial(\rho Z_1 H_\mathrm{H})/\partial\rho$ follows from Eq.(1):

$$\frac{\partial (\rho Z_1 H_\mathrm{H})}{\partial \rho} = Z_1 H_0 \left(\frac{1}{K_1(2\sqrt{\iota\rho_0})} \frac{\partial[\rho K_1(2\sqrt{\iota\rho})]}{\partial \rho} + \frac{\partial[\rho w(\rho,\theta)]}{\partial \rho} \right) \tag{4}$$

In a table of integrals[1] we find:

$$\frac{1}{z}\frac{d}{dz}[zK_1(z)] = -K_0(z), \quad \frac{d}{d\rho}\left[\rho K_1(2\sqrt{\iota\rho})\right] = -2\sqrt{\iota}\rho K_0(2\sqrt{\iota\rho}) \tag{5}$$

Plots of $K_1(2\sqrt{\iota\rho})$ and $K_0(2\sqrt{\iota\rho})$ are shown in Fig.3.3-1.

The derivative $\partial[\rho w(\rho,\theta)]/\partial\rho$ in Eq.(4) follows from Eq.(3.2-52). We note that only the term $J_1(\omega\rho)$ contains the variable ρ:

[1] Gradshteyn and Ryzhik 1980, p. 970, 8.486/14

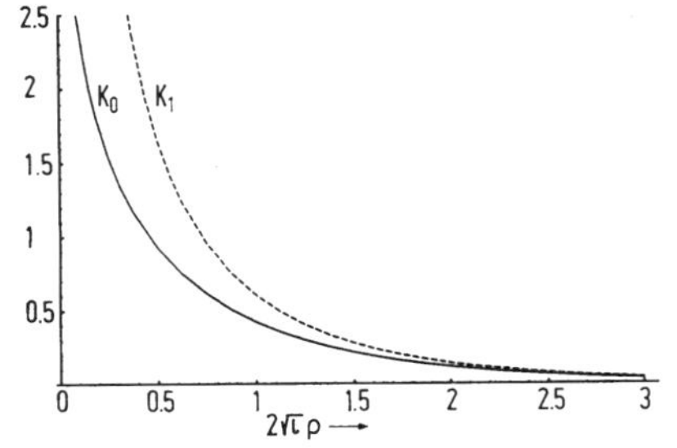

FIG.3.3-1. Plots of the modified Bessel functions $K_1(2\sqrt{\iota}\rho)$ and $K_0(2\sqrt{\iota}\rho)$ as functions of $2\sqrt{\iota}\rho$.

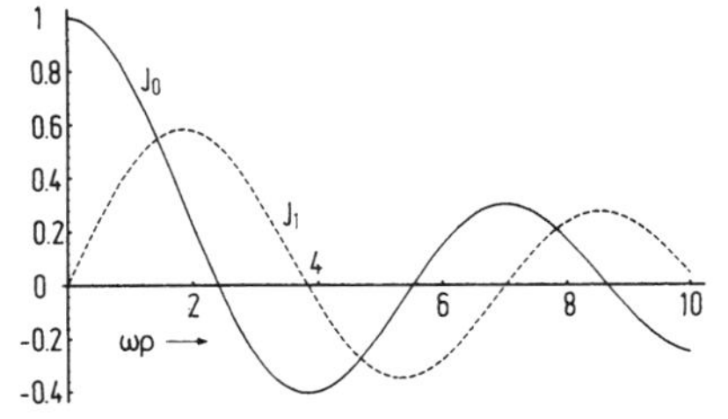

FIG.3.3-2. Plots of the Bessel functions $J_1(\omega\rho)$ and $J_0(\omega\rho)$ as functions of $\omega\rho$.

$$
\begin{aligned}
\frac{\partial(\rho w)}{\partial\rho} = \frac{e^{-a\theta}}{2\sqrt{\iota}K_1(2\sqrt{\iota}\rho_0)} \Bigg(\int_0^{\omega_0} \bigg\{ \operatorname{ch}\Big[(a^2-b^2)^{1/2}\theta\Big] \\
+ \frac{a\operatorname{sh}\big[(a^2-b^2)^{1/2}\theta\big]}{(a^2-b^2)^{1/2}} \bigg\} \frac{\omega^2}{\omega^2+4\iota} \frac{d[\rho J_1(\omega\rho)]}{d\rho}\, d\omega \\
+ \int_{\omega_0}^{\infty} \bigg\{ \cos\Big[(b^2-a^2)^{1/2}\theta\Big] + \frac{a\sin\big[(b^2-a^2)^{1/2}\theta\big]}{(b^2-a^2)^{1/2}} \bigg\} \frac{\omega^2}{\omega^2+4\iota} \frac{d[\rho J_1(\omega\rho)]}{d\rho}\, d\omega \\
a = \iota+1,\ b^2 = \omega^2+4\iota,\ \omega_0 = |1-\iota|
\end{aligned} \tag{6}
$$

With the help of a table[2] we obtain:

$$
\frac{1}{z}\frac{d}{dz}[zJ_1(z)] = J_0(z), \quad \frac{d}{d\rho}[\rho J_1(\omega\rho)] = \omega\rho J_0(\omega\rho) \tag{7}
$$

Plots of $J_0(\omega\rho)$ and $J_1(\omega\rho)$ are shown in Fig.3.3-2.

The terms $d[\rho J_1(\omega\rho)]/d\rho$ in Eq.(6) are replaced by Eq.(7) and the following relation is obtained:

[2] Gradshteyn and Ryzhik 1980, p. 968, 8.471/3

$$\frac{\partial[\rho w(\rho,\theta)]}{\partial\rho} = \frac{e^{-a\theta}}{2\sqrt{\iota}K_1(2\sqrt{\iota\rho_0})}\left(\int_0^{\omega_0}\left\{\mathrm{ch}\left[(a^2-b^2)^{1/2}\theta\right]\right.\right.$$

$$\left.+\frac{a\,\mathrm{sh}\left[(a^2-b^2)^{1/2}\theta\right]}{(a^2-b^2)^{1/2}}\right\}\frac{\omega^3\rho J_0(\omega\rho)}{\omega^2+4\iota}\,d\omega$$

$$\left.+\int_{\omega_0}^{\infty}\left\{\cos\left[(b^2-a^2)^{1/2}\theta\right]+\frac{a\sin\left[(b^2-a^2)^{1/2}\theta\right]}{(b^2-a^2)^{1/2}}\right\}\frac{\omega^3\rho J_0(\omega\rho)}{\omega^2+4\iota}\,d\omega\right) \quad (8)$$

Equation (3) is transformed into the following form with the help of Eqs.(4) and (5):

$$E_{\mathrm{H}}(\rho,\theta) =$$

$$e^{-2\theta}\left[\frac{Z_1H_0}{\rho}\int\left(-\frac{2\sqrt{\iota\rho}K_0(2\sqrt{\iota\rho})}{K_1(2\sqrt{\iota\rho_0})}+\frac{\partial[\rho w(\rho,\theta)]}{\partial\rho}\right)e^{2\theta}d\theta+E_{\mathrm{r}2}(\rho)\right] \quad (9)$$

The integration over θ can be done analytically:

$$e^{-2\theta}\int e^{2\theta}d\theta=\frac{1}{2} \quad (10)$$

$$I_{11}(\omega,\theta)=e^{-2\theta}\int e^{-a\theta}\,\mathrm{ch}\left[(a^2-b^2)^{1/2}\theta\right]e^{2\theta}d\theta$$

$$=\frac{e^{-a\theta}}{(2-a)^2-(a^2-b^2)}\left\{(2-a)\,\mathrm{ch}\left[(a^2-b^2)^{1/2}\theta\right]\right.$$

$$\left.-(a^2-b^2)^{1/2}\,\mathrm{sh}\left[(a^2-b^2)^{1/2}\theta\right]\right\} \quad (11)$$

$$I_{12}(\omega,\theta)=e^{-2\theta}\int e^{-a\theta}\,\mathrm{sh}\left[(a^2-b^2)^{1/2}\theta\right]e^{2\theta}d\theta$$

$$=\frac{e^{-a\theta}}{(2-a)^2-(a^2-b^2)}\left\{(2-a)\,\mathrm{sh}\left[(a^2-b^2)^{1/2}\theta\right]\right.$$

$$\left.-(a^2-b^2)^{1/2}\,\mathrm{ch}\left[(a^2-b^2)^{1/2}\theta\right]\right\} \quad (12)$$

$$I_{13}(\omega,\theta)=e^{-2\theta}\int e^{-a\theta}\cos\left[(b^2-a^2)^{1/2}\theta\right]e^{2\theta}d\theta$$

$$=\frac{e^{-a\theta}}{(2-a)^2+(b^2-a^2)}\left\{(2-a)\cos\left[(b^2-a^2)^{1/2}\theta\right]\right.$$

$$\left.+(b^2-a^2)^{1/2}\sin\left[(b^2-a^2)^{1/2}\theta\right]\right\} \quad (13)$$

$$\begin{aligned} I_{14}(\omega,\theta) &= e^{-2\theta}\int e^{-a\theta}\sin\left[(b^2-a^2)^{1/2}\theta\right]e^{2\theta}d\theta \\ &= \frac{e^{-a\theta}}{(2-a)^2+(b^2-a^2)}\Big\{(2-a)\sin\left[(b^2-a^2)^{1/2}\theta\right] \\ &\qquad -(b^2-a^2)^{1/2}\cos\left[(b^2-a^2)^{1/2}\theta\right]\Big\} \end{aligned} \tag{14}$$

Equation (9) may now be rewritten into the following form with the help of Eqs.(10)–(14):

$$\begin{aligned} E_{\mathrm{H}}(\rho,\theta) &= \frac{Z_1H_0}{2\sqrt{\iota}K_1(2\sqrt{\iota\rho_0})}\Bigg\{-2\iota K_0(2\sqrt{\iota\rho}) \\ &\quad +\Bigg[\int_0^{\omega_0}\left(I_{11}(\omega,\theta)+\frac{aI_{12}(\omega,\theta)}{(a^2-b^2)^{1/2}}\right)\frac{\omega^3J_0(\omega\rho)}{\omega^2+4\iota}d\omega \\ &\quad +\int_{\omega_0}^{\infty}\left(I_{13}(\omega,\theta)+\frac{aI_{14}(\omega,\theta)}{(b^2-a^2)^{1/2}}\right)\frac{\omega^3J_0(\omega\rho)}{\omega^2+4\iota}d\omega\Bigg]\Bigg\}+E_{\mathrm{r2}}(\rho)e^{-2\theta} \\ &= -Z_1H_0\rho_0\Bigg\{2\iota K_0(2\sqrt{\iota\rho}) \\ &\quad +e^{-a\theta}\Bigg[\int_0^{\omega_0}\left(2\iota\,\mathrm{ch}[(a^2-b^2)^{1/2}\theta]+\frac{(2\iota a-b^2)\,\mathrm{sh}[(a^2-b^2)^{1/2}\theta]}{(a^2-b^2)^{1/2}}\right)\frac{\omega J_0(\rho\omega)}{\omega^2+4\iota}d\omega \\ &\quad +\int_{\omega_0}^{\infty}\left(2\iota\cos[(b^2-a^2)^{1/2}\theta]+\frac{(2\iota a-b^2)\sin[(b^2-a^2)^{1/2}\theta]}{(b^2-a^2)^{1/2}}\right)\frac{\omega J_0(\rho\omega)}{\omega^2+4\iota}d\omega\Bigg]\Bigg\} \\ &\quad +e^{-2\theta}E_{\mathrm{r2}}(\rho) \end{aligned}$$

$$(2-a)^2-a^2+b^2=\omega^2,\ a=\iota+1,\ b^2=\omega^2+4\iota$$

$$K_1(2\sqrt{\iota\rho_0})\approx 1/2\sqrt{\iota\rho_0}\ \text{for}\ 2\sqrt{\iota\rho_0}\ll 1 \tag{15}$$

We have succeeded in rewriting Eq.(3) as a single integral of the variable ω rather than a double integral of the variables θ and ω. Equation (2) is a double integral of the variables ρ and ω and we must rewrite it into a single integral of the variable ω too. To this end we calculate $\partial(Z_1H_{\mathrm{H}})/\partial\theta$ with the help of Eqs.(1) and (3.2-52):

$$\frac{\partial(Z_1 H_{\mathrm{H}})}{\partial\theta} = Z_1 H_0 \frac{\partial w(\rho,\theta)}{\partial\theta} = Z_1 H_0 \Bigg[- a w(\rho,\theta) - \frac{e^{-a\theta}}{2\sqrt{\iota}K_1(2\sqrt{\iota}\rho_0)} \Bigg(\int_0^{\omega_0} \Big\{ (a^2-b^2)^{1/2} \,\mathrm{sh}\left[(a^2-b^2)^{1/2}\theta\right] + a\,\mathrm{ch}\left[(a^2-b^2)^{1/2}\theta\right] \Big\} \frac{\omega^2 J_1(\omega\rho)}{\omega^2+4\iota}\,d\omega - \int_{\omega_0}^{\infty} \Big\{ (b^2-a^2)^{1/2}\sin\left[(b^2-a^2)^{1/2}\theta\right] - a\cos\left[(b^2-a^2)^{1/2}\theta\right] \Big\} \frac{\omega^2 J_1(\omega\rho)}{\omega^2+4\iota}\,d\omega \Bigg) \Bigg] \quad (16)$$

The term $2\iota(Z_1 H_{\mathrm{H}})$ in Eq.(2) is written in the form

$$2\iota(Z_1 H_{\mathrm{H}}) = 2\iota Z_1 H_0 \left[\frac{K_1(2\sqrt{\iota}\rho)}{K_1(2\sqrt{\iota}\rho_0)} + w(\rho,\theta) \right] \quad (17)$$

and Eq.(2) becomes:

$$E_{\mathrm{H}} = Z_1 H_0 \int \Bigg[\frac{2\iota K_1(2\iota\rho)}{K_1(2\sqrt{\iota}\rho_0)} - \frac{(2\iota - a)e^{-a\theta}}{2\sqrt{\iota}K_1(2\sqrt{\iota}\rho_0)} \Bigg(\int_0^{\omega_0} \Big\{ \mathrm{ch}\left[(a^2-b^2)^{1/2}\theta\right] + \frac{a\,\mathrm{sh}\left[(a^2-b^2)^{1/2}\theta\right]}{(a^2-b^2)^{1/2}} \Big\} \frac{\omega^2 J_1(\omega\rho)}{\omega^2+4\iota}\,d\omega + \int_{\omega_0}^{\infty} \Big\{ \cos\left[(b^2-a^2)^{1/2}\theta\right] + \frac{a\sin\left[(b^2-a^2)^{1/2}\theta\right]}{(b^2-a^2)^{1/2}} \Big\} \frac{\omega^2 J_1(\omega\rho)}{\omega^2+4\iota}\,d\omega \Bigg) - \frac{e^{-a\theta}}{2\sqrt{\iota}K_1(2\sqrt{\iota}\rho_0)} \Bigg(\int_0^{\omega_0} \Big\{ (a^2-b^2)^{1/2}\,\mathrm{sh}\left[(a^2-b^2)^{1/2}\theta\right] + a\,\mathrm{ch}\left[(a^2-b^2)^{1/2}\theta\right] \Big\} \frac{\omega^2 J_1(\omega\rho)}{\omega^2+4\iota}\,d\omega - \int_{\omega_0}^{\infty} \Big\{ (b^2-a^2)^{1/2}\sin\left[(b^2-a^2)^{1/2}\theta\right] - a\cos\left[(b^2-a^2)^{1/2}\theta\right] \frac{\omega^2 J_1(\omega\rho)}{\omega^2+4\iota}\,d\omega \Bigg) \Bigg] d\rho + E_{\mathrm{r}1}(\theta) \quad (18)$$

Only the two integrals $\int K_1(2\sqrt{\iota}\rho)d\rho$ and $\int J_1(\omega\rho)\,d\rho$ of the variable ρ occur. They can both be integrated analytically[3]

[3] Gradshteyn and Ryzhik, 1980 p. 634, 6.56/1

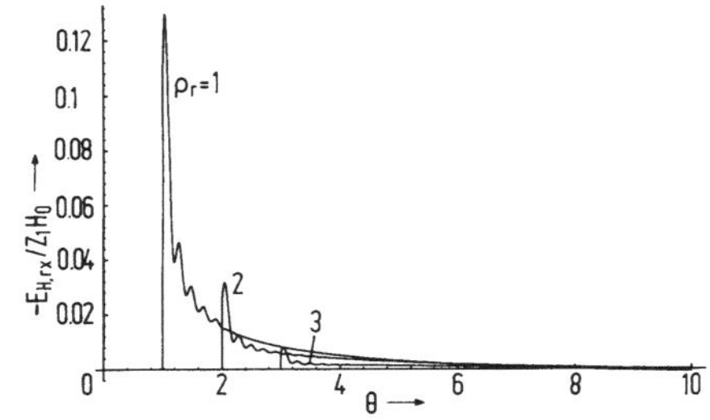

FIG.3.3-3. Plots of $E_{H,rx}/Z_1H_0$ according to Eq.(21) for $E_{r1}(\theta) = 0$, $\rho_0 = 0.1$, $\iota = 0.01$, and $\rho_r = 1, 2, 3$.

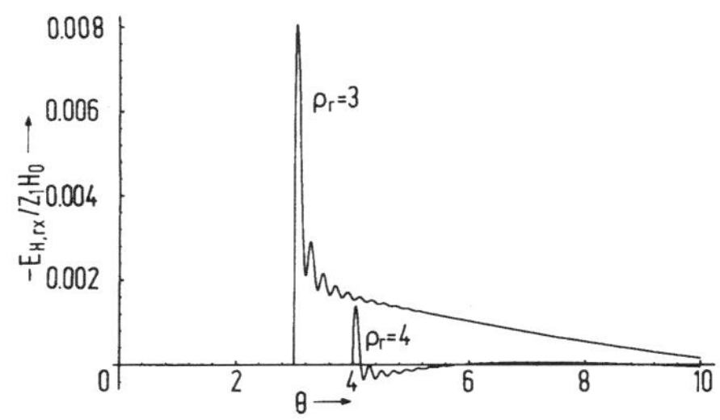

FIG.3.3-4. Plots of $E_{H,rx}/Z_1H_0$ according to Eq.(21) for $E_{r1} = 0$, $\rho_0 = 0.1$, $\iota = 0.01$, and $\rho_r = 3, 4$.

$$\int K_1(2\sqrt{\iota}\rho)d\rho = -\frac{1}{2\sqrt{\iota}}K_0(2\sqrt{\iota}\rho) \tag{19}$$

$$\int J_1(\omega\rho)d\rho = -\frac{1}{\omega}J_0(\omega\rho) \tag{20}$$

and Eq.(18) can be written in the following form with the subscripts r and x reinstated:

$$E_{H,rx}(\rho_r,\theta) = Z_1H_0F_{H,rx}(\rho_r,\theta) = -Z_1H_0\rho_0\Bigg\{2\iota K_0(2\sqrt{\iota}\rho_r)$$

$$+ e^{-a\theta}\left[\int_0^{\omega_0}\left(2\iota\,\mathrm{ch}\left[(a^2-b^2)^{1/2}\theta\right] + \frac{(2\iota a - b^2)\,\mathrm{sh}\left[(a^2-b^2)^{1/2}\theta\right]}{(a^2-b^2)^{1/2}}\right)\frac{\omega J_0(\rho_r\omega)}{\omega^2+4\iota}\,d\omega\right.$$

$$\left.+\int_{\omega_0}^{\infty}\left(2\iota\cos\left[(b^2-a^2)^{1/2}\theta\right] + \frac{(2\iota a - b^2)\sin\left[(b^2-a^2)^{1/2}\theta\right]}{(b^2-a^2)^{1/2}}\right)\frac{\omega J_0(\rho_r\omega)}{\omega^2+4\iota}\,d\omega\right]\Bigg\}$$

$$+ E_{r1}(\theta)$$

$$a = \iota + 1,\ b^2 = \omega^2 + 4\iota,\ E_{H,rx} = E_H,\ \rho_r = \rho$$

$$K_1(2\sqrt{\iota}\rho_0) \approx 1/2\sqrt{\iota}\rho_0 \text{ for } 2\sqrt{\iota} \ll 1 \tag{21}$$

Since Eqs.(15) and (21) must be equal we get the following condition for the integration constants $E_{r1}(\theta)$ and $E_{r2}(\rho) = E_{r2}(\rho_r)$:

$$e^{-2\theta} E_{r2}(\rho_r) = E_{r1}(\theta) \tag{22}$$

The left side cannot be a function of ρ_r and $E_{r2}(\rho_r)$ must be a constant. If we choose the initial field strength $E_{r1}(0)$ equal to zero we get $E_{r2} = E_{r1} = 0$.

To check the correctness of Eq.(21) we produce plots as shown in Fig.3.3-3 and 3.3-4. The plots should be similar to those in Fig.3.2-9. We recognize that the similarity exists but that it is difficult to obtain infinite or large peaks as well as smooth curves from the integral of Eq.(21) just as in the case of Fig.3.2-10. A close look at Fig.3.3-3 reveals that the plot for $\rho_r = 3$ drops below zero at $\theta = 10$. This deficiency is removed when the larger scale of Fig.3.3-4 is used. However, the precision of the computation is insufficient for the plot $\rho_r = 4$.

3.4 Electric Excitation Force

For signal solutions excited by an electric force function we must use Eqs.(3.1-39):

$$\frac{\partial^2 E_{rx}}{\partial \rho_r^2} - \frac{\partial^2 E_{rx}}{\partial \theta^2} - 2(\iota + 1)\frac{\partial E_{rx}}{\partial \theta} + \frac{1}{\rho_r}\frac{\partial E_{rx}}{\partial \rho_r} - 4\iota E_{rx} = 0 \tag{1}$$

For the excitation force at the location $\rho_r = 0$ we assume an electric step function $E_0 S(\theta)$:

$$\begin{aligned} E_{rx}(0,\theta) = E_{ix} = E_0 S(\theta) &= 0 \quad \text{for } \theta < 0 \\ &= E_0 \quad \text{for } \theta \geq 0 \end{aligned} \tag{2}$$

For great distances, $\rho_r \to \infty$, we have as a second boundary condition the requirement that the electric field strength must be finite:

$$E_{rx}(\rho_r \to \infty, \theta) = \text{finite} \tag{3}$$

At the time $\theta = 0$ both E_{rx} and $H_{r\varphi}$ shall be zero for $\rho_r > 0$, which implies the initial conditions[1]

$$E_{rx}(\rho_r, 0) = H_{r\varphi}(\rho_r, 0) = 0 \tag{4}$$

If $E_{rx}(\rho_r, 0)$ and $H_{r\varphi}(\rho_r, 0)$ are zero for $\rho_r > 0$, their derivatives with respect to ρ_r must vanish:

$$\partial E_{rx}(\rho_r, 0)/\partial \rho_r = \partial H_{r\varphi}(\rho_r, 0)/\partial \rho_r = 0 \tag{5}$$

Due to Eqs.(3.1-33) and (3.1-34) the initial conditions of Eqs.(4) and (5) imply a further initial condition for $\rho_r > 0$ and $\theta = 0$:

[1] As in Section 3.2 this initial condition holds for $\theta \leq 0$ too since the excitation force $E_{rx}(0,\theta)$ of Eq.(2) is zero for $\theta < 0$.

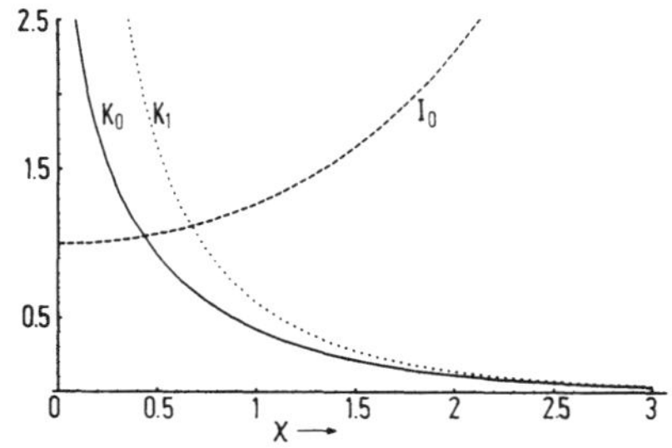

FIG.3.4-1. Plots of the modified Bessel functions $K_0(\chi)$, $I_0(\chi)$, and $K_1(\chi)$; $\chi = 2\sqrt{\iota\rho}$.

$$\partial E_{\mathrm{r}x}(\rho_\mathrm{r}, \theta)/\partial\theta = \partial H_{\mathrm{r}\varphi}(\rho_\mathrm{r}, \theta)/\partial\theta = 0 \tag{6}$$

We try to find a sufficiently general solution of Eq.(1) by the method of *steady state solution plus deviation from it* used in Section 3.2. In analogy to Eq.(3.2-6) we try the following solution:

$$E_{\mathrm{r}x}(\rho_\mathrm{r}, \theta) = E_{\mathrm{E,r}x}(\rho_\mathrm{r}, \theta) = E_\mathrm{E}(\rho, \theta) = E_0[w(\rho, \theta) + F(\rho)] \tag{7}$$

The subscript E indicates the electric field strength is due to an electric force function according to Eq.(2) rather than a magnetic force function according to Eq.(3.2-1). The subscripts r and x are dropped temporarily to simplify writing.

Substitution of $E_0F(\rho)$ into Eq.(1) yields the differential equation for the steady state:

$$\frac{d^2F}{d\rho^2} + \frac{1}{\rho}\frac{dF}{d\rho} - 4\iota F = 0 \tag{8}$$

The substitution $\chi = 2\sqrt{\iota\rho}$ yields the differential equation of the modified Bessel functions $K_0(\chi)$ and $I_0(\chi)$:

$$\chi^2\frac{d^2F}{d\chi^2} + \chi\frac{dF}{d\chi} - \chi^2 F = 0, \quad \chi = 2\sqrt{\iota\rho} \tag{9}$$

$$F(\chi) = A_{00}K_0(\chi) + A_{01}I_0(\chi) \tag{10}$$

Figure 3.4-1 shows plots of $K_0(\chi)$, $I_0(\chi)$, as well as of $K_1(\chi)$. The function $I_0(\chi)$ is not usable since it is not finite for $\rho \to \infty$ but $K_0(\chi)$ looks very similar to $K_1(\chi)$ which has yielded a good result before. According to Fig.3.2-2 we assume that the exciting field strength in a point y is not given in the line interval $y-\chi_0 < y < y+\chi_0$ but on the half-circle with radius χ_0. In analogy to Eqs.(3.2-11)–(3.2-14) and using the boundary condition of Eq.(2) for $\chi \to 0$, $\rho \to 0$ we get:

$$E_{\rm E}(\chi_0 \rightarrow 0, \theta) = E_0[w(\chi_0, \theta) + A_{00}K_0(\chi_0)] = E_0 \qquad \text{for } \theta \geq 0,\ \chi \geq \chi_0 = 2\sqrt{\iota}\rho_0 \tag{11}$$

$$A_{00} = 1/K_0(\chi_0) \tag{12}$$

$$w(\chi_0, \theta) = 0 \tag{13}$$

$$F(\chi) = K_0(\chi)/K_0(\chi_0) \tag{14}$$

We substitute now $E_0 w(\rho, \theta)$ of Eq.(7) into Eq.(1) and obtain the same partial differential equation with $E_{\rm rx}$ replaced by w:

$$\frac{\partial^2 w}{\partial \rho^2} - \frac{\partial^2 w}{\partial \theta^2} - (2\iota + 1)\frac{\partial w}{\partial \theta} + \frac{1}{\rho}\frac{\partial w}{\partial \rho} - 4\iota w = 0 \tag{15}$$

Again we look for a particular solution $w_\omega(\rho, \theta)$ obtained by Bernoulli's product method

$$w_\omega(\rho, \theta) = u(\rho)v(\theta) \tag{16}$$

$$\frac{1}{u}\frac{\partial^2 u}{\partial \rho^2} + \frac{1}{\rho}\frac{1}{u}\frac{\partial u}{\partial \rho} = \frac{1}{v}\frac{\partial^2 v}{\partial \theta^2} + 2(\iota + 1)\frac{1}{v}\frac{\partial v}{\partial \theta} + 4\iota = -\omega^2 \tag{17}$$

that yields two ordinary differential equations

$$\frac{d^2 u}{d\rho^2} + \frac{1}{\rho}\frac{du}{d\rho} + \omega^2 u = 0 \tag{18}$$

and

$$\frac{d^2 v}{d\theta^2} + 2(\iota + 1)\frac{dv}{d\theta} + (\omega^2 + 4\iota)v = 0 \tag{19}$$

Equation (19) is the same as Eq.(3.2-23) while Eq.(18) is transformed by the substitution $\psi = \omega\rho$ into the differential equation of the Bessel function $J_0(\psi)$ and the Neumann function $N_0(\psi) = Y_0(\psi)$:

$$\psi^2\frac{d^2 u}{d\psi^2} + \psi\frac{du}{d\psi} + \psi^2 u = 0, \quad \psi = \omega\rho \tag{20}$$

$$u(\psi) = A_{20}J_0(\psi) + A_{21}Y_0(\psi) \tag{21}$$

Figure 3.4-2 shows plots of $J_0(\psi)$ and $Y_0(\psi)$. Neither function can satisfy the condition of Eq.(13) for $\chi_0 \rightarrow 0$, but both behave acceptably for $\chi \rightarrow \infty$. The plot of the sum $J_0(\psi) + Y_0(\psi)$ in Fig.3.4-2 shows how to proceed. We can produce a function $u(\chi)$ that is zero for as small a value $\chi = \chi_0$ as we want if we choose one constant in Eq.(21) properly. We use a strange notation in order to connect to published results:

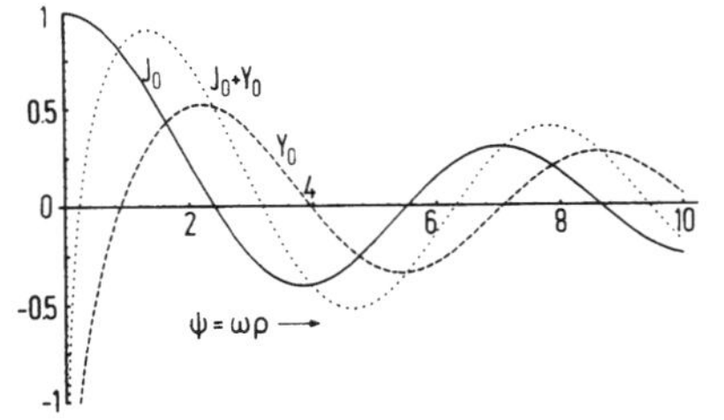

FIG.3.4-2 Plots of the Bessel functions $J_0(\psi)$ and $Y_0(\psi)$ as well as of their sum $J_0(\psi)+Y_0(\psi)$; $\psi = \omega\rho$.

$$u(\psi) = A_0[J_0(\psi)\cos\alpha_0 + Y_0(\psi)\sin\alpha_0] = A_0 Z_0(\psi) \tag{22}$$

The function $Z_0(\psi)$ can be made zero for any value $\psi = \psi_0 \ll 1$ by choosing α_0 according to the following relation:

$$J_0(\psi_0)\cos\alpha_0 + Y_0(\psi_0)\sin\alpha_0 = 0$$
$$\frac{\sin\alpha_0}{\cos\alpha_0} = \mathrm{tg}\,\alpha_0 \approx \sin\alpha_0 \approx \alpha_0 = -\frac{J_0(\psi_0)}{Y_0(\psi_0)} \approx -\frac{1}{Y_0(\psi_0)}$$
$$\psi_0 \ll 1,\ 0 < \alpha_0 \ll 1 \tag{24}$$

We consider the following relation known in the theory of Bessel functions[2] that we will use to find a Fourier-Bessel-Neumann transform pair for the function $Z_0(\psi)$ in Eq.(22):

$$\int_0^\infty \omega\,d\omega \int_0^\infty g(\rho)Z_\nu(\omega\rho)Z_\nu(\omega\lambda)\rho\,d\rho = \frac{1}{2}A_0^2[g(\rho+0)+g(\rho-0)]$$
$$-\frac{2A_0^2\sin\alpha_0\sin(\alpha_0+\nu\pi)}{\pi\sin\nu\pi}\int_0^\infty \frac{\rho^{2\nu}-\lambda^{2\nu}}{\rho^{\nu-1}\lambda^\nu(\rho^2-\lambda^2)}g(\rho)\,d\rho$$
$$0 \le \nu \le 1/2 \tag{25}$$

The functions $Z_\nu(\omega\rho)$ and $Z_\nu(\omega\lambda)$ are more general than our function $Z_0(\psi)$ with $\psi = \omega\rho = \omega\lambda$ but the substitutions $\nu = 0$ and $\rho = \lambda$ make the term on the right side of Eq.(25) undetermined. We must calculate the two limits $\nu \rightarrow 0$ and $\lambda \rightarrow \rho$. First we write:

$$\rho^{2\nu} = e^{2\nu\ln\rho} = 1 + 2\nu\ln\rho + \ldots,\ \nu \ll 1$$
$$\lambda^{2\nu} = 1 + 2\nu\ln\lambda + \ldots$$
$$\rho^{2\nu} - \lambda^{2\nu} \approx 2\nu(\ln\rho - \ln\lambda) = 2\nu\ln\frac{\rho}{\lambda} \tag{26}$$

[2] Watson 1966, p. 468

The variable λ is rewritten:

$$\lambda = \rho(1+\delta), \ \delta \ll 1 \tag{27}$$

$$\rho^2 - \lambda^2 = \rho^2 - \rho^2(1+2\delta) = -2\delta\rho^2 \tag{28}$$

$$\ln\frac{\rho}{\lambda} = \ln(1-\delta) \approx -\delta$$

$$\frac{\rho^{2\nu} - \lambda^{2\nu}}{\rho^2 - \lambda^2} = \frac{-2\nu\delta}{-2\delta\rho^2} = \frac{\nu}{\rho^2} \tag{29}$$

The last term on the right side of Eq.(25) may now be written as follows for $\nu \ll 1$, $\delta \ll 1$:

$$-\frac{2A_0^2\sin^2\alpha_0}{\nu\pi^2}\int_0^\infty \frac{\nu}{\rho^2}\rho g(\rho)\,d\rho$$

$$= -\frac{2A_0^2\sin^2\alpha_0}{\pi^2}\int_0^\infty \frac{g(\rho)}{\rho}\,d\rho = -2\left(\frac{A_0}{\pi Y_0(\psi_0)}\right)^2\int_0^\infty \frac{g(\rho)}{\rho}\,d\rho \tag{30}$$

This expression becomes arbitrarily small for sufficiently small values of ψ_0 provided the integral exists and has a finite value.

For non-discontinuous points of $g(\rho)$ we may write $g(\rho)$ instead of the average $[g(\rho+0)+g(\rho-0)]/2$. The constant A_0 is chosen to be 1. Equation (25) may then be written as follows:

$$f(\omega) = \int_0^\infty g(\rho)Z_0(\omega\rho)\rho\,d\rho \tag{31}$$

$$g(\rho) = \int_0^\infty f(\omega)Z_0(\omega\rho)\omega\,d\omega \tag{32}$$

$$Z_0(\omega\rho) = J_0(\omega\rho) - Y_0(\omega\rho)/Y_0(\psi_0) \tag{33}$$

$$\psi_0 \ll 1, \ \int_0^\infty \rho^{-1}g(\rho)\,d\rho = \text{finite} \tag{34}$$

The similarity of Eqs.(31) and (32) to Eqs.(3.2-37) and (3.2-38) suggests the name Fourier-Bessel-Neumann transform pair. We note that Eqs.(31) and (32) hold for $n = 0$ only rather than for $n = 0, 1, 2, \ldots$.

We return to the particular solution $w_\omega(\rho,\theta)$ of Eq.(16). The function $v(\theta)$ is defined by Eqs.(3.2-24)–(3.2-26). With the help of Eqs.(22) and (33) we get:

$$w_\omega(\rho,\theta) = \left(A_1e^{\gamma_1\theta} + A_2e^{\gamma_2\theta}\right)\left(J_0(\omega\rho) - \frac{Y_0(\omega\rho)}{Y_0(\psi_0)}\right), \ \psi_0 \ll 1, \ \psi_0 \le \omega\rho \tag{35}$$

The particular solution is generalized in the usual way by making A_1 and A_2 functions of ω and integrating over all values of ω:

$$w(\rho,\theta) = \int_{\psi_0/\rho}^{\infty} \left[A_1(\omega)e^{\gamma_1\theta} + A_2(\omega)e^{\gamma_2\theta}\right]\left(J_0(\omega\rho) - \frac{Y_0(\omega\rho)}{Y_0(\psi_0)}\right)d\omega \tag{36}$$

The time derivative of $w(\rho,\theta)$ is needed too:

$$\frac{\partial w}{\partial\theta} = \int_{\psi_0/\rho}^{\infty} \left[A_1(\omega)\gamma_1 e^{\gamma_1\theta} + A_2(\omega)\gamma_2 e^{\gamma_2\theta}\right]\left(J_0(\omega\rho) - \frac{Y_0(\omega\rho)}{Y_0(\psi_0)}\right)d\omega \tag{37}$$

The initial conditions at the time $\theta = 0$ of Eqs.(4) and (6) impose the following conditions according to Eqs.(11) and (14):

$$\int_{\psi_0/\rho}^{\infty} \left[A_1(\omega) + A_2(\omega)\right]\left(J_0(\omega\rho) - \frac{Y_0(\omega\rho)}{Y_0(\psi_0)}\right)d\omega = -\frac{K_0(2\sqrt{\iota\rho})}{K_0(\chi_0)} \tag{38}$$

$$\int_{\psi_0/\rho}^{\infty} \left[A_1(\omega)\gamma_1 + A_2(\omega)\gamma_2\right]\left(J_0(\omega\rho) - \frac{Y_0(\omega\rho)}{Y_0(\psi_0)}\right)d\omega = 0 \tag{39}$$

The two equations are rewritten in the form of Eq.(32):

$$-\frac{K_0(2\sqrt{\iota\rho})}{K_0(\chi_0)} = \int_{\psi_0/\rho}^{\infty} \frac{1}{\omega}\left[A_1(\omega) + A_2(\omega)\right]\left(J_0(\omega\rho) - \frac{Y_0(\omega\rho)}{Y_0(\psi_0)}\right)\omega\, d\omega \tag{40}$$

$$0 = \int_{\psi_0/\rho}^{\infty} \frac{1}{\omega}\left[A_1(\omega)\gamma_1 + A_2(\omega)\gamma_2\right]\left(J_0(\omega\rho) - \frac{Y_0(\omega\rho)}{Y_0(\psi_0)}\right)\omega\, d\omega \tag{41}$$

We can use Eq.(31) to obtain explicit equations for $A_1(\omega)$ and $A_2(\omega)$ provided either the lower limit $\omega = \psi_0/\rho$ in Eqs.(40) and (41) can be replaced by $\omega = 0$ or the lower limit $\omega = 0$ in Eqs.(31) and (32) can be replaced by $\omega = \psi_0/\rho$ for sufficiently small values of ψ_0. In order to find out we replace $\omega = 0$ by $\omega = \psi_0/\rho$ in Eq.(32) and $\rho = 0$ by $\rho = \psi_0/\omega$ in Eq.(31), and apply these modified equations to Eqs.(40) and (41):

$$\frac{1}{\omega}\left[A_1(\omega) + A_2(\omega)\right] = -\frac{1}{K_0(\chi_0)}\int_{\psi_0/\omega}^{\infty} K_0(2\sqrt{\iota\rho})\left(J_0(\omega\rho) - \frac{Y_0(\omega\rho)}{Y_0(\psi_0)}\right)\rho\, d\rho \tag{42}$$

$$\frac{1}{\omega}[A_1(\omega)\gamma_1 + A_2(\omega)\gamma_2 = 0 \tag{43}$$

$$\psi_0/\omega \ll 1,\ \psi_0/\rho \ll 1 \tag{44}$$

One obtains:

$$A_1(\omega) = \frac{\omega\gamma_2 G(\omega)}{\gamma_1 - \gamma_2}, \quad A_2(\omega) = \frac{\omega\gamma_1 G(\omega)}{\gamma_2 - \gamma_1}$$

$$G(\omega) = \frac{1}{K_0(\chi_0)}\Big(\int_{\psi_0/\omega}^{\infty} K_0(2\sqrt{\iota\rho})J_0(\omega\rho)\rho\,d\rho$$

$$- \frac{1}{Y_0(\psi_0)}\int_{\psi_0/\omega}^{\infty} K_0(2\sqrt{\iota\rho})Y_0(\omega\rho)\rho\,d\rho\Big)$$

$$2\sqrt{\iota\rho} \geq \chi_0, \ \omega\rho \geq \psi_0, \ \chi_0 \ll 1, \ \psi_0 \ll 1 \tag{45}$$

In order to evaluate the integrals of $G(\omega)$ we need $K_0(2\sqrt{\iota\rho})$ and—later—$K_1(2\sqrt{\iota\rho})$ expressed in terms of the functions $J_n(jz)$ and $Y_n(jz)$ with $z = 2\sqrt{\iota\rho}$:

$$K_0(z) = j\frac{\pi}{2}\left[J_0(jz) + jY_0(jz)\right], \quad K_1(z) = -\frac{\pi}{2}\left[J_1(jz) + jY_1(jz)\right] \tag{46}$$

$$G(\omega) = \frac{j\pi}{2K_0(\chi_0)}\Big(\int_{\psi_0/\omega}^{\infty} [J_0(2j\sqrt{\iota\rho}) + jY_0(2j\sqrt{\iota\rho})]J_0(\omega\rho)\rho\,d\rho$$

$$- \frac{1}{Y_0(\psi_0)}\int_{\psi_0/\omega}^{\infty} [J_0(2j\sqrt{\iota\rho}) + jY_0(2j\sqrt{\iota\rho})]Y_0(\omega\rho)\rho\,d\rho\Big) \tag{47}$$

The following indefinite integral is listed[3]; the functions Z_n and $\overline{Z}_n$ stand for any Bessel function J_n or Y_n:

$$\int Z_n(\eta\rho)\overline{Z}_n(\omega\rho)\rho\,d\rho = \rho\,\frac{\omega Z_n(\eta\rho)\overline{Z}_{n-1}(\omega\rho) - \eta Z_{n-1}(\eta\rho)\overline{Z}_n(\omega\rho)}{\eta^2 - \omega^2} \tag{48}$$

We obtain the following four integrals of $G(\omega)$:

$$\int J_0(2j\sqrt{\iota\rho})J_0(\omega\rho)\rho\,d\rho = \rho\,\frac{\omega J_0(2j\sqrt{\iota\rho})J_1(\omega\rho) - 2j\sqrt{\iota}J_1(2j\sqrt{\iota\rho})J_0(\omega\rho)}{\omega^2 + 4\iota} \tag{49}$$

$$\int Y_0(2j\sqrt{\iota\rho})J_0(\omega\rho)r\,d\rho = \rho\,\frac{\omega Y_0(2j\sqrt{\iota\rho})J_1(\omega\rho) - 2j\sqrt{\iota}Y_1(2j\sqrt{\iota\rho})J_0(\omega\rho)}{\omega^2 + 4\iota} \tag{50}$$

[3]Gradshteyn and Ryzhik 1980, p. 634, 5.54/1

$$\int J_0(2j\sqrt{\iota\rho})Y_0(\omega\rho)\rho\,d\rho = \rho\,\frac{\omega J_0(2j\sqrt{\iota\rho})Y_1(\omega\rho) - 2j\sqrt{\iota}J_1(2j\sqrt{\iota\rho})Y_0(\omega\rho)}{\omega^2+4\iota} \tag{51}$$

$$\int Y_0(2j\sqrt{\iota\rho})Y_0(\omega\rho)\rho\,d\rho = \rho\,\frac{\omega Y_0(2j\sqrt{\iota\rho})Y_1(\omega\rho) - 2j\sqrt{\iota}Y_1(2j\sqrt{\iota\rho})Y_0(\omega\rho)}{\omega^2+4\iota} \tag{52}$$

Using the relations of Eq.(46) we may write $G(\omega)$ as follows:

$$\begin{aligned} G(\omega) = \Bigg\{\frac{\rho}{\omega^2+4\iota}\Bigg[\omega\,\frac{K_0(2\sqrt{\iota\rho})}{K_0(\chi_0)}\left(J_1(\omega\rho) - \frac{Y_1(\omega\rho)}{Y_0(\psi_0)}\right) \\ - 2\sqrt{\iota}\,\frac{K_1(2\sqrt{\iota\rho})}{K_0(\chi_0)}\left(J_0(\omega\rho) - \frac{Y_0(\omega\rho)}{Y_0(\psi_0)}\right)\Bigg]\Bigg\}_{\rho=\psi_0/\omega}^{\infty} \end{aligned} \tag{53}$$

For the determination of the upper limit $\rho \to \infty$ we need the following asymptotic expansions[4]:

$$J_0(\omega\rho) \approx \sqrt{\frac{2}{\pi\omega\rho}}\cos\left(\omega\rho - \frac{\pi}{4}\right), \quad J_1(\omega\rho) \approx \sqrt{\frac{2}{\pi\omega\rho}}\sin\left(\omega\rho - \frac{\pi}{4}\right) \tag{54}$$

$$Y_0(\omega\rho) \approx \sqrt{\frac{2}{\pi\omega\rho}}\sin\left(\omega\rho - \frac{\pi}{4}\right), \quad Y_1(\omega\rho) \approx -\sqrt{\frac{2}{\pi\omega\rho}}\cos\left(\omega\rho - \frac{\pi}{4}\right) \tag{55}$$

$$K_0(2\sqrt{\iota\rho}) \approx \sqrt{\frac{\pi}{4\sqrt{\iota\rho}}}e^{-2\sqrt{\iota\rho}}, \qquad K_1(2\sqrt{\iota\rho}) \approx \sqrt{\frac{\pi}{4\sqrt{\iota\rho}}}e^{-2\sqrt{\iota\rho}} \tag{56}$$

The exponent $e^{-2\sqrt{\iota\rho}}$ of K_0 and K_1 makes the following four products zero:

$$\lim_{\rho\to\infty}\rho K_0(2\sqrt{\iota\rho})J_1(\omega\rho) = 0, \quad \lim_{\rho\to\infty}\rho K_0(2\sqrt{\iota\rho})Y_1(\omega\rho) = 0 \tag{57}$$

$$\lim_{\rho\to\infty}\rho K_1(2\sqrt{\iota\rho})J_0(\omega\rho) = 0, \quad \lim_{\rho\to\infty}\rho K_1(2\sqrt{\iota\rho})Y_0(\omega\rho) = 0 \tag{58}$$

The function $G(\omega)$ of Eq.(53) assumes the following form for the lower limit $\rho = \psi_0/\omega$:

$$\begin{aligned} G(\omega) = \frac{\psi_0}{\omega(\omega^2+4\iota)}\Bigg[2\sqrt{\iota}\,\frac{K_1(2\sqrt{\iota\psi_0/\omega})}{K_0(\chi_0)}[J_0(\psi_0)-1] \\ - \omega\,\frac{K_0(2\sqrt{\iota\psi_0/\omega})}{K_0(\chi_0)}\left(J_1(\psi_0) - \frac{Y_1(\psi_0)}{Y_0(\psi_0)}\right)\Bigg] \end{aligned} \tag{59}$$

[4] Gradshteyn and Ryzhik 1980, pp. 961–963

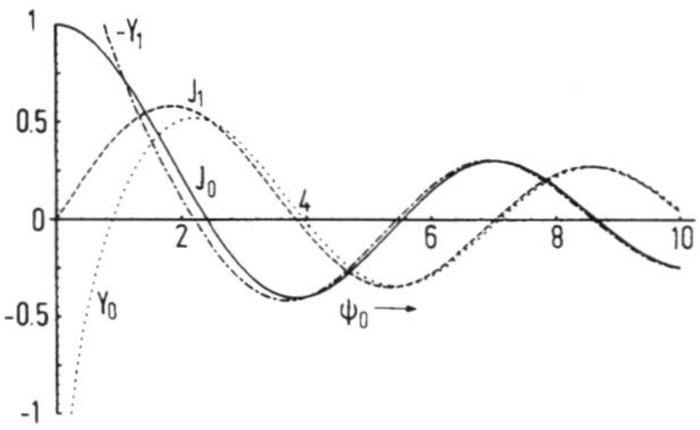

FIG.3.4-3 Plots of the Bessel functions $J_0(\psi_0)$, $J_1(\psi_0)$ and the Neumann functions $Y_0(\psi_0)$, $-Y_1(\psi_0)$ in the interval $0 \leq \psi_0 \leq 10$.

The modified Bessel functions K_0 and K_1 are plotted for the variable $\chi = 2\sqrt{\iota}\psi_0/\omega$ in Fig.3.4-1. The Bessel functions $J_0(\psi_0)$, $J_1(\psi_0)$ and the Neumann functions $Y_0(\psi_0)$, $-Y_1(\psi_0)$ are shown in Fig.3.4-3.

Since $G(\omega)$ contains three parameters ι, ψ_0, and χ_0 we simplify it for $\psi_0 \ll 1$, $\chi_0 \ll 1$. This calls for the following ascending series expansions[5]:

$$J_0(\psi_0) \approx 1 - \frac{1}{4}\psi_0^2, \quad J_1(\psi_0) \approx \frac{1}{2}\psi_0\left(1 - \frac{1}{8}\psi_0^2\right) \tag{60}$$

$$Y_0(\psi_0) \approx \frac{2}{\pi}\left(1 - \frac{1}{4}\psi_0^2\right)\ln\left(\frac{\psi_0}{2}e^C\right) + \frac{1}{2\pi}\psi_0^2 \tag{61}$$

$$Y_1(\psi_0) \approx -\frac{1}{\pi}\left[\frac{2}{\psi_0} - \psi_0\ln\left(\frac{\psi_0}{2}e^C\right) + \frac{1}{2}\psi_0\right] \tag{62}$$

$$I_0(\chi_0) \approx 1 + \frac{1}{4}\chi_0^2, \quad I_0(2\sqrt{\iota}\psi_0/\omega) \approx 1 + \frac{\iota}{\omega^2}\psi_0^2 \tag{63}$$

$$I_1(2\sqrt{\iota}\psi_0/\omega) \approx \frac{\sqrt{\iota}\psi_0}{\omega}\left(1 + \frac{\iota}{2\omega^2}\psi_0^2\right) \tag{64}$$

$$K_0(\chi_0) \approx -I_0(\chi_0)\ln\left(\frac{\chi_0}{2}e^C\right) \approx -\left(1 + \frac{1}{4}\chi_0^2\right)\ln\left(\frac{\chi_0}{2}e^C\right) \tag{65}$$

$$K_0(2\sqrt{\iota}\psi_0/\omega) \approx -\left(1 + \frac{\iota}{\omega^2}\psi_0^2\right)\ln\left(\frac{\sqrt{\iota}\psi_0}{\omega}e^C\right) \tag{66}$$

$$\begin{aligned} K_1(2\sqrt{\iota}\psi_0/\omega) &\approx \frac{\omega}{2\sqrt{\iota}\psi_0} + I_1(2\sqrt{\iota}\psi_0/\omega)\ln\left(\frac{\sqrt{\iota}\psi_0}{\omega}e^C\right) \\ &\approx \frac{\omega}{2\sqrt{\iota}\psi_0} + \frac{\sqrt{\iota}\psi_0}{\omega}\ln\left(\frac{\sqrt{\iota}\psi_0}{\omega}e^C\right) \end{aligned} \tag{67}$$

$$C \approx 0.577\,216 \text{ (Euler's constant)}, \; \psi_0 \ll 1, \; \chi_0 \ll 1$$

Using these approximations we obtain for certain terms in Eq.(59) simpler relations

$$\psi_0 K_1\left(\frac{2\sqrt{\iota}\psi_0}{\omega}\right)[J_0(\psi_0) - 1] \approx -\frac{1}{4}\psi_0^2\left[\frac{\omega}{2\sqrt{\iota}} + \frac{\sqrt{\iota}\psi_0^2}{\omega}\ln\left(\frac{\sqrt{\iota}\psi_0}{\omega}e^C\right)\right] \tag{68}$$

[5] Gradshteyn and Ryzhik 1980, pp. 959–961

$$\psi_0 K_0\left(\frac{2\sqrt{\iota}\psi_0}{\omega}\right)J_1(\psi_0)\approx-\frac{1}{2}\psi_0^2\left(1+\frac{\sqrt{\iota}\psi_0^2}{\omega}\right)\ln\left(\frac{\sqrt{\iota}\psi_0}{\omega}e^C\right) \tag{69}$$

$$\frac{1}{Y_0(\psi_0)}\approx\frac{\pi}{2\ln(\psi_0 e^C/2)}\left(1+\frac{1}{4}\psi_0^2\right) \tag{70}$$

$$\psi_0 K_0\left(\frac{2\sqrt{\iota}\psi_0}{\omega}\right)\frac{Y_1(\psi_0)}{Y_0(\psi_0)}\approx\left(1+\frac{\iota\psi_0^2}{\omega^2}\right)\frac{\ln\left(\sqrt{\iota}\psi_0 e^C/\omega\right)}{\ln\left(\psi_0 e^C/2\right)} \tag{71}$$

and $G(\omega)$ becomes:

$$\begin{aligned}G(\omega)&\approx\frac{1}{\omega^2+4\iota}\frac{\ln\left(\sqrt{\iota}\psi_0 e^C/\omega\right)}{K_0(\chi_0)\ln\left(\psi_0 e^C/2\right)}\\&\approx-\frac{1}{\omega^2+4\iota}\frac{\ln\left(\sqrt{\iota}\psi_0 e^C/\omega\right)}{\ln\left(\chi_0 e^C/2\right)\ln\left(\psi_0 e^C/2\right)}\end{aligned} \tag{72}$$

From Eq.(45) we introduce the notation

$$\begin{gathered}2\sqrt{\iota}\rho_0=\chi_0,\quad \omega\rho_0=\psi_0\\ \rho_0=\frac{\psi_0}{\omega}=\frac{\chi_0}{2\sqrt{\iota}},\quad \frac{\sqrt{\iota}\psi_0}{\omega}=\frac{\chi_0}{2}\end{gathered} \tag{73}$$

and $G(\omega)$ becomes:

$$G(\omega)\approx=\frac{1}{\omega^2+4\iota}\frac{1}{\ln(1/\psi_0)},\quad \psi_0\ll 1 \tag{74}$$

The functions $A_1(\omega)$ and $A_2(\omega)$ of Eq.(45) may be written as follows with the help of Eqs.(3.2-26) and (74):

$$(\iota+1)^2>\omega^2+4\iota$$

$$A_1(\omega)=-\frac{1}{2\ln(1/\psi_0)}\frac{\omega}{\omega^2+4\iota}\left(1+\frac{\iota+1}{[(\iota+1)^2-\omega^2-4\iota]^{1/2}}\right) \tag{75}$$

$$A_2(\omega)=-\frac{1}{2\ln(1/\psi_0)}\frac{\omega}{\omega^2+4\iota}\left(1-\frac{\iota+1}{[(\iota+1)^2-\omega^2-4\iota]^{1/2}}\right) \tag{76}$$

$$(\iota+1)^2<\omega^2+4\iota$$

$$A_1(\omega)=-\frac{1}{2\ln(1/\psi_0)}\frac{\omega}{\omega^2+4\iota}\left(1-j\frac{\iota+1}{[\omega^2+4\iota-(\iota+1)^2]^{1/2}}\right) \tag{77}$$

$$A_2(\omega)=-\frac{1}{2\ln(1/\psi_0)}\frac{\omega}{\omega^2+4\iota}\left(1+j\frac{\iota+1}{[\omega^2+4\iota-(\iota+1)^2]^{1/2}}\right) \tag{78}$$

We substitute Eqs.(75) and (76) for $(\iota+1)^2 > \omega^2 + 4\iota$ or $\omega^2 < \omega_0^2 = (\iota-1)^2$ into Eq.(36) while Eqs.(77) and (78) are substituted for $\omega > \omega_0$:

$$w(\rho,\theta) = -\frac{e^{-a\theta}}{\ln(1/\psi_0)}\Bigg[\int_{\psi_0/\rho}^{\omega_0}\left(\operatorname{ch}[(a^2-b^2)^{1/2}\theta] + \frac{a\operatorname{sh}[(a^2-b^2)^{1/2}\theta]}{(a^2-b^2)^{1/2}}\right) \times \frac{\omega}{\omega^2+4\iota}\left(J_0(\omega\rho) - \frac{Y_0(\omega\rho)}{Y_0(\psi_0)}\right)d\omega$$
$$+\int_{\omega_0}^{\infty}\left(\cos[(b^2-a^2)^{1/2}\theta] + \frac{a\sin[(b^2-a^2)^{1/2}\theta]}{(b^2-a^2)^{1/2}}\right) \times \frac{\omega}{\omega^2+4\iota}\left(J_0(\omega\rho) - \frac{Y_0(\omega\rho)}{Y_0(\psi_0)}\right)d\omega\Bigg]$$
$$a = \iota+1,\ b^2 = \omega^2+4\iota,\ \omega_0 = |1-\iota| \tag{79}$$

The electric field strength $E_{\mathrm{E,rx}}(\rho_{\mathrm{r}},\theta)$ of Eq.(7) becomes with the help of Eqs.(14) and (79):

$$E_{\mathrm{E}}(\rho_{\mathrm{r}},\theta) = E_{\mathrm{E,rx}}(\rho_{\mathrm{r}},\theta) = E_0\left(\frac{K_0(2\sqrt{\iota\rho_{\mathrm{r}}})}{K_0(2\sqrt{\iota\rho_0})} + w(\rho_{\mathrm{r}},\theta)\right)$$
$$\rho_{\mathrm{r}} = \rho \geq \rho_0 > 0,\ \chi_0 = 2\sqrt{\iota\rho_0} \tag{80}$$

In order to obtain some understanding of the physical content of Eqs.(79) and (80) we investigate the limit

$$s \to 0,\ \iota = 0,\ \omega_0 = 1,\ a = 1,\ b = \omega, \tag{81}$$

return to nonnormalized notation

$$\theta = \sigma_1 t/2\epsilon_1,\ \rho_{\mathrm{r}} = \sigma_1 r_{\mathrm{r}}/2\epsilon_1 c_1,\ \rho_0 = \sigma_1 r_0/2\epsilon_1 c_1 \tag{82}$$

and introduce certain definitions used for previously published results[6]

$$\sigma_1/2\epsilon_1 = Z_1 c_1 \sigma_1/2 = \alpha,\ (1-\omega^2)^{1/2}\sigma_1/2\epsilon_1 = (\alpha^2 - \beta^2 c^2)^{1/2},$$
$$\alpha\omega/c = \beta,\ \rho_{\mathrm{r}} = \alpha r_{\mathrm{r}}/c_1,\ 1/\rho_{\mathrm{r}} = c_1/\alpha r_{\mathrm{r}},\ Z_1 = Z,\ c_1 = c,\ \sigma_1 = \sigma, \tag{83}$$

to rewrite Eq.(79) into the following form:

[6] Harmuth 1986, pp. 53–55

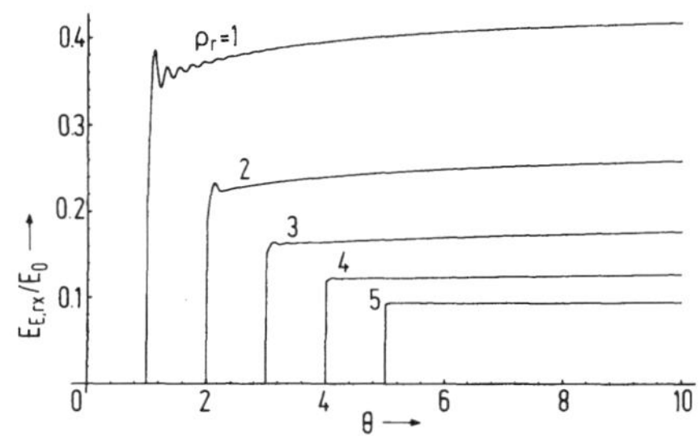

FIG.3.4-4. Plots of $E_{\mathrm{E,rs}}/E_0$ according to Eq.(80) for $\psi_0 = 0.001$, $\iota = 0.01$, $\rho_0 = 0.1$, and $\rho_{\mathrm{r}} = 1, 2, 3, 4, 5$.

$$\begin{aligned}
w(r_{\mathrm{r}}, t) = -\frac{e^{-\alpha t}}{\ln(1/\psi_0)} \Bigg[\int\limits_{\psi_0/r_{\mathrm{r}}}^{Z\sigma/2} & \left(\mathrm{ch}(\alpha^2 - \beta^2 c^2)^{1/2} t + \frac{\alpha\, \mathrm{sh}(\alpha^2 - \beta^2 c^2) t}{(\alpha^2 - \beta^2 c^2)^{1/2}} \right) \\
& \times \left(J_0(\beta r_{\mathrm{r}}) - \frac{Y_0(\beta r_{\mathrm{r}})}{Y_0(\psi_0)} \right) \frac{d\beta}{\beta} \\
+ \int\limits_{Z\sigma/2}^{\infty} & \left(\cos(\beta^2 c^2 - \alpha^2)^{1/2} t + \frac{\alpha \sin(\beta^2 c^2 - \alpha^2)^{1/2} t}{(\beta^2 c^2 - \alpha^2)^{1/2}} \right) \\
& \times \left(J_0(\beta r_{\mathrm{r}}) - \frac{Y_0(\beta r_{\mathrm{r}})}{Y_0(\psi_0)} \right) \frac{d\beta}{\beta} \Bigg]
\end{aligned} \tag{84}$$

For the smallest permissible value of σ we get

$$Z\sigma/2 = \psi_0/r_{\mathrm{r}} = \alpha/c, \ \alpha = c\psi_0/r_{\mathrm{r}} \ll 1 \tag{85}$$

and

$$w(r_{\mathrm{r}}, t) = -\frac{1}{\ln(1/\psi_0)} \int\limits_{\psi_0/r_{\mathrm{r}}}^{\infty} \cos(\beta c t) \left(J_0(\beta r_{\mathrm{r}}) - \frac{Y_0(\beta r_{\mathrm{r}})}{Y_0(\psi_0)} \right) \frac{d\beta}{\beta} \tag{86}$$

For the first term in Eq.(80) we get for $\iota \to 0$ and finite values of ρ and ρ_0 with the help of Eq.(65):

$$\frac{K_0(2\sqrt{\iota}\rho)}{K_0(2\sqrt{\iota}\rho_0)} \approx \frac{\ln(\sqrt{\iota}\rho)}{\ln(\sqrt{\iota}\rho_0)} = \frac{\ln\sqrt{\iota} + \ln\rho}{\ln\sqrt{\iota} + \ln\rho_0} \approx \frac{\ln\sqrt{\iota}}{\ln\sqrt{\iota}} = 1 \tag{87}$$

Equations (80) and (86) yield $E_{\mathrm{E,r}x}(r_{\mathrm{r}}, t)$ for $\iota \to 0$:

$$\begin{aligned}
E_{\mathrm{E,r}x}(r_{\mathrm{r}}, t) &= E_{\mathrm{E,r}x}(r_{\mathrm{r}}/r_0, ct/r_0) \\
&= E_0 \left[1 - \frac{1}{\ln(1/\psi_0)} \int\limits_{\frac{\psi_0}{r_{\mathrm{r}}/r_0}}^{\infty} \cos\left(\beta \frac{ct}{r_0}\right) \left(J_0(\beta r_{\mathrm{r}}/r_0) - \frac{Y_0(\beta r_{\mathrm{r}}/r_0)}{Y_0(\psi_0)} \right) \frac{d\beta}{\beta} \right]
\end{aligned} \tag{88}$$

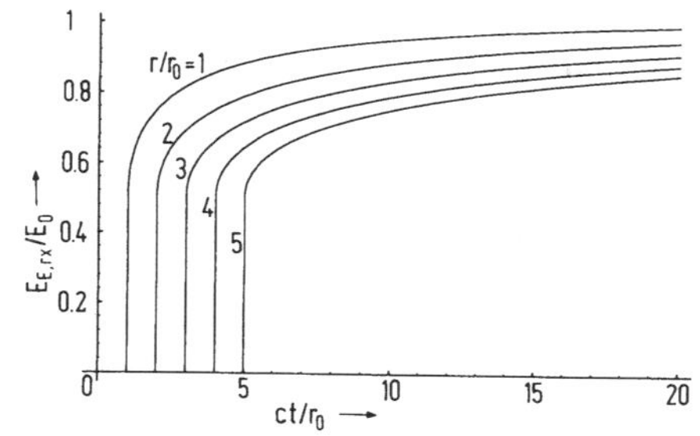

FIG.3.4-5. Plots of $E_{\mathrm{E,rx}}/E_0$ according to Eq.(88) for $\psi_0 = 0.01$ and $r_\mathrm{r}/r_0 = 1, 2, 3, 4, 5$.

Plots of the electric field strength according to Eq.(80) are shown in Fig.3.4-4. These plots are remarkably good if one considers the complexity of Eq.(80). The decaying oscillations shown clearly by the plot $\rho_\mathrm{r} = 1$ are probably not a computer artifact since such oscillations have been found and studied previously for planar waves[7].

Figure 3.4-5 shows plots of the electric field strength for $\iota = 0$ according to Eq.(88).

3.5 Associated Magnetic Field Strength

The electric field strength $E_{\mathrm{E,r}x}$ excited by an electric step function at the surface of a cylinder with radius ρ_0 is defined by Eq.(3.4-80). The associated magnetic field strength $H_{\mathrm{r}\varphi}$ is determined by Eqs.(3.1-40) and (3.1-41). We write $H_{\mathrm{r}\varphi} = H_{\mathrm{E,r}\varphi} = H_\mathrm{E}$ and $E_{\mathrm{E,r}x} = E_{\mathrm{r}x} = E_\mathrm{E}$ to emphasize that the field strengths are due to electric excitation. The subscript r is temporarily left out to simplify writing:

$$Z_1 H_\mathrm{E} = \frac{1}{\rho} \int \rho \left(\frac{\partial E_\mathrm{E}}{\partial \theta} + E_\mathrm{E} \right) d\rho + Z_1 H_{\mathrm{r}1}(\theta), \quad \rho = \rho_\mathrm{r} \tag{1}$$

$$Z_1 H_\mathrm{E} = e^{-2\iota\theta} \left(\int \frac{\partial E_\mathrm{E}}{\partial \rho} e^{2\iota\theta} d\theta + Z_1 H_{\mathrm{r}2}(\rho) \right) \tag{2}$$

We consider first Eq.(2). The term $\partial E_\mathrm{E}/\partial\rho$ follows from Eq.(3.4-80):

$$\frac{\partial E_\mathrm{E}}{\partial \rho} = E_0 \left(\frac{1}{K_0(2\sqrt{\iota}\rho_0)} \frac{dK_0(2\sqrt{\iota}\rho)}{d\rho} + \frac{\partial w}{\partial \rho} \right) \tag{3}$$

We find in a table[1]:

$$\frac{dK_0(2\sqrt{\iota}\rho)}{d\rho} = -2\sqrt{\iota} K_1(2\sqrt{\iota}\rho) \tag{4}$$

The derivative $\partial w/\partial\rho$ in Eq.(3) follows from Eq.(3.4-79). The terms $J_0(\omega\rho)$ and $Y_0(\omega\rho)$ contain the variable ρ but ρ also occurs in the lower integral limit. Using the

[7] Harmuth and Hussain 1994a, pp. 6, 7

[1] Gradshteyn and Ryzhik 1980, p. 970, 4.486/18

general formula for the differentiation of an integral with the variable in the limits as well as in the kernel and formulas for the derivatives of J_0 and Y_0 we get[2]

$$\frac{d}{d\rho}\int_{\psi_0/\rho}^{\omega_0}\left(J_0(\omega\rho)-\frac{Y_0(\omega\rho)}{Y_0(\psi_0)}\right)d\omega$$
$$=\frac{\psi_0}{\rho^2}[J_0(\psi_0)-1]+\int_{\psi_0/\rho}^{\omega_0}\left(\frac{dJ_0(\omega\rho)}{d\rho}-\frac{1}{Y_0(\psi_0)}\frac{dY_0(\omega\rho)}{d\rho}\right)d\omega \quad (5)$$

$$\frac{dJ_0(\omega\rho)}{d\rho}=-\omega J_1(\omega\rho),\quad \frac{dY_0(\omega\rho)}{d\rho}=-\omega Y_1(\omega\rho) \quad (6)$$

and the derivative $\partial w/\partial\rho$ of w in Eq.(3.4-79) becomes:

$$\begin{aligned}\frac{\partial w}{\partial\rho}=\frac{e^{-a\theta}}{\ln(1/\psi_0)}\Bigg[&-\frac{\psi_0}{\rho^2}\left(\mathrm{ch}[(a^2-b^2)^{1/2}\theta]+\frac{a\,\mathrm{sh}[(a^2-b^2)^{1/2}\theta]}{(a^2-b^2)^{1/2}}\right)\\&\times\frac{\omega}{\omega^2+4\iota}[J_0(\psi_0)-1]\\&+\int_{\psi_0/\rho}^{\omega_0}\left(\mathrm{ch}[(a^2-b^2)^{1/2}\theta]+\frac{a\,\mathrm{sh}[(a^2-b^2)^{1/2}\theta]}{(a^2-b^2)^{1/2}}\right)\\&\times\frac{\omega^2}{\omega^2+4\iota}\left(J_1(\omega\rho)-\frac{Y_1(\omega\rho)}{Y_0(\psi_0)}\right)d\omega\\&+\int_{\omega_0}^{\infty}\left(\cos[(b^2-a^2)^{1/2}\theta]+\frac{a\sin[(b^2-a^2)^{1/2}\theta]}{(b^2-a^2)^{1/2}}\right)\\&\times\frac{\omega^2}{\omega^2+4\iota}\left(J_1(\omega\rho)-\frac{Y_1(\omega\rho)}{Y_0(\psi_0)}\right)d\omega\Bigg]\end{aligned} \quad (7)$$

Equation (2) is transformed with the help of Eqs.(3) and (4) into the following form:

$$Z_1H_{\mathrm{E}}(\rho,\theta)=e^{-2\iota\theta}\left[E_0\int\left(-\frac{2\sqrt{\iota}K_1(2\sqrt{\iota\rho})}{K_0(2\sqrt{\iota\rho_0})}+\frac{\partial w}{\partial\rho}\right)e^{2\iota\theta}d\theta+Z_1H_{\mathrm{r}2}(\rho)\right] \quad (8)$$

The integration over θ can be done analytically[3]:

[2] Gradshteyn and Ryzhik 1980, p. 18, 0.410; p. 968, 8.473/4, 8.473/5
[3] Gradshteyn and Ryzhik 1980, p. 128, 2.481/2, 2.481/1; p. 196, 2.663/3, 2.663/1

$$e^{-2\iota\theta}\int e^{2\iota\theta}d\theta = \frac{1}{2\iota} \tag{9}$$

$$\begin{aligned} I_{21}(\omega,\theta) &= e^{-2\iota\theta}\int e^{-a\theta}\,\mathrm{ch}[(a^2-b^2)^{1/2}\theta]e^{2\iota\theta}d\theta \\ &= \frac{e^{-a\theta}}{(2\iota-a)^2-(a^2-b^2)}\{(2\iota-a)\,\mathrm{ch}[(a^2-b^2)^{1/2}\theta] \\ &\qquad -(a^2-b^2)^{1/2}\,\mathrm{sh}[(a^2-b^2)^{1/2}\theta]\} \end{aligned} \tag{10}$$

$$\begin{aligned} I_{22}(\omega,\theta) &= e^{-2\iota\theta}\int e^{-a\theta}\,\mathrm{sh}[(a^2-b^2)^{1/2}\theta]e^{2\iota\theta}d\theta \\ &= \frac{e^{-a\theta}}{(2\iota-a)^2-(a^2-b^2)}\{(2\iota-a)\,\mathrm{sh}[(a^2-b^2)^{1/2}\theta] \\ &\qquad -(a^2-b^2)^{1/2}\,\mathrm{ch}[(a^2-b^2)^{1/2}\theta]\} \end{aligned} \tag{11}$$

$$\begin{aligned} I_{23}(\omega,\theta) &= e^{-2\iota\theta}\int e^{-a\theta}\cos[(b^2-a^2)^{1/2}\theta]e^{2\iota\theta}d\theta \\ &= \frac{e^{-a\theta}}{(2\iota-a)^2+(b^2-a^2)}\{(2\iota-a)\cos[(b^2-a^2)^{1/2}\theta] \\ &\qquad +(b^2-a^2)^{1/2}\sin[(b^2-a^2)^{1/2}\theta]\} \end{aligned} \tag{12}$$

$$\begin{aligned} I_{24}(\omega,\theta) &= e^{-2\iota\theta}\int e^{-a\theta}\sin[(b^2-a^2)^{1/2}\theta]e^{2\iota\theta}d\theta \\ &= \frac{e^{-a\theta}}{(2\iota-a)^2+(b^2-a^2)}\{(2\iota-a)\sin[(b^2-a^2)^{1/2}\theta] \\ &\qquad -(b^2-a^2)^{1/2}\cos[(b^2-a^2)^{1/2}\theta]\} \end{aligned} \tag{13}$$

Equation (8) is rewritten:

$$\begin{aligned} Z_1H_{\mathrm{E}}(\rho,\theta) = E_0\Bigg\{ &-\frac{K_1(2\sqrt{\iota}\rho)}{\sqrt{\iota}K_0(2\sqrt{\iota}\rho_0)} \\ &+\frac{1}{\ln(1/\psi_0)}\Bigg[-\frac{\psi_0}{\rho^2}\left(I_{21}(\omega,\theta)+\frac{aI_{22}(\omega,\theta)}{(a^2-b^2)^{1/2}}\right)\frac{\omega}{\omega^2+4\iota}[J_0(\psi_0)-1] \\ &+\int_{\psi_0/\rho}^{\omega_0}\left(I_{21}(\omega,\theta)+\frac{aI_{22}(\omega,\theta)}{(a^2-b^2)^{1/2}}\right)\frac{\omega^2}{\omega^2+4\iota}\left(J_1(\omega\rho)-\frac{Y_1(\omega\rho)}{Y_0(\psi_0)}\right)d\omega \\ &+\int_{\omega_0}^{\infty}\left(I_{23}(\omega,\theta)+\frac{aI_{24}(\omega,\theta)}{(b^2-a^2)^{1/2}}\right)\frac{\omega^2}{\omega^2+4\iota}\left(J_1(\omega\rho)-\frac{Y_1(\omega\rho)}{Y_0(\psi_0)}\right)d\omega\Bigg]\Bigg\} \\ &+Z_1H_{\mathrm{r}2}(\rho)e^{-2\iota\theta} \end{aligned}$$

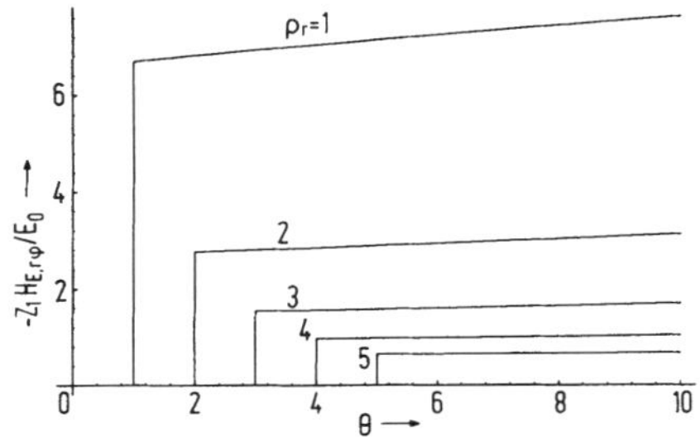

FIG.3.5-1. Plots of $Z_1 H_{E,r\varphi}/E_0$ according to Eq.(15) for $H_{r2}(\rho) = 0$, $\psi_0 = 0.001$, $\iota = 0.01$, $\rho_0 = 0.1$, and $\rho = \rho_r = 1, 2, 3, 4, 5$.

$$(2\iota - a)^2 - a^2 + b^2 = \omega^2,\ a = \iota + 1,\ a^2 - b^2 = (\iota+1)^2 - \omega^2 \tag{14}$$

Substitution of the terms I_{21} to I_{24} brings with $H_E(\rho,\theta) = H_{E,r\varphi}(\rho,\theta)$:

$$\begin{aligned}
Z_1 H_{E,r\varphi}(\rho,\theta) = -E_0 \Bigg\{ & \frac{K_1(2\sqrt{\iota\rho})}{\sqrt{\iota}K_0(2\sqrt{\iota\rho_0})} + \frac{e^{-a\theta}}{\ln(1/\psi_0)} \Bigg[\frac{[1 - J_0(\psi_0)]\psi_0}{\rho^2\omega(\omega^2+4\iota)} \\
& \times \left(2\operatorname{ch}[(a^2-b^2)^{1/2}\theta] - \frac{[\omega^2 - 2(1-\iota)]\operatorname{sh}[(a^2-b^2)^{1/2}\theta]}{(a^2-b^2)^{1/2}} \right) \\
& + \int_{\psi_0/\rho}^{\omega_0} \left(2\operatorname{ch}[(a^2-b^2)^{1/2}\theta] - \frac{[\omega^2 - 2(1-\iota)]\operatorname{sh}[(a^2-b^2)^{1/2}\theta]}{(a^2-b^2)^{1/2}} \right) \\
& \qquad \times \frac{1}{\omega^2+4\iota}\left(J_1(\omega\rho) - \frac{Y_1(\omega\rho)}{Y_0(\psi_0)} \right) d\omega \\
& + \int_{\omega_0}^{\infty} \left(2\cos[(b^2-a^2)^{1/2}\theta] - \frac{[\omega^2 - 2(1-\iota)]\sin[(b^2-a^2)^{1/2}\theta]}{(b^2-a^2)^{1/2}} \right) \\
& \qquad \times \frac{1}{\omega^2+4\iota}\left(J_1(\omega\rho) - \frac{Y_1(\omega\rho)}{Y_0(\psi_0)} \right) d\omega \Bigg] \Bigg\} + Z_1 H_{r2}(\rho) e^{-2\iota\theta}
\end{aligned}$$

$$(2\iota - a)^2 - a^2 + b^2 = \omega^2,\ a(\iota - 1) - a^2 + b^2 = \omega^2 - 2(1-\iota),\ \rho = \rho_r \tag{15}$$

We note the relation

$$[1 - J_0(\psi_0)]\psi_0 \approx \frac{1}{4}\psi_0^3,\quad \psi_0 \ll 1 \tag{16}$$

which makes the second term in Eq.(15) that is multiplied by $[1 - J_0(\psi_0)]\psi_0$ very small.

Equation (15) represents $H_E(\rho,\theta) = H_{E,r\varphi}(\rho_r,\theta)$ by a single integral of the variable ω, which is much easier to evaluate numerically by computer than the double integral of the variables θ and ω implied by Eq.(2). The integration constant $H_{r2}(\rho_r)$ is zero if there is initially no magnetic field strength at the time $\theta = 0$.

We may now turn to Eq.(1) to see whether it yields Eq.(15) too, but this attempt runs into a mathematical difficulty. Instead of needing the derivative of an integral with the variable both in the kernel and one limit as shown by Eq.(5) we must now face the integral of that expression

$$\int \left[\int_{\psi_0/\rho}^{\omega_0} \left(J_0(\omega\rho) - \frac{Y_0(\omega\rho)}{Y_0(\psi_0)} \right) d\omega \right] d\rho$$

and we shall pass up this challenge.

Plots of the magnetic field strength according to Eq.(15) are shown in Fig.3.5-1.

3.6 Transmission of Signals Into Medium 2

For the investigation of the transmission of an incident, perpendicularly polarized wave from medium 1 into medium 2 according to Fig.2.2-7 we proceed in analogy to Section 3.2. We must now use Eqs.(3.1-46)–(3.1-48) or (3.1-49)–(3.1-51). Let us start with Eq.(3.1-46) and assume a magnetic excitation step function that is the same as in Eq.(3.2-1) but we must write the index t for transmitted rather than r for reflected. A constant H_1 rather than H_0 is used in order to be able to satisfy the boundary conditions of Eqs.(2.1-53) and (2.1-54) at the very end of our calculations:

$$\begin{aligned} H_{\mathrm{t}\varphi}(0,\theta) = H_{\mathrm{i}y} = H_1 S(\theta) &= 0 \quad \text{for } \theta < 0 \\ &= H_1 \quad \text{for } \theta \geq 0 \end{aligned} \tag{1}$$

The distance ρ_r is replaced by ρ_t and Eqs.(3.2-2)–(3.2-5) are rewritten:

$$\rho_\mathrm{r} \rightarrow \rho_\mathrm{t}$$

$$H_{\mathrm{t}\varphi} = \text{finite} \tag{2}$$

$$H_{\mathrm{t}\varphi}(\rho_\mathrm{t},0) = E_{\mathrm{t}x}(\rho_\mathrm{t},0) = 0 \quad \text{for } \rho_\mathrm{t} > 0 \tag{3}$$

$$\partial H_{\mathrm{t}\varphi}(\rho_\mathrm{t},0)/\partial\rho_\mathrm{t} = \partial E_{\mathrm{t}x}(\rho_\mathrm{t},0)/\partial\rho_\mathrm{t} = 0 \tag{4}$$

$$\partial H_{\mathrm{t}\varphi}(\rho_\mathrm{t},\theta)/\partial\theta = \partial E_{\mathrm{t}x}(\rho_\mathrm{t},\theta)/\partial\theta = 0 \quad \text{for } \rho_\mathrm{t} > 0,\ \theta = 0 \tag{5}$$

Equations (3.1-46)–(3.1-48) are rewritten into the following form:

$$\begin{aligned} \frac{\partial^2(Z_2H_{\mathrm{t}\varphi})}{\partial\rho_\mathrm{t}^2} - \frac{\partial^2(Z_2H_{\mathrm{t}\varphi})}{\partial(\theta\gamma_\mu/\gamma_\sigma)^2} - 2\left(\iota\frac{\gamma_\mu\gamma_\sigma}{\gamma_\epsilon\gamma_\mathrm{s}} + 1\right)\frac{\partial(Z_2H_{\mathrm{t}\varphi})}{\partial(\theta\gamma_\epsilon/\gamma_\sigma)} \\ + \frac{1}{\rho_\mathrm{t}}\frac{\partial(Z_2H_{\mathrm{t}\varphi})}{\partial\rho_\mathrm{t}} - \left(\frac{1}{\rho_\mathrm{t}^2} + 4\iota\frac{\gamma_\mu\gamma_\sigma}{\gamma_\epsilon\gamma_\mathrm{s}}\right)(Z_2H_{\mathrm{t}\varphi}) = 0 \end{aligned} \tag{6}$$

$$E_{\mathrm{t}x} = \int\left(\frac{\partial(Z_2H_{\mathrm{r}\varphi})}{\partial(\theta\gamma_\epsilon/\gamma_\sigma)} + 2\iota\frac{\gamma_\mu\gamma_\sigma}{\gamma_\epsilon\gamma_\mathrm{s}}(Z_2H_{\mathrm{t}\varphi})\right)d\rho_\mathrm{t} + E_{\mathrm{t}1}(\theta\gamma_\epsilon/\gamma_\sigma) \tag{7}$$

$$E_{\mathrm{t}x} = e^{-2(\theta\gamma_\epsilon/\gamma_\sigma)}\left(\frac{1}{\rho_{\mathrm{t}}}\int \frac{\partial(\rho_{\mathrm{t}} Z_2 H_{\mathrm{t}\varphi})}{\partial\rho_{\mathrm{t}}} e^{2(\theta\gamma_\epsilon/\gamma_\sigma)} d(\theta\gamma_\epsilon/\gamma_\sigma) + E_{\mathrm{t}2}(\rho_{\mathrm{t}})\right) \tag{8}$$

Equations (3.1-36)–(3.1-38) are transformed into Eqs.(6)–(8) if we make the following substitutions:

$$\begin{aligned}
&Z_1 \to Z_2, \quad \rho_{\mathrm{r}} \to \rho_{\mathrm{t}}, \quad \theta \to \theta\gamma_\epsilon/\gamma_\sigma, \quad \iota \to \iota\gamma_\mu\gamma_\sigma/\gamma_\epsilon\gamma_{\mathrm{s}} = \iota' \\
&H_{\mathrm{r}\varphi} \to H_{\mathrm{t}\varphi}, \quad E_{\mathrm{r}x} \to E_{\mathrm{t}x}, \quad E_{\mathrm{r}1} \to E_{\mathrm{t}1}, \quad E_{\mathrm{r}2} \to E_{\mathrm{t}2}
\end{aligned} \tag{9}$$

The magnetic field strength $H_{\mathrm{t}\varphi} = H_{\mathrm{H,t}\varphi}$ follows with the same substitutions from Eq.(3.2-53):

$$\begin{aligned}
&H_{\mathrm{H,t}\varphi}(\rho_{\mathrm{t}}, \theta\gamma_\epsilon/\gamma_\sigma) = H_{\mathrm{H,t}\varphi}(\rho_{\mathrm{t}}, \theta) = H_1\left(\frac{K_1(2\sqrt{\iota'}\rho_{\mathrm{t}})}{K_1(2\sqrt{\iota'}\rho_0)} + w'(\rho_{\mathrm{t}}, \theta)\right) \\
&\iota' = \iota\gamma_\mu\gamma_\sigma/\gamma_\epsilon\gamma_{\mathrm{s}}, \ \rho_{\mathrm{t}} \ge \rho_0 > 0
\end{aligned} \tag{10}$$

The function $w'(\rho_{\mathrm{t}}, \theta)$ follows from Eq.(3.2-52) with changed values for θ, a, b, ω_0, and ι:

$$\begin{aligned}
w'(\rho_{\mathrm{t}}, \theta) = &-\frac{e^{-a'\theta\gamma_\epsilon/\gamma_\sigma}}{2\sqrt{\iota'}K_1(2\sqrt{\iota'}\rho_0)}\left[\int_0^{\omega_0'}\left(\mathrm{ch}[(a'^2 - b'^2)^{1/2}\theta\gamma_\epsilon/\gamma_\sigma]\right.\right. \\
&\left. + \frac{a'\,\mathrm{sh}[(a'^2 - b'^2)^{1/2}\theta\gamma_\epsilon/\gamma_\sigma]}{(a'^2 - b'^2)^{1/2}}\right)\frac{\omega^2}{\omega^2 + 4\iota'} J_1(\omega\rho_{\mathrm{t}})\,d\omega \\
&+ \int_{\omega_0'}^{\infty}\left(\cos[(b'^2 - a'^2)^{1/2}\theta\gamma_\epsilon/\gamma_\sigma] + \frac{a'\sin[(b'^2 - a'^2)^{1/2}\theta\gamma_\epsilon/\gamma_\sigma]}{(b'^2 - a'^2)^{1/2}}\right) \\
&\left.\times \frac{\omega^2}{\omega^2 + 4\iota'} J_1(\omega\rho_{\mathrm{t}})\,d\omega\right] \\
&a' = \iota' + 1, \ b'^2 = \omega^2 + 4\iota', \ \omega_0' = |1 - \iota'|, \ \iota' = \iota\gamma_\mu\gamma_\sigma/\gamma_\epsilon\gamma_{\mathrm{s}}
\end{aligned} \tag{11}$$

In addition to the excited magnetic field strength we need the associated electric field strength. This electric field strength $E_{\mathrm{t}x} = E_{\mathrm{H,t}x}$ was derived in Section 3.3 and is defined by Eq.(3.3-21):

$$E_{\mathrm{H,t}x}(\rho_{\mathrm{t}},\theta) = Z_2 H_1 F_{\mathrm{H,t}x}(\rho_{\mathrm{t}},\theta) = -Z_2 H_1 \rho_0 \Bigg\{ 2\iota' K_0(2\sqrt{\iota'}\rho_{\mathrm{t}}) + e^{-a'\theta\gamma_\epsilon/\gamma_\sigma}$$
$$\times \Bigg[\int_0^{\omega_0'} \left(2\iota' \operatorname{ch}[(a'^2 - b'^2)^{1/2}\theta\gamma_\epsilon/\gamma_\sigma] + \frac{(2\iota' a' - b'^2)\operatorname{sh}[(a'^2 - b'^2)^{1/2}\theta\gamma_\epsilon/\gamma_\sigma]}{(a'^2 - b'^2)^{1/2}} \right)$$
$$\times \frac{\omega J_0(\rho_{\mathrm{t}}\omega)}{\omega^2 + 4\iota'}\, d\omega$$
$$+ \int_{\omega_0'}^{\infty} \left(2\iota' \cos[(b'^2 - a'^2)^{1/2}\theta\gamma_\epsilon/\gamma_\sigma] + \frac{(2\iota' a' - b'^2)\sin[(b'^2 - a'^2)^{1/2}\theta\gamma_\epsilon/\gamma_\sigma]}{(b'^2 - a'^2)^{1/2}} \right)$$
$$\times \frac{\omega J_0(\rho_{\mathrm{t}}\omega)}{\omega^2 + 4\iota'}\, d\omega \Bigg] \Bigg\} + E_{\mathrm{t}1} \quad (12)$$

The 'integration constant' $E_{\mathrm{t}1}$ is indeed a constant according to Eq.(3.3-22) and the following text. If the initial field strength $E_{\mathrm{H,t}x}(\rho_{\mathrm{t}},0)$ for $\rho_{\mathrm{t}} > 0$ is zero we get $E_{\mathrm{t}1} = 0$.

Turning from a magnetic excitation force to an electric one we rewrite Eqs.(3.1-49)–(3.1-51) in analogy to Eqs.(6)–(7):

$$\frac{\partial^2 E_{\mathrm{t}x}}{\partial\rho_{\mathrm{t}}^2} - \frac{\partial^2 E_{\mathrm{t}x}}{\partial(\theta\gamma_\epsilon/\gamma_\sigma)^2} - 2\left(\iota\frac{\gamma_\mu\gamma_\sigma}{\gamma_\epsilon\gamma_{\mathrm{s}}} + 1\right)\frac{\partial E_{\mathrm{t}x}}{\partial(\theta\gamma_\epsilon/\gamma_\sigma)}$$
$$+ \frac{1}{\rho_{\mathrm{t}}}\frac{\partial E_{\mathrm{t}x}}{\partial\rho_{\mathrm{t}}} - 4\iota\frac{\gamma_\mu\gamma_\sigma}{\gamma_\epsilon\gamma_{\mathrm{s}}}E_{\mathrm{t}x} = 0 \quad (13)$$

$$Z_2 H_{\mathrm{t}\varphi} = \frac{1}{\rho_{\mathrm{t}}}\int \rho_{\mathrm{t}}\left(\frac{\partial E_{\mathrm{t}x}}{\partial(\theta\gamma_\epsilon/\gamma_\sigma)} + 2E_{\mathrm{t}x}\right) d\rho_{\mathrm{t}} + Z_2 H_{\mathrm{t}1}(\theta\gamma_\epsilon/\gamma_\sigma) \quad (14)$$
$$Z_2 H_{\mathrm{t}\varphi} = e^{-2\iota'(\theta\gamma_\epsilon/\gamma_\sigma)}$$
$$\times \left(\int \frac{\partial E_{\mathrm{t}x}}{\partial\rho_{\mathrm{t}}} e^{2\iota'(\theta\gamma_\epsilon/\gamma_\sigma)} d(\theta\gamma_\epsilon/\gamma_\sigma) + Z_2 H_{\mathrm{t}2}(\theta\gamma_\epsilon/\gamma_\sigma)\right) \quad (15)$$
$$\iota' = \iota\gamma_\mu\gamma_\sigma/\gamma_\epsilon\gamma_{\mathrm{s}}$$

The substitutions of Eq.(9) again transform Eqs.(3.1-39)–(3.1-41) into Eqs.(13) and (14). The electric field strength $E_{\mathrm{E,t}x}(\rho_{\mathrm{t}},\theta)$ of the transmitted wave follows thus readily from Eqs.(3.4-80) and (3.4-79) that hold for the reflected wave. The constant E_0 is replaced by a new constant E_1 so that the boundary conditions of Eqs.(2.1-53) and (2.1-54) can be satisfied:

$$E_{\mathrm{E,t}x}(\rho_{\mathrm{t}},\theta) = E_1\left(\frac{K_0(\sqrt{\iota'}\rho_{\mathrm{t}})}{K_0(2\sqrt{\iota'}\rho_0)} + w(\rho_{\mathrm{t}},\theta)\right)$$
$$\rho_{\mathrm{t}} \geq \rho_0 > 0,\ \chi_0 = 2\sqrt{\iota'}\rho_0 \quad (16)$$

$$w(\rho_t, \theta) = -\frac{e^{-a'\theta\gamma_\epsilon/\gamma_\sigma}}{\ln(1/\psi_0)}$$
$$\times \left[\int_{\psi_0/\rho_t}^{\omega_0'} \left(\mathrm{ch}[(a'^2 - b'^2)^{1/2}\theta\gamma_\epsilon/\gamma_\sigma] + \frac{a'\,\mathrm{sh}[(a'^2 - b'^2)^{1/2}\theta\gamma_\epsilon/\gamma_\sigma]}{(a'^2 - b'^2)^{1/2}} \right) \right.$$
$$\times \frac{\omega}{\omega^2 + 4\iota'} \left(J_0(\omega\rho_t) - \frac{Y_0(\omega\rho_t)}{Y_0(\psi_0)} \right) d\omega$$
$$+ \int_{\omega_0'}^{\infty} \left(\cos[(b'^2 - a'^2)^{1/2}\theta\gamma_\epsilon/\gamma_\sigma] + \frac{a'\sin[(b'^2 - a'^2)^{1/2}\theta\gamma_\epsilon/\gamma_\sigma]}{(b'^2 - a'^2)^{1/2}} \right)$$
$$\left. \times \frac{\omega}{\omega^2 + 4\iota'} \left(J_0(\omega\rho_t) - \frac{Y_0(\omega\rho_t)}{Y_0(\psi_0)} \right) d\omega \right]$$
$$a' = \iota' + 1,\ b'^2 = \omega'^2 = 4\iota',\ \omega_0' = |1 - \iota'|,\ \iota' = \iota\gamma_\mu\gamma_\sigma/\gamma_\epsilon\gamma_s \tag{17}$$

The associated magnetic field strength $Z_2 H_{E,t\varphi}(\rho_t, \theta)$ is obtained by substituting ρ_t, $\theta\gamma_\epsilon/\gamma_\sigma$, a', b', ω_0', ι' for ρ, θ, a, b, ω_0, ι in Eq.(3.5-15).

Let us turn to the boundary conditions of Eqs.(2.1-53) and (2.1-54). First we rewrite them for cylinder coordinates and a certain point y on the boundary plane. The index H is added to emphasize that the calculation refers to a magnetic excitation force. The electric excitation force $E_{ix}(y, z = 0, t)$ in Eq.(2.1-53) is in this case zero by definition:

$$0 = E_{H,rx}(\rho_0, \theta) + E_{H,tx}(\rho_0, \theta) \tag{18}$$
$$-H_{H,i\varphi}(\rho_0, \theta) = H_{H,r\varphi}(\rho_0, \theta) + H_{H,t\varphi}(\rho_0, \theta) \tag{19}$$

Equation (3.2-53) yields for $\rho_r = \rho_0$:

$$H_{H,r\varphi}(\rho_0, \theta) = H_0 w(\rho_0, \theta) \tag{20}$$

In order to avoid writing the long Eq.(3.3-21) we substitute $F_{H,rx}(\rho_r, \theta)$ for the right side of this equation excepting the constants Z_1 and H_0:

$$E_{H,rx}(\rho_0, \theta) = Z_1 H_0 F_{H,rx}(\rho_0, \theta) \tag{21}$$

From Eqs.(10) and (12) we obtain in analogy

$$H_{H,t\varphi}(\rho_0, \theta) = H_1 w'(\rho_0, \theta) \tag{22}$$

and

$$E_{H,tx}(\rho_0, \theta) = Z_2 H_1 F_{H,tx}(\rho_0, \theta) \tag{23}$$

Substituting Eqs.(21) and (23) into Eq.(18) we get

$$0 = Z_1 H_0 F_{\mathrm{H,rz}}(\rho_0, \theta) + Z_1 H_1 F_{\mathrm{H,tz}}(\rho_0, \theta) \tag{24}$$

while the substitution of Eqs.(20) and (22) into Eq.(19) yields:

$$-H_{\mathrm{H,i}\varphi}(\rho_0, \theta) = H_0 w(\rho_0, \theta) + H_1 w'(\rho_0, \theta) \tag{25}$$

Since we assumed a step function excitation in Eqs.(3.2-1) and (3.6-1) the functions with the time variable θ in Eqs.(24) and (25) are constants for $\theta > 0$. We indicate this by suppressing θ in these equations:

$$Z_1 H_0 F_{\mathrm{H,rz}}(\rho_0) + Z_2 H_1 F_{\mathrm{H,tz}}(\rho_0) = 0 \tag{26}$$

$$H_0 w(\rho_0) + H_1 w'(\rho_0) = -H_{\mathrm{H,i}\varphi}(\rho_0) \tag{27}$$

Only H_0 and H_1 are unknowns in these equations:

$$H_0 = \frac{Z_2 F_{\mathrm{H,tz}}(\rho_0)}{Z_1 F_{\mathrm{H,rz}}(\rho_0) w'(\rho_0) - Z_2 F_{\mathrm{H,tz}}(\rho_0) w(\rho_0)} H_{\mathrm{H,i}\varphi}(\rho_0) \tag{28}$$

$$H_1 = \frac{Z_1 F_{\mathrm{H,rz}}(\rho_0)}{Z_2 F_{\mathrm{H,tz}}(\rho_0) w(\rho_0) - Z_1 F_{\mathrm{H,rz}}(\rho_0) w'(\rho_0)} H_{\mathrm{H,i}\varphi}(\rho_0) \tag{29}$$

4 Wave Theory of Electromagnetic Missiles

4.1 Line Array of Radiators

In Section 1.7 we discussed electromagnetic missiles or *electromagnetic waves with moving focal point* in terms of ray optics. We want to extend the analysis here to the wave theory

Figure 4.1-1 shows a line array with $2m + 1$ radiators uniformly spaced along the y-axis. The array axis in the direction of x. The distance between adjacent radiators is d. The array length is denoted D:

$$D = 2md \tag{1}$$

It is desired to concentrate or focus the electromagnetic energy radiated by the array elements at the point $P(L, 0)$ at the distance $x = L$ on the array axis $y = 0$. One may use the term *beam forming with a focused array* for this process. Beam forming is usually discussed for waves with sinusoidal time variation that have no beginning and no end; their energy is necessarily infinite. Beam forming with a focused array is more readily discussed for waves with finite duration and finite energy. The simplest such wave has the time variation $R(t)$ of a rectangular pulse with duration Δt. Since the solutions of the partial differential equations describing the propagation of a wave that begins at a certain time are most readily obtained for excitation forces having the time variation of a step function or an exponential function resembling a step function, we represent a rectangular pulse with amplitude 1 by means of two unit step functions:

$$\begin{aligned} R(t, \Delta t) &= S(t) - S(t - \Delta t) \\ S(t) &= 0 \qquad \text{for } t < 0 \\ &= 1 \qquad \text{for } t \geq 0 \end{aligned} \tag{2}$$

In order to achieve focusing of the radiated energy at the point $P(L, 0)$ at the distance $x = L$ on the array axis $y = 0$ in Fig.4.1-1, one must set a proper delay time for each radiating array element so that the leading edges of the radiated pulses arrived simultaneously at the point $P(L, 0)$. Since the distance from the array center $i = 0$ to the focus point $P(L, 0)$ equals L, the required delay time t_i for each element i is given by

$$t_i = L/c - \left[L^2 + (id)^2\right]^{1/2}/c, \quad i = 0, \pm 1, \ldots, \pm m \tag{3}$$

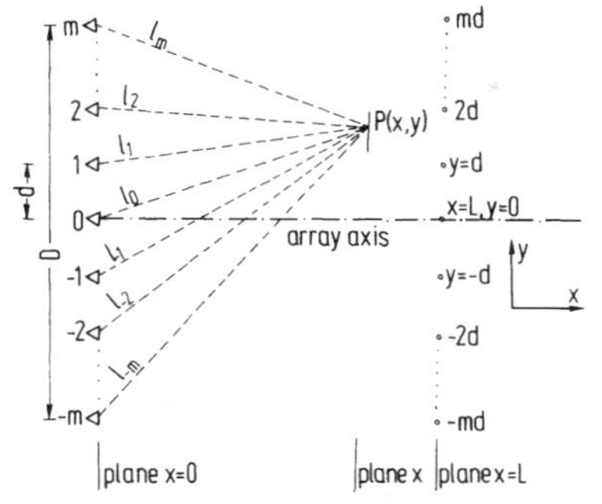

FIG.4.1-1. Line array with $2m+1$ radiators in the plane $x=0$ and the array axis in the direction of x.

where c is the velocity of light. The times t_i are negative for $i \neq 0$ since the waves must be radiated earlier from a radiator $i \neq 0$ than from the radiator $i=0$ at the center of the array..

According to Fig.4.1-1 the distance $l_i(x,y)$ from the radiator i to a point $P(x,y)$ equals

$$l_i(x,y) = \left[x^2 + (y-id)^2\right]^{1/2} \tag{4}$$

and the propagation time from the radiators i to the point $P(x,y)$ becomes $l_i(x,y)/c$.

Consider the sum of all electric field strengths in the point $x=L$ on the array axis produced by the $2m+1$ radiators. The distance to this point from the radiator m equals

$$l_m(L,0) = \left[L^2 + (D/2)^2\right]^{1/2} = L\left[1 + (D/2L)^2\right]^{1/2} \tag{5}$$

For a spherical radiator the field strength produced at a certain distance varies inversely with the distance. Hence, the amplitude produced by the $2m+1$ radiators will depend on the distance of each radiator. However, for a distance $L = 10D$ we get from Eq.(5)

$$l_m(L,0) = L\left[1 + (1/20)^2\right]^{1/2} \approx L(1 + 0.00125) \tag{6}$$

and the variation of the amplitudes produced by the radiators $i = 0, \pm1, \ldots, \pm m$ becomes very small. We shall ignore this effect. If every radiator produces a rectangular pulse according to Eq.(2) at a point with the distance $l_0(x,y)$ from the center of the array we get the sum $S_0(t,x,y)$ produced by $2m+1$ spherical radiators:

$$\begin{aligned}
S_0(t,x,y) &= \frac{1}{l_0(x,y)} \sum_{i=-m}^{m} R[t - \tau_i(x,y), \Delta t] \\
\tau_i(x,y) &= t_i + l_i(x,y)/c \\
&= \left\{L - \left[L^2 + (id)^2\right]^{1/2} + \left[x^2 + (y-id)^2\right]^{1/2}\right\}/c
\end{aligned} \tag{7}$$

The total delay $\tau_i(x,y)$ is the sum of the focusing delay t_i of Eq.(3) and the propagation delay $l_i(x,y)/c$ of Eq.(4).

The minimum time required for a wave radiated at the plane $x = 0$ in Fig.4.1-1 to reach the plane $x = L$ is L/c. We use this minimum time for normalization of $\tau_i(x,y)$:

$$\theta_i(x,y) = \frac{\tau_i(x,y)}{L/c} = 1 - \left[1 + \left(\frac{2md}{L}\right)^2 \left(\frac{i}{2m}\right)^2\right]^{1/2} + \left[\left(\frac{x}{L}\right)^2 + \left(\frac{y}{L} - \frac{2md}{L}\frac{i}{2m}\right)^2\right]^{1/2}$$

$$= 1 - \left[1 + \frac{1}{q^2}\left(\frac{i}{2m}\right)^2\right]^{1/2} + \left[\left(\frac{x}{L}\right)^2 + \left(\frac{y}{L} - \frac{1}{q}\frac{i}{2m}\right)^2\right]^{1/2}$$

$$q = L/2md = L/D \tag{8}$$

The normalized pulse duration Δt used in Eq.(2) is denoted $\Delta\theta$:

$$\Delta\theta = \frac{\Delta t}{L/c} = \frac{c\Delta t}{L} \tag{9}$$

4.2 Pulse Shape Along the Array Axis

Let the radiator array of Fig.4.1-1 be timed so that a pulse with rectangular time variation is produced at the focusing point $x = L$, $y = 0$. We want to show the time variation of the pulse for shorter distances $x < L$ and for longer distances $x > L$ along the array axis $y = 0$. Equation (4.1-8) yields for $y = 0$

$$\theta_i(x,0) = 1 - \left[1 + \frac{1}{q^2}\left(\frac{i}{2m}\right)^2\right]^{1/2} + \left[\left(\frac{x}{L}\right)^2 + \frac{1}{q^2}\left(\frac{i}{2m}\right)^2\right]^{1/2} \tag{1}$$

and Eq.(4.1-7) assumes the form

$$S_1(\theta,x,0) = S_0(\theta,x,0)l_0(x,0) = \sum_{i=-m}^{m} R[\theta - \theta_i(x,0), \Delta\theta]$$

$$\theta = ct/L,\ \Delta\theta = c\Delta t/L \tag{2}$$

By using the function $S_1(\theta,x,0)$ rather than $S_0(\theta,x,0)$ we eliminate the decrease of amplitude with distance and concentrate on the variation of the shape of the pulse.

For large values of m we may use the substitutions $p = i/2m$ and $\theta_i(x,0) = \theta_p(x,0)$ to approximate the sum in Eq.(2) by an integral over the interval $-1/2 < p < +1/2$:

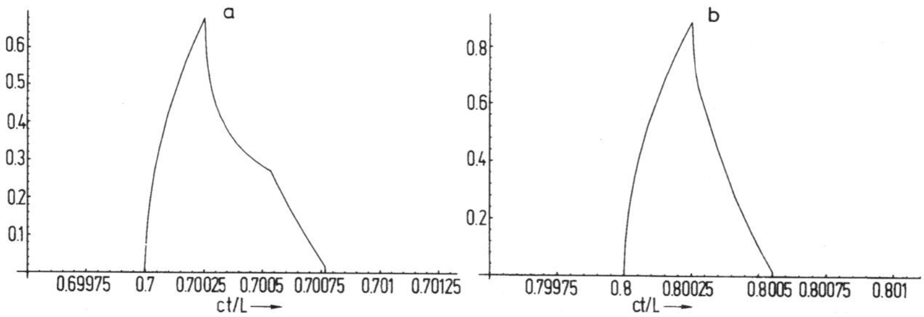

FIG.4.2-1. Distortion of a rectangular pulse on the array axis for $c\Delta t/L = 0.00025$ and $L/D = 10$ at a distance $x/L = 0.7$ (a) as well as $x/L = 0.8$ (b).

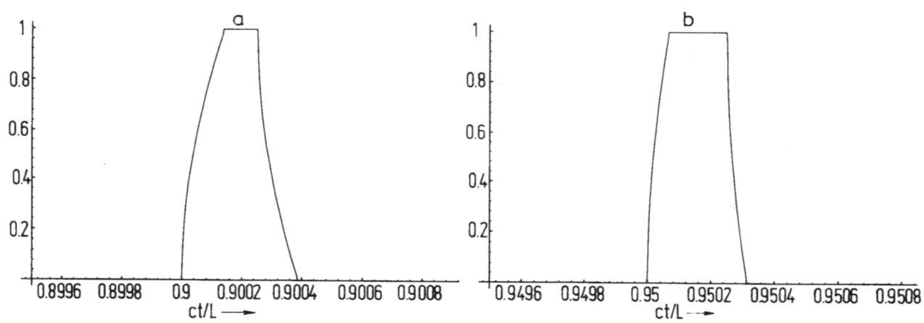

FIG.4.2-2. Distortion of a rectangular pulse on the array axis for $c\Delta t/L = 0.00025$ and $L/D = 10$ at a distance $x/L = 0.9$ (a) as well as $x/L = 0.95$ (b).

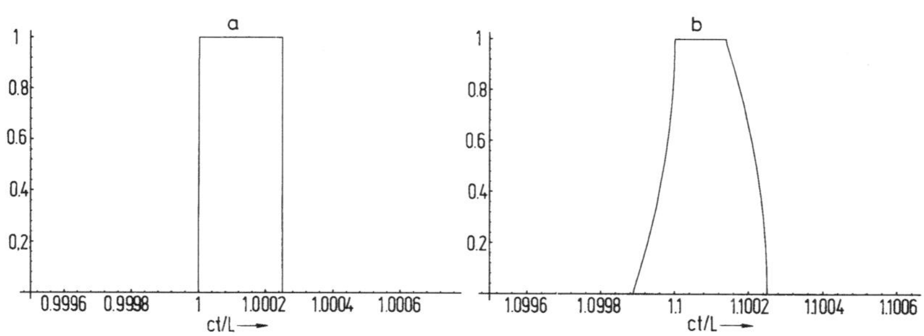

FIG.4.2-3. Distortion of a rectangular pulse on the array axis for $c\Delta t/L = 0.00025$ and $L/D = 10$ at a distance $x/L = 1$ (a) as well as $x/L = 1.1$ (b).

$$S_1(\theta, x, 0) = 2m \int_{-1/2}^{1/2} R[\theta - \theta_p(x,0), \Delta\theta]dp, \quad p = i/2m, \ m \gg 1$$

$$\theta_p(x,0) = 1 - (1 + p^2/q^2)^{1/2} + \left[(x/L)^2 + p^2/q^2\right]^{1/2} \tag{3}$$

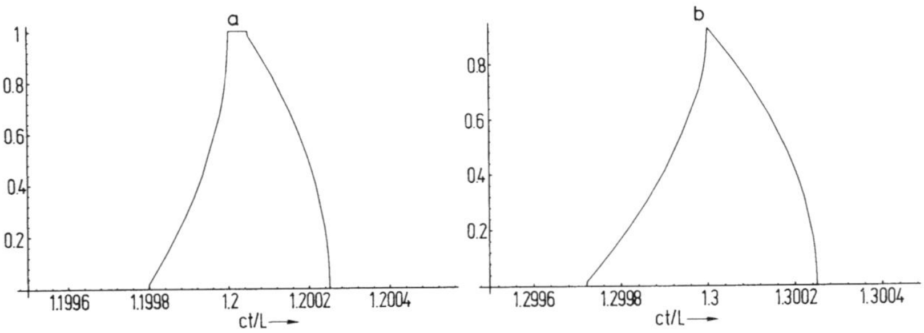

FIG.4.2-4. Distortion of a rectangular pulse on the array axis for $c\Delta t/L = 0.00025$ and $L/D = 10$ at a distance $x/L = 1.2$ (a) as well as $x/L = 1.3$ (b).

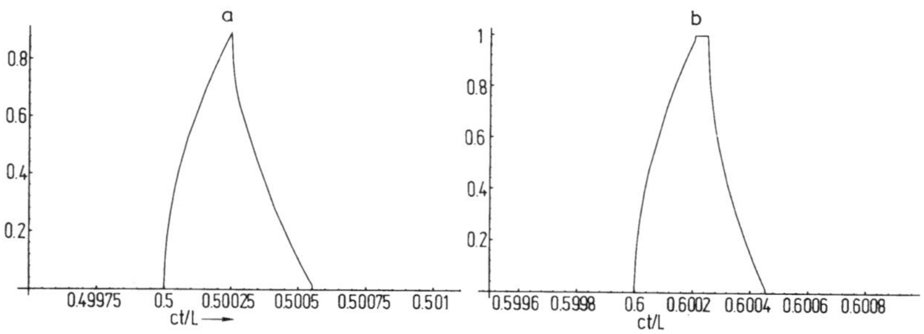

FIG.4.2-5. Distortion of a rectangular pulse on the array axis for $c\Delta t/L = 0.00025$ and $L/D = 20$ at a distance $x/L = 0.5$ (a) as well as $x/L = 0.6$ (b).

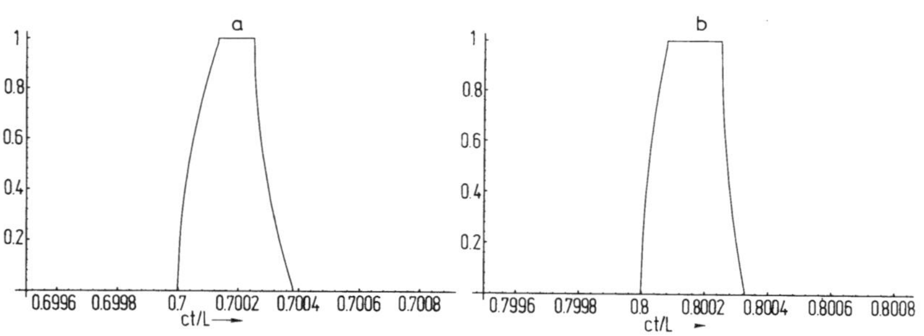

FIG.4.2-6. Distortion of a rectangular pulse on the array axis for $c\Delta t/L = 0.00025$ and $L/D = 20$ at a distance $x/L = 0.7$ (a) as well as $x/L = 0.8$ (b).

Plots of the normalized integral $S_1(\theta, x, 0)/2m$ as function of $\theta = ct/L$ are shown in Figs.4.2-1 to 4.2-15 for a normalized pulse duration $\Delta\theta = c\Delta t/L = 0.00025$. Figures 4.2-1 to 4.2-4 hold for a distance $L = 10D$ to the focusing point. The time variation of the nominally rectangular pulse is shown for the distances $x = 0.7L = 7D$, $8D$, $9D$, $9.5D$, $10D$, $11D$,$12D$, and $13D$. The rectangular pulse is obtained in Fig.4.2-3a for $x/L = x/10D = 1$. For shorter distances $x/L < 1$ and longer distances $x/L > 1$ we recognize significantly distorted pulses.

Figures 4.2-5 to 4.2-9 hold for a focusing distance $L = 20D$. The time variation

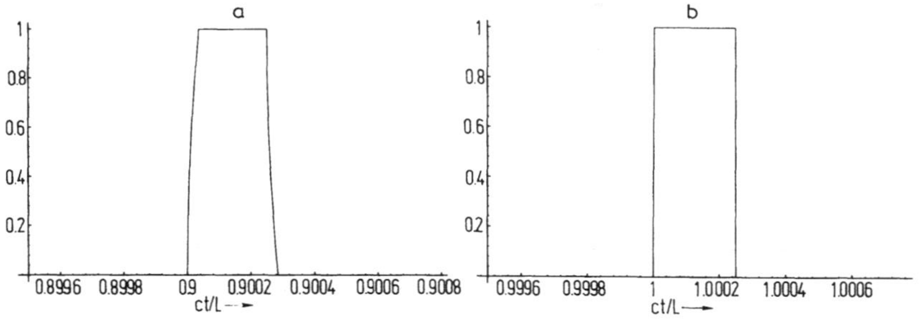

FIG.4.2-7. Distortion of a rectangular pulse on the array axis for $c\Delta t/L = 0.00025$ and $L/D = 20$ at a distance $x/L = 0.9$ (a) as well as $x/L = 1$ (b).

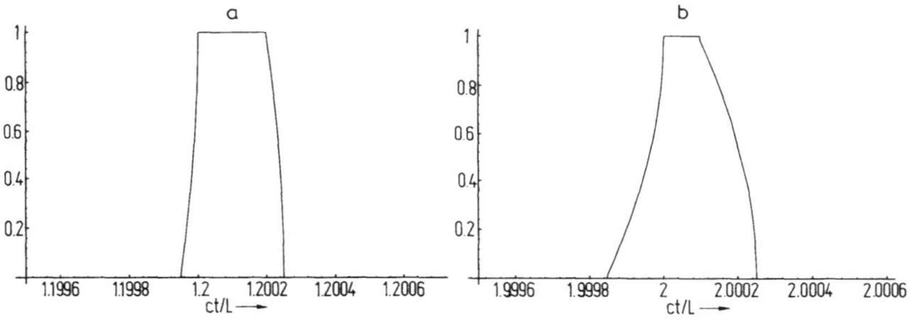

FIG.4.2-8. Distortion of a rectangular pulse on the array axis for $c\Delta t/L = 0.00025$ and $L/D = 20$ at a distance $x/L = 1.2$ (a) as well as $x/L = 2$ (b).

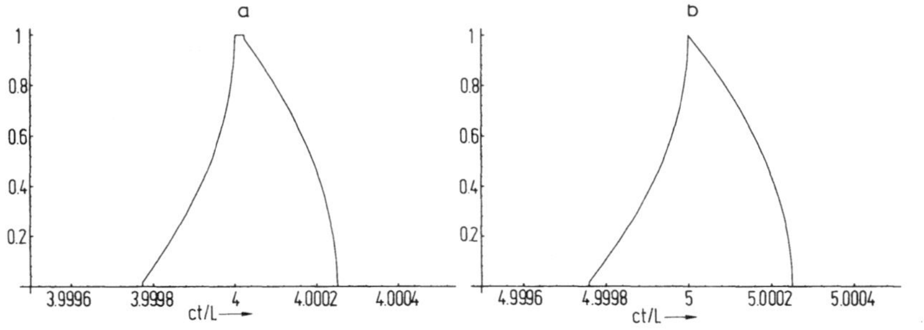

FIG.4.2-9. Distortion of a rectangular pulse on the array axis for $c\Delta t/L = 0.00025$ and $L/D = 20$ at a distance $x/L = 4$ (a) as well as $x/L = 5$ (b).

of the pulses on the array axis is shown for distances $x/L = x/20D = 0.5$, 0.6, 0.7, 0.8, 0.9, 1, 1.2, 2, 4, and 10. Again there is an undistorted pulse for $x/L = 1$ in Fig.4.2-7b while distorted pulses are shown for shorter and longer distances x/L.

Figures 4.2-10 to 4.2-15 hold for a focusing distance $L = 50D$. The time variation on the array axis is shown for the distances $x/L = x/50D = 0.1$, 0.15, 0.2, 0.3, 0.5, 0.7, 1, 5, 7, 10, 20, and 40. The undistorted pulse for $x/L = 1$ is shown in

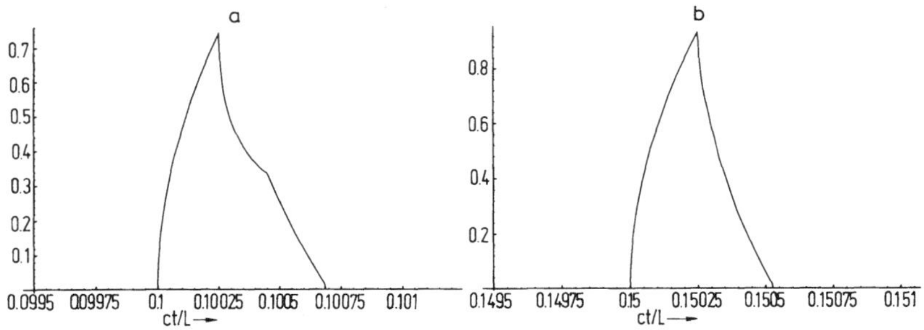

FIG.4.2-10. Distortion of a rectangular pulse on the array axis for $c\Delta t/L = 0.00025$ and $L/D = 50$ at a distance $x/L = 0.1$ (a) as well as $x/L = 0.15$ (b).

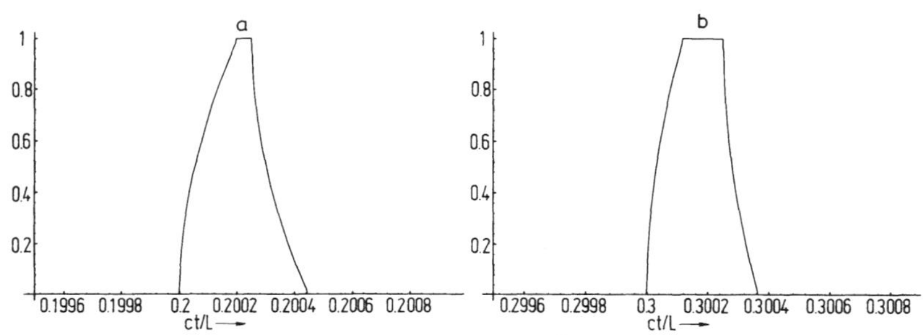

FIG.4.2-11. Distortion of a rectangular pulse on the array axis for $c\Delta t/L = 0.00025$ and $L/D = 50$ at a distance $x/L = 0.2$ (a) as well as $x/L = 0.3$ (b).

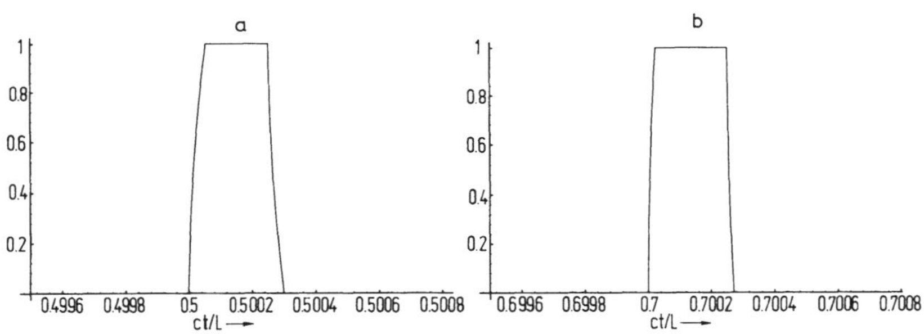

FIG.4.2-12. Distortion of a rectangular pulse on the array axis for $c\Delta t/L = 0.00025$ and $L/D = 50$ at a distance $x/L = 0.5$ (a) as well as $x/L = 0.7$ (b).

Fig.4.2-13a. We recognize that there is now a fairly large range of distances from $x/L = 0.7$ in Fig.4.2-12b to $x/L = 5$ in Fig.4.2-13b in which the deviation from the rectangular time variation is quite small. The smallness of the distortions is particularly conspicuous for large distances $5 \leq x/L \leq 40$. The increase of the *range of small distortions* for increasing focusing distances $L = 10D$, $20D$, $50D$ corresponds to the increase of the depth of field—or range of good focusing—with distance familiar from photography.

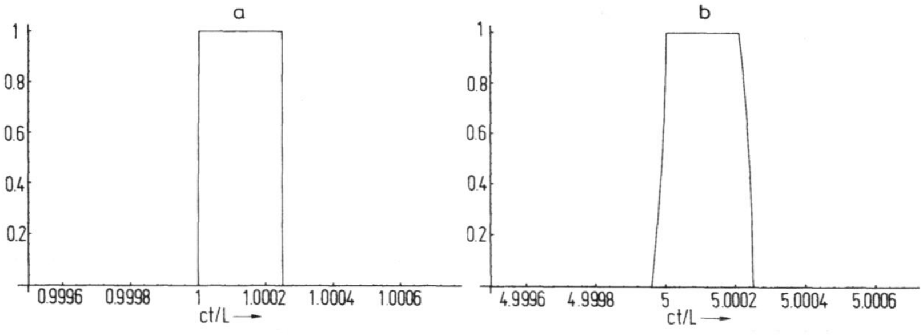

FIG.4.2-13. Distortion of a rectangular pulse on the array axis for $c\Delta t/L = 0.00025$ and $L/D = 50$ at a distance $x/L = 1$ (a) as well as $x/L = 5$ (b).

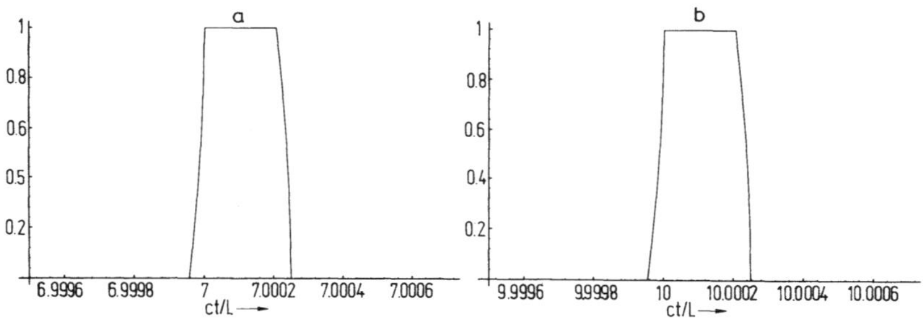

FIG.4.2-14. Distortion of a rectangular pulse on the array axis for $c\Delta t/L = 0.00025$ and $L/D = 50$ at a distance $x/L = 7$ (a) as well as $x/L = 10$ (b).

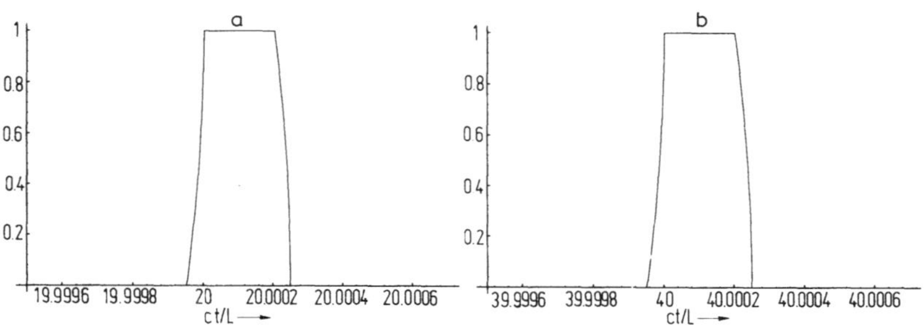

FIG.4.2-15. Distortion of a rectangular pulse on the array axis for $c\Delta t/L = 0.00025$ and $L/D = 50$ at a distance $x/L = 20$ (a) as well as $x/L = 40$ (b).

4.3 Pulse Shape Perpendicular to the Array Axis

We turn to the time variation of pulses in the focusing plane $x = L$ perpendicular to the array axis $x = 0$ in Fig.4.1-1. Equation (4.1-8) yields for $x = L$:

$$\theta_i(L,y) = \frac{\tau_i(L,y)}{L/c} = 1 - \left[1 + \frac{1}{q^2}\left(\frac{i}{2m}\right)^2\right]^{1/2} + \left[1 + \left(\frac{y}{L} - \frac{1}{q}\frac{i}{2m}\right)^2\right]^{1/2} \quad (1)$$

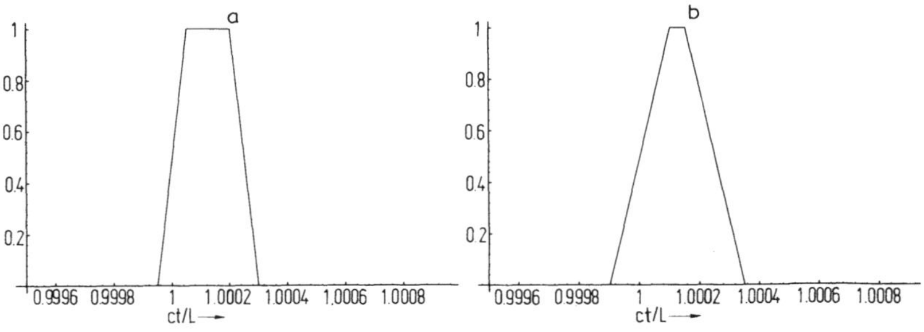

FIG.4.3-1. Distortion of a rectangular pulse at the distance $x/L = 1$ for $c\Delta t/L = 0.00025$ and $L/D = 10$ for distances from the array axis $y/L = \pm 0.001$ (a) as well as $y/L = \pm 0.002$ (b).

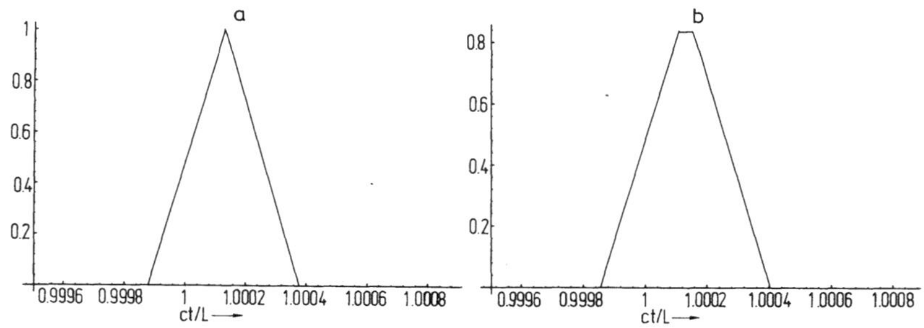

FIG.4.3-2. Distortion of a rectangular pulse at the distance $x/L = 1$ for $c\Delta t/L = 0.00025$ and $L/D = 10$ for distances from the array axis $y/L = \pm 0.0025$ (a) as well as $y/L = \pm 0.003$ (b).

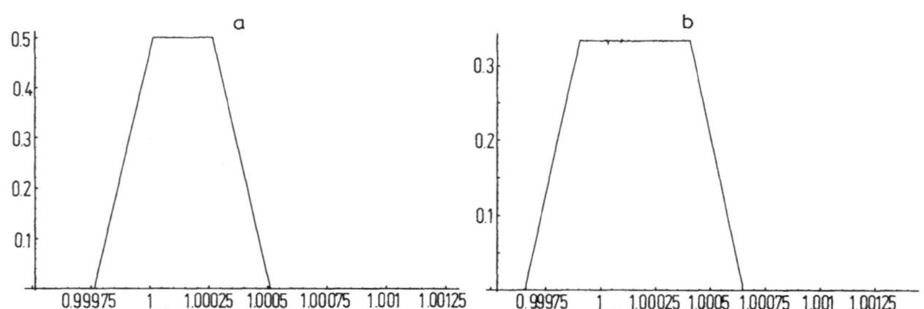

FIG.4.3-3. Distortion of a rectangular pulse at the distance $x/L = 1$ for $c\Delta t/L = 0.00025$ and $L/D = 10$ for distances from the array axis $y/L = \pm 0.005$ (a) as well as $y/L = \pm 0.0075$ (b).

In analogy to Eq.(4.2-1) we obtain the function $S_1(\theta, L, y)$:

$$S_1(\theta, L, y) = S_0(\theta, L, y) l_0(L, y) \sum_{i=-m}^{m} R[\theta - \theta_i(L, y), \Delta\theta] \tag{2}$$

For large values of m we may again use the substitution $p = i/2m$ and approximate the sum in Eq.(2) by an integral over the interval $-1/2 < p < +1/2$:

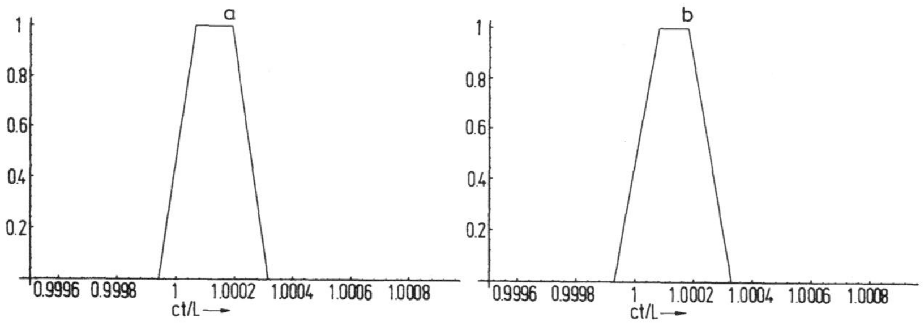

FIG.4.3-4. Distortion of a rectangular pulse at the distance $x/L = 1$ for $c\Delta t/L = 0.00025$ and $L/D = 20$ for distances from the array axis $y/L = \pm 0.0025$ (a) as well as $y/L = \pm 0.003$ (b).

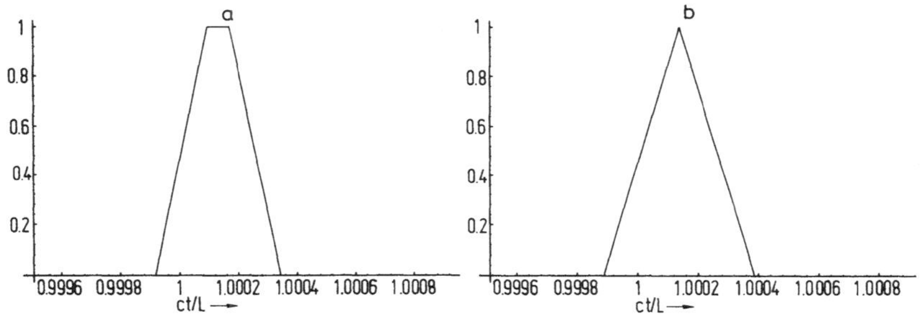

FIG.4.3-5. Distortion of a rectangular pulse at the distance $x/L = 1$ for $c\Delta t/L = 0.00025$ and $L/D = 20$ for distances from the array axis $y/L = \pm 0.0035$ (a) as well as $y/L = \pm 0.005$ (b).

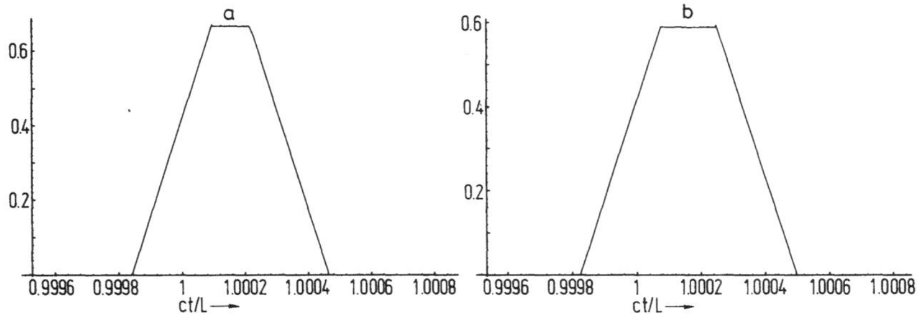

FIG.4.3-6. Distortion of a rectangular pulse at the distance $x/L = 1$ for $c\Delta t/L = 0.00025$ and $L/D = 20$ for distances from the array axis $y/L = \pm 0.0075$ (a) as well as $y/L = \pm 0.0085$ (b).

$$S_1(\theta, L, y) = 2m \int_{-1/2}^{1/2} R[\theta - \theta_p(L, y), \Delta\theta]dp$$

$$\theta_p(L, y) = 1 - (1 + p^2/q^2)^{1/2} + [1 + (y/L - p/q)^2]^{1/2}$$

$$p = i/2m, \quad -1/2 < p < +1/2, \quad m \gg 1 \tag{3}$$

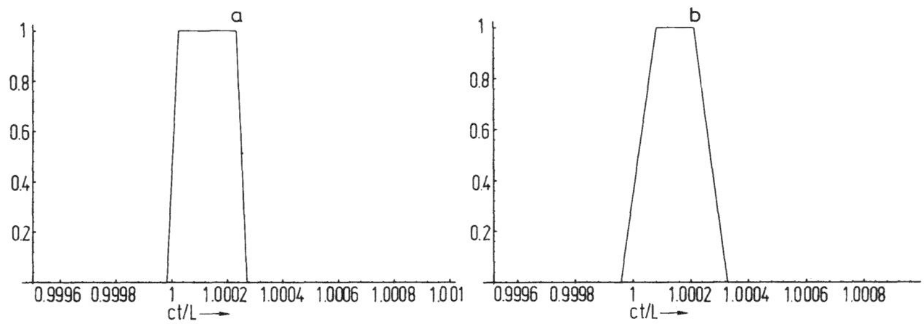

FIG.4.3-7. Distortion of a rectangular pulse at the distance $x/L = 1$ for $c\Delta t/L = 0.00025$ and $L/D = 50$ for distances from the array axis $y/L = \pm 0.002$ (a) as well as $y/L = \pm 0.006$ (b).

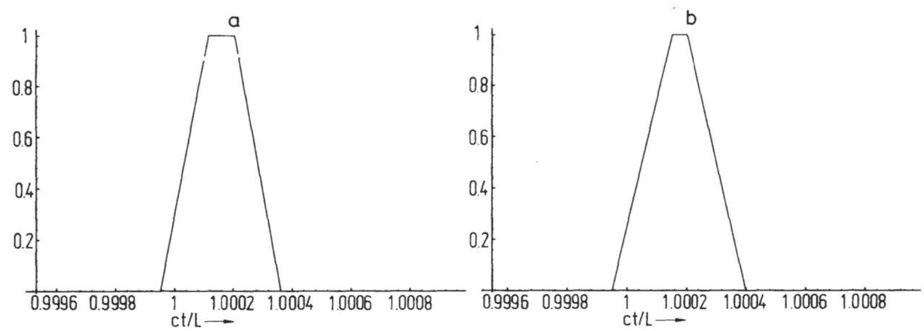

FIG.4.3-8. Distortion of a rectangular pulse at the distance $x/L = 1$ for $c\Delta t/L = 0.00025$ and $L/D = 50$ for distances from the array axis $y/L = \pm 0.008$ (a) as well as $y/L = \pm 0.01$ (b).

FIG.4.3-9. Distortion of a rectangular pulse at the distance $x/L = 1$ for $c\Delta t/L = 0.00025$ and $L/D = 50$ for distances from the array axis $y/L = \pm 0.0125$ (a) as well as $y/L = \pm 0.015$ (b).

Plots of the normalized integral $S_1(\theta, L, y)/2m$ are shown in Figs.4.3-1 to 4.3-10 for a nominal pulse duration $\Delta\theta = c\Delta t/L = 0.00025$. Figures 4.3-1 to 4.3-3 hold for a distance of $L = 10D$ according to the definitions of Fig.4.1-1. The distances from the array axis x range from $y = \pm 0.001L$ in Fig.4.3-1a to $y = \pm 0.0075L$ in Fig.4.3-3b. The rectangular pulse for $y = 0$ is not shown since it is the same as in Fig.4.2-3a. We see that the rectangular pulse becomes a trapezoidal pulse for $|y| < 0.0025L$ in Figs.4.3-1a and b. For $|y| = 0.0025L$ we obtain the triangular pulse

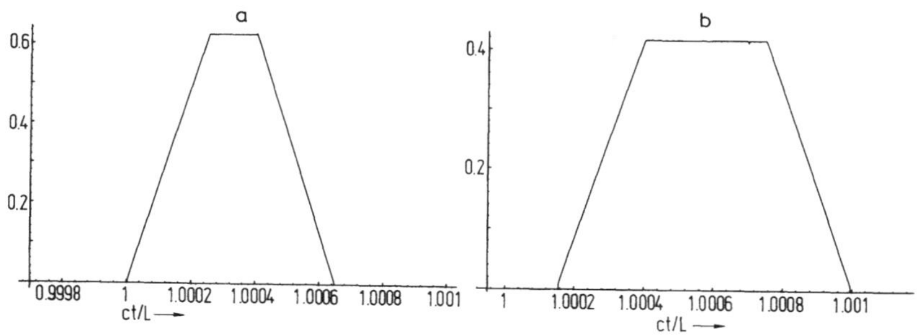

FIG.4.3-10. Distortion of a rectangular pulse at the distance $x/L = 1$ for $c\Delta t/L = 0.00025$ and $L/D = 50$ for distances from the array axis $y/L = \pm 0.02$ (a) as well as $y/L = \pm 0.03$ (b).

of Fig.4.3-2a. Larger values $|y| > 0.0025L$ yield again trapezoidal pulses but with decreasing amplitude as shown in Figs.4.3-2b to 4.3-3b.

The larger distance $L = 20D$ to the focusing plane $x = L$ yields the plots of Figs.4.3-4 to 4.3-6 for $S_1(\theta, Ly)$ according to Eq.(3). The rectangular pulse at $y = 0$ is again not shown since it is the same as in Fig.4.2-7b. For $|y| < 0.005L$ we obtain the trapezoidal pulses of Figs.4.3-4a to 4.3-5a. A triangular pulse is shown for $y = \pm 0.005L$ in Fig.4.3-5b. Trapezoidal pulses with reduced amplitude are obtained for $|y| > 0.005L$ and shown in Figs.4.3-6a and b.

The distance $L = 50D$ to the focusing plane yields the plots of Figs.4.3-7 to 4.3-10 for $S_1(\theta, L, y)$. The rectangular pulse for $y = 0$ is shown in Fig.4.2-13a. Trapezoidal pulses with the same amplitude are obtained for $|y| < 0.015L$ and shown in Figs.4.3-7a to 4.3-8b. A triangular pulse is shown for $y = \pm 0.015L$ in Fig.4.3-9a. Larger values $|y| > 0.015L$ yield the trapezoidal pulses with reduced amplitude in Figs.4.3-9b to 4.3-10b.

We note that all the distorted pulses in the focusing plane $x = L$ have either trapezoidal or triangular shape while the distorted pulses along the array axis x in Section 4.2 had curved leading and trailing edges.

4.4 Effect of Pulse Duration on Pulse Shape

In Sections 4.2 and 4.3 we investigated the distortion of pulses with the fixed duration $\Delta\theta = c\Delta t/L = 0.00025$ for certain distances L of the focusing plane from the radiator array and the distances x from the array or y from the intersection of the array axis with the focusing plane. We extend now the investigation to pulses with the durations $\Delta\theta = c\Delta t/L = 0.0005$ and 0.001.

Figures 4.4-1 to 4.4-6 show plots of $S_1(\theta, x, 0)/2m$ defined by Eq.(4.2-3) for $\Delta\theta = 0.0005$ or $\Delta t = 0.0005L/c$ and $L = 20D$. The distances x from the array vary from $x = 0.2L = 4D$ in Fig.4.4-1a to $x = 10L = 200D$ in Fig.4.4-6b. For $x = L$ we obtain the undistorted rectangular pulse of Fig.4.4-4b. Shorter distances $x < L$ and larger distances $x > L$ yield the pulses of Figs.4.4-1a to 4.4-4a and Figs.4.4-5a to 4.4-6b with curved leading and trailing edges. Figures 4.4-1 to 4.4-6 should be compared with Figs.4.2-5 to 4.2-9 that also hold for $L = 20D$ but use the pulse duration $\Delta t = 0.00025L/c$.

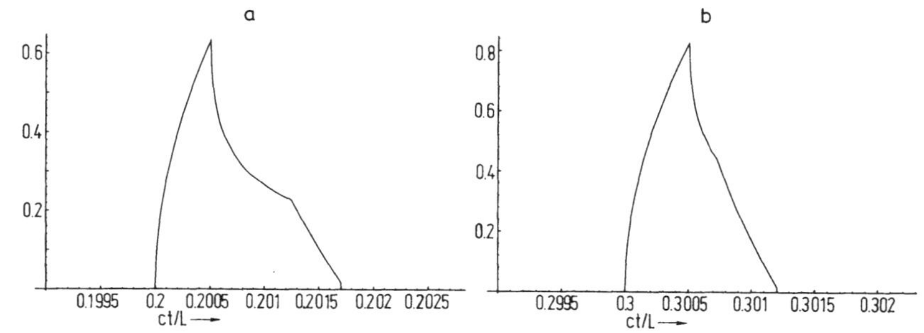

FIG.4.4-1. Distortion of a rectangular pulse on the array axis for $c\Delta t/L = 0.0005$ and $L/D = 20$ at a distance $x/L = 0.2$ (a) as well as $x/L = 0.3$ (b).

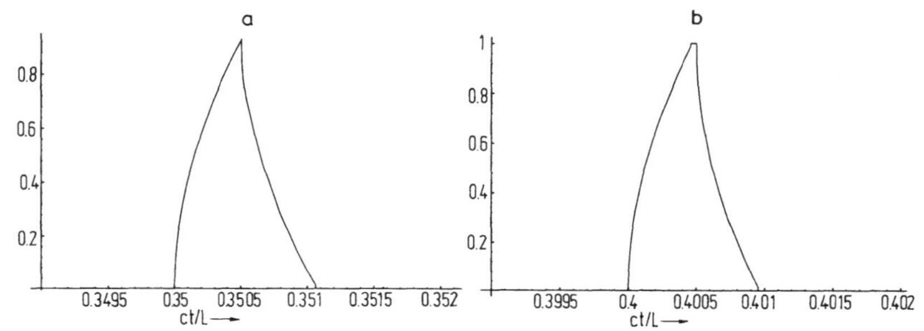

FIG.4.4-2. Distortion of a rectangular pulse on the array axis for $c\Delta t/L = 0.0005$ and $L/D = 20$ at a distance $x/L = 0.35$ (a) as well as $x/L = 0.4$ (b).

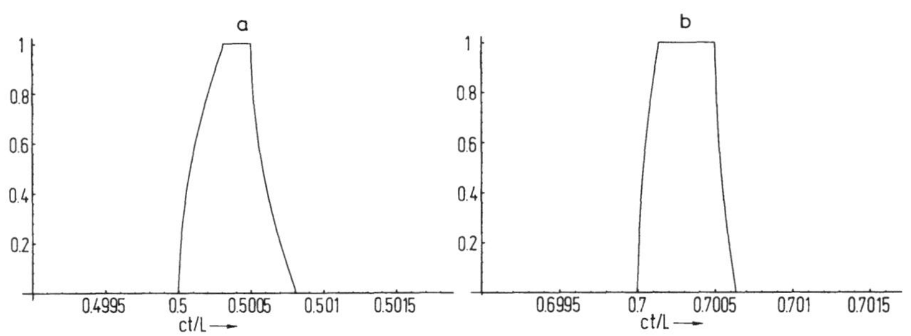

FIG.4.4-3. Distortion of a rectangular pulse on the array axis for $c\Delta t/L = 0.0005$ and $L/D = 20$ at a distance $x/L = 0.5$ (a) as well as $x/L = 0.7$ (b).

Figures 4.4-7 to 4.4-9 show plots of $S_1(\theta, L, y)/2m$ defined by Eq.(4.3-3) for $\Delta t = 0.0005L/c$ and $L = 20D$. Starting with the undistorted rectangular pulse for $x = 0$ in Fig.4.4-7a we obtain trapezoidal pulses with the same amplitude for $y = \pm 0.005L$ and $y = \pm 0.0075L$ in Figs.4.4-7b and 4.4-8a. A triangular pulse is obtained for $y = \pm 0.01L$ and shown in Fig.4.4-8b. For larger values of $|y|$ we obtain the trapezoidal pulses with reduced amplitude in Fig.4.4-9. Figures 4.4-7 to 4.4-9 should be compared with Figs.4.3-4 to 4.3-6 holding for $\Delta t = 0.00025L/c$ and

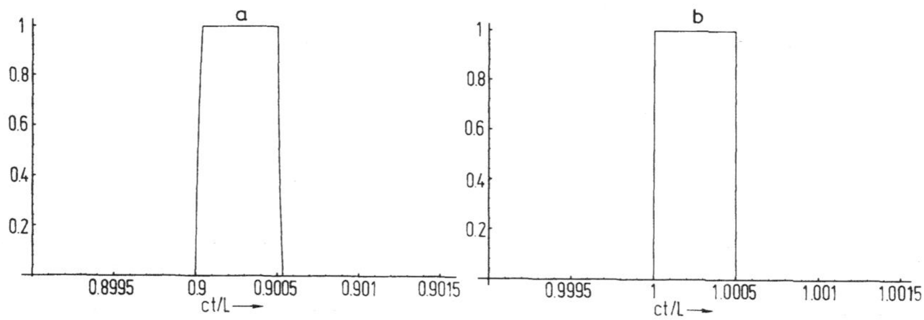

FIG.4.4-4. Distortion of a rectangular pulse on the array axis for $c\Delta t/L = 0.0005$ and $L/D = 20$ at a distance $x/L = 0.9$ (a) as well as $x/L = 1$ (b).

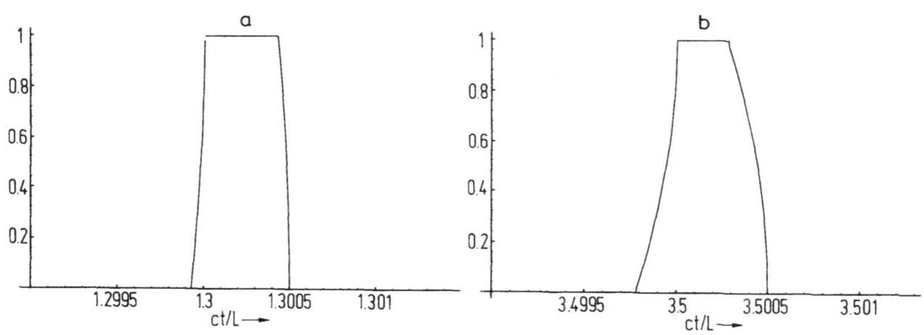

FIG.4.4-5. Distortion of a rectangular pulse on the array axis for $c\Delta t/L = 0.0005$ and $L/D = 20$ at a distance $x/L = 1.3$ (a) as well as $x/L = 3.5$ (b).

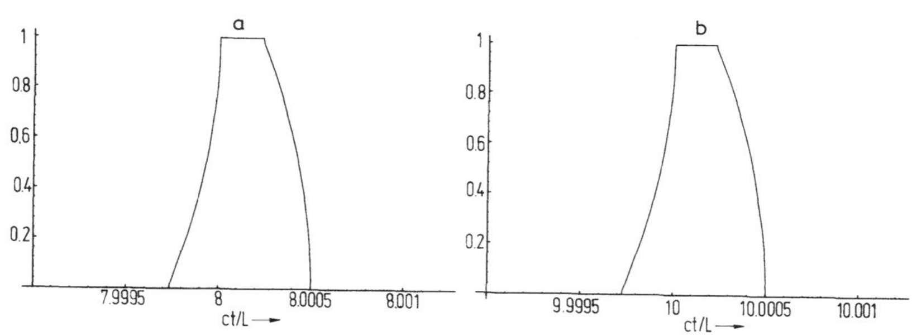

FIG.4.4-6. Distortion of a rectangular pulse on the array axis for $c\Delta t/L = 0.0005$ and $L/D = 20$ at a distance $x/L = 8$ (a) as well as $x/L = 10$ (b).

$L = 20D$.

Figures 4.4-10 to 4.4-16 hold for $\Delta t = 0.001L/c$ and $L = 50D$. The distorted pulses along the array axis according to $S_1(\theta, x, 0)/2m$ of Eq.(4.2-3) are shown for $x < L$ in Figs.4.4-10a to Fig.4.4-12a. In Fig.4.4-12b we see the undistorted rectangular pulse for $x = L$. Figures 4.4-13a and b for the larger distances $x = 10L$ and $x = 200L$ show essentially the undistorted rectangular pulse too. This is different from the pulses in Figs.4.2-13b to 4.2-15b that hold for the shorter pulse

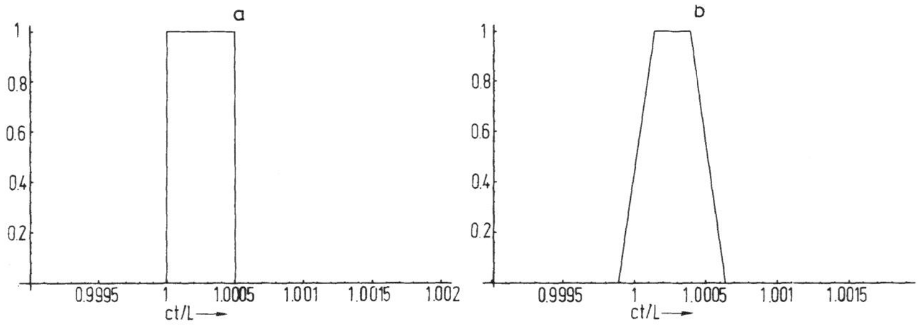

FIG.4.4-7. Distortion of a rectangular pulse at the distance $x/L = 1$ for $c\Delta t/L = 0.0005$ and $L/D = 20$ for distances from the array axis $y/L = 0$ (a) as well as $y/L = \pm 0.0005$ (b).

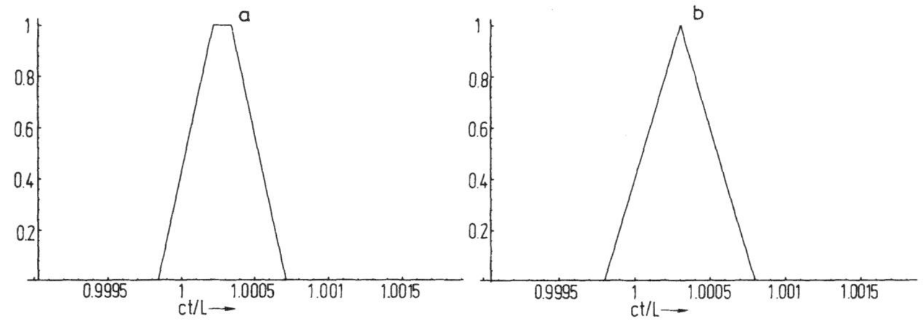

FIG.4.4-8. Distortion of a rectangular pulse at the distance $x/L = 1$ for $c\Delta t/L = 0.0005$ and $L/D = 20$ for distances from the array axis $y/L = \pm 0.0075$ (a) as well as $y/L = \pm 0.01$ (b).

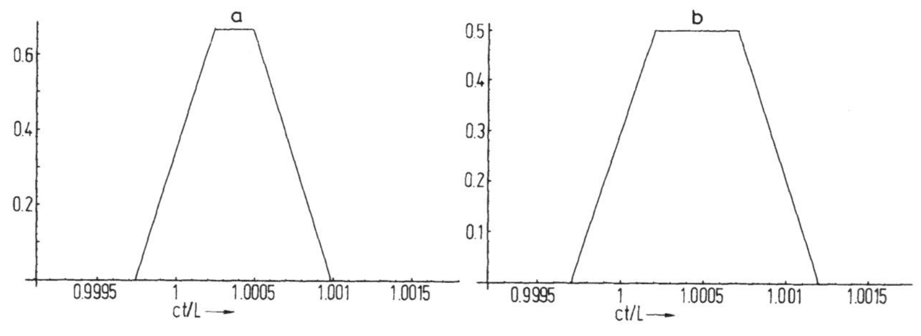

FIG.4.4-9. Distortion of a rectangular pulse at the distance $x/L = 1$ for $c\Delta t/L = 0.0005$ and $L/D = 20$ for distances from the array axis $y/L = \pm 0.015$ (a) as well as $y/L = \pm 0.02$ (b).

duration $\Delta t = 0.00025L/c$.

Distorted pulses for $\Delta t = 0.001L/c$ and $L = 50D$ according to the normalized integral $S_1(\theta, L, y)/2m$ of Eq.(4.3-3) at locations perpendicular to the array axis are shown in Figs.4.4-14 to 4.4-16. We recognize trapezoidal pulses with amplitude 1 for $|y| < 0.05$ in Figs.4.4-14a to 4.4-15a, the triangular pulse for $y = \pm 0.05$ in Fig.4.4-

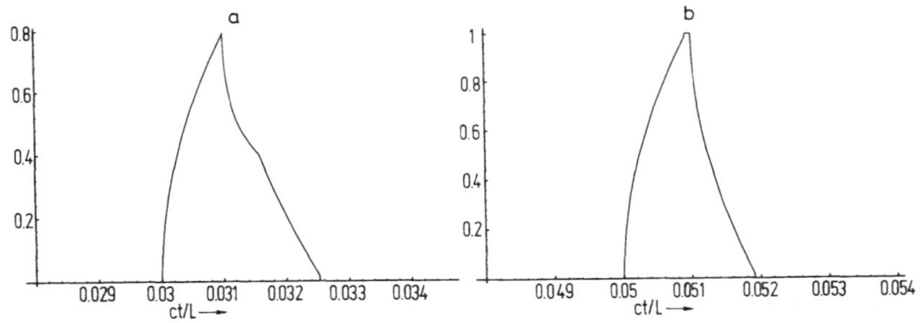

FIG.4.4-10. Distortion of a rectangular pulse on the array axis for $c\Delta t/L = 0.001$ and $L/D = 50$ at a distance $x/L = 0.03$ (a) as well as $x/L = 0.05$ (b).

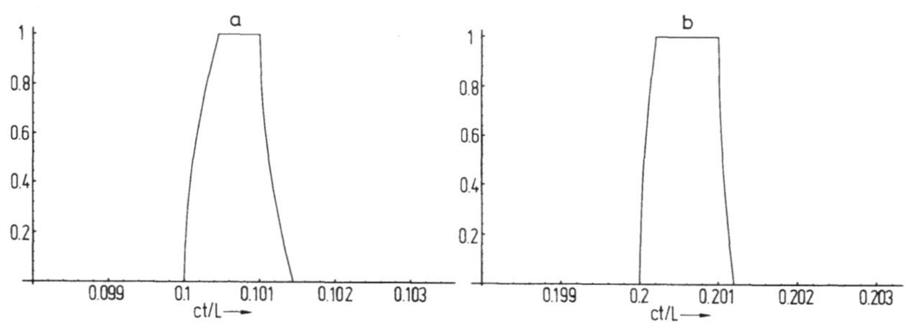

FIG.4.4-11. Distortion of a rectangular pulse on the array axis for $c\Delta t/L = 0.001$ and $L/D = 50$ at a distance $x/L = 0.1$ (a) as well as $x/L = 0.2$ (b).

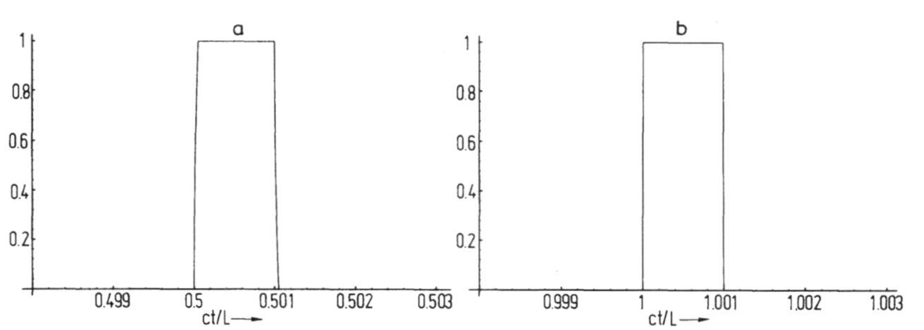

FIG.4.4-12. Distortion of a rectangular pulse on the array axis for $c\Delta t/L = 0.001$ and $L/D = 50$ at a distance $x/L = 0.5$ (a) as well as $x/L = 1$ (b).

15b, and trapezoidal pulses with reduced amplitude for $|y| > 0.05$ in Fig.4.4-16. The comparable pulses for $\Delta t = 0.00025L/c$ and $L = 50D$ are shown in Figs.4.3-7 to 4.3-10.

the focusing of periodic sinusoidal waves one knows that only amplitude and phase can vary since the sum of sinusoidal functions with the same frequency will always yield a sinusoidal function with that frequency. The situation is different for signals since the sum of signals with the same time variation but various am-

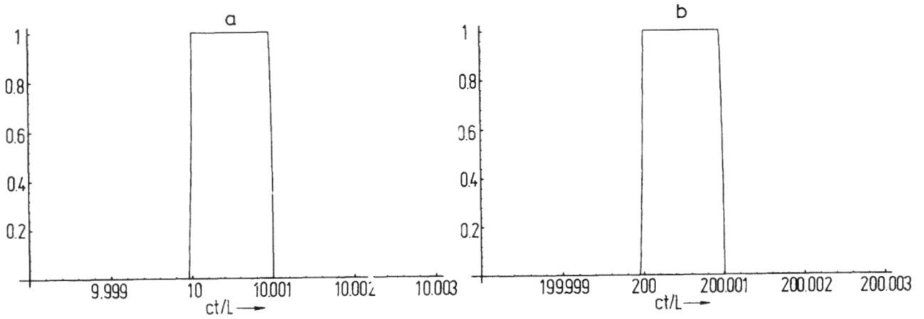

FIG.4.4-13. Distortion of a rectangular pulse on the array axis for $c\Delta t/L = 0.001$ and $L/D = 50$ at a distance $x/L = 10$ (a) as well as $x/L = 200$ (b).

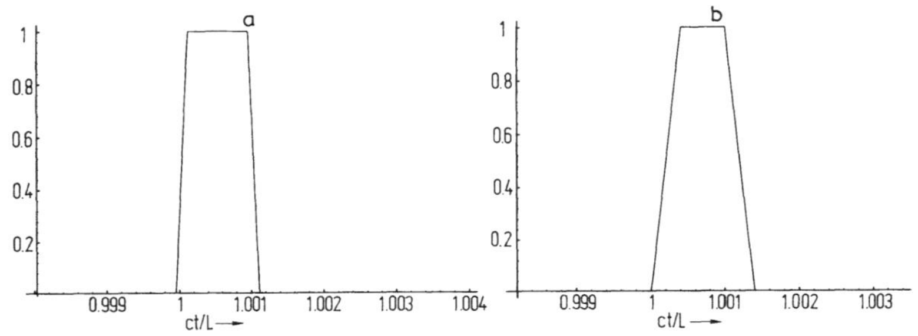

FIG.4.4-14. Distortion of a rectangular pulse at the distance $x/L = 1$ for $c\Delta t/L = 0.001$ and $L/D = 50$ for distances from the array axis $y/L = \pm 0.008$ (a) as well as $y/L = \pm 0.02$ (b).

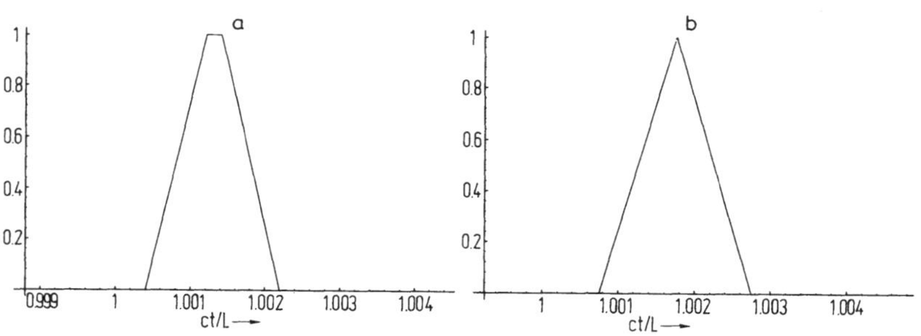

FIG.4.4-15. Distortion of a rectangular pulse at the distance $x/L = 1$ for $c\Delta t/L = 0.001$ and $L/D = 50$ for distances from the array axis $y/L = \pm 0.04$ (a) as well as $y/L = \pm 0.05$ (b).

plitudes and time delays may create signals with completely different time variation. One may see in this effect just one more unwanted complication. But the change of the time variation represents information that periodic sinusoidal functions cannot provide and one may see an opportunity to make use of this information.

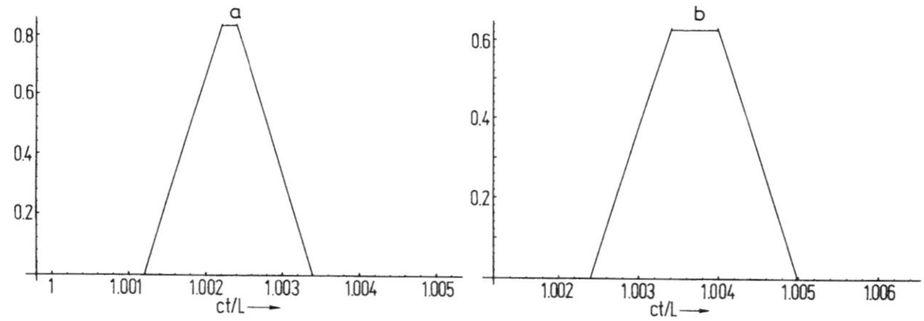

FIG.4.4-16. Distortion of a rectangular pulse at the distance $x/L = 1$ for $c\Delta t/L = 0.001$ and $L/D = 50$ for distances from the array axis $y/L = \pm 0.06$ (a) as well as $y/L = \pm 0.08$ (b).

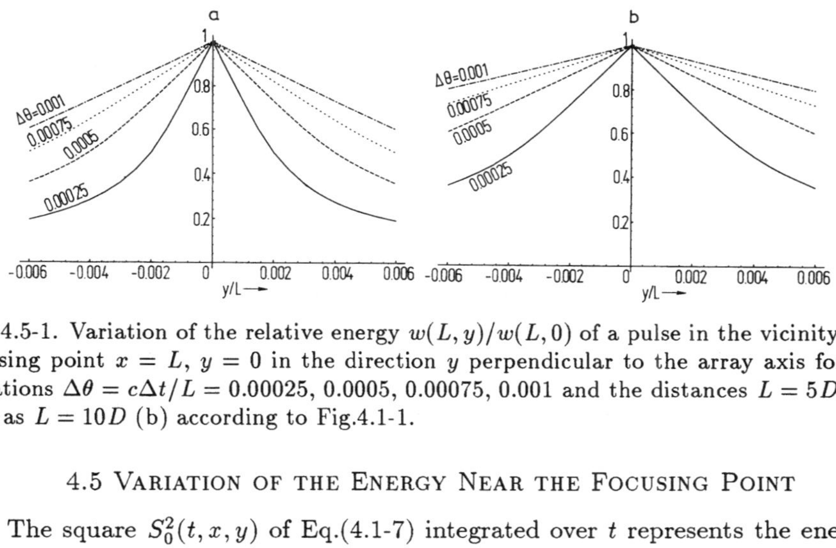

FIG.4.5-1. Variation of the relative energy $w(L,y)/w(L,0)$ of a pulse in the vicinity of the focusing point $x = L$, $y = 0$ in the direction y perpendicular to the array axis for pulse durations $\Delta\theta = c\Delta t/L = 0.00025, 0.0005, 0.00075, 0.001$ and the distances $L = 5D$ (a) as well as $L = 10D$ (b) according to Fig.4.1-1.

4.5 VARIATION OF THE ENERGY NEAR THE FOCUSING POINT

The square $S_0^2(t,x,y)$ of Eq.(4.1-7) integrated over t represents the energy of a pulse in the point x, y. In order to represent the effect of the pulse shape rather than that of the distance $l_0(x,y)$ on the energy, we compute

$$w(x,y) = \int_{-\infty}^{\infty} S_1^2(\theta,x,y)d\theta = \int_{-\infty}^{\infty} S_0^2(\theta,x,y)l_0^2(x,y)d\theta \tag{1}$$

where the integration goes from beginning to end of the pulses in Section 4.2 to 4.4. The largest energy is obtained for the focusing point $x = L$, $y = 0$:

$$w(L,0) = \int_{-\infty}^{\infty} S_1^2(\theta,L,0)d\theta \tag{2}$$

The function $S_1(\theta,x,y)$ can be approximated for a large number $2m+1$ of radiators according to Eq.(4.3-3):

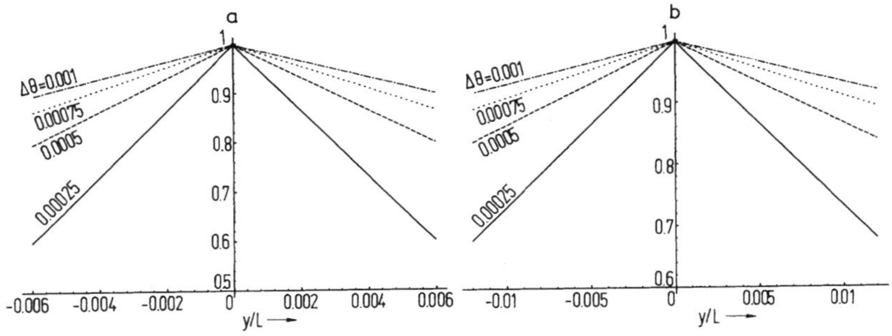

FIG.4.5-2. Variation of the relative energy $w(L,y)/w(L,0)$ of a pulse in the vicinity of the focusing point $x = L$, $y = 0$ in the direction y perpendicular to the array axis for pulse durations $\Delta\theta = c\Delta t/L = 0.00025$, 0.0005, 0.00075, 0.001 and the distances $L = 20D$ (a) as well as $L = 50D$ (b) according to Fig.4.1-1.

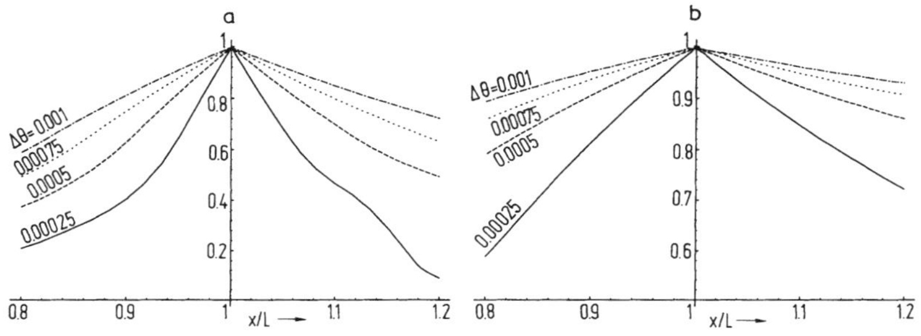

FIG.4.5-3. Variation of the relative energy $w(x,0)/w(L,0)$ of a pulse in the vicinity of the focusing point $x = L$, $y = 0$ in the direction x of the array axis for pulse durations $\Delta\theta = c\Delta t/L = 0.00025$, 0.0005, 0.00075, 0.001 and the distances $L = 5D$ (a) as well as $L = 10D$ (b) according to Fig.4.1-1.

$$S_1(\theta, x, y) = 2m \int_{-1/2}^{1/2} R[\theta - \theta_p(x,y), \Delta\theta]dp \qquad (3)$$

Plots of the normalized energy $w(L,y)/w(L,0)$ are shown in Figs.4.5-1 and 4.5-2 for distances $L = 5D$, $10D$, $20D$, $50D$ and pulse durations $\Delta\theta = c\Delta t/L = 0.00025$, 0.0005, 0.00075, 0.001. All plots are are symmetric with respect to $y/L = 0$.

Figures 4.5-3 and 4.5-4 show plots of $w(x,0)/w(L,0)$ again for the distances $L = 5D$, $10D$, $20D$, $50D$ and pulse durations $\Delta\theta = 0.00025$, 0.0005, 0.00075, 0.001. These plots are not symmetric with respect to $x/L = 1$.

As one would expect the drop of the relative energy from 1 at the focusing point is fastest for the shortest pulses with $\Delta\theta = 0.00025$ represented by the solid lines in Figs.4.5-1 to 4.5-4 and less fast for increasing values of $\Delta\theta$. There are other features that the interested reader will recognize.

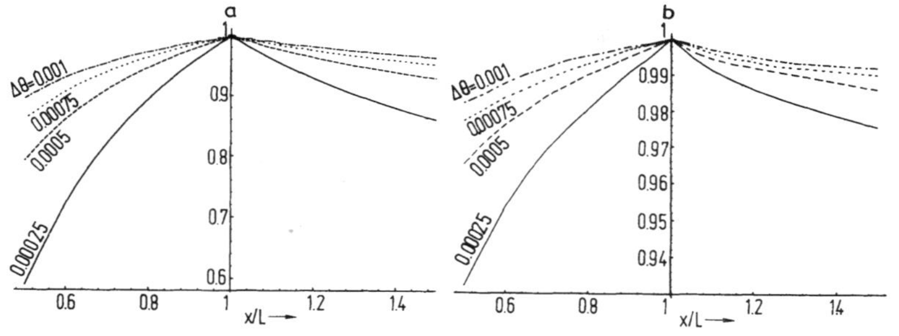

FIG.4.5-4. Variation of the relative energy $w(x,0)/w(L,0)$ of a pulse in the vicinity of the focusing point $x = L$, $y = 0$ in the direction x of the array axis for pulse durations $\Delta\theta = c\Delta t/L = 0.00025, 0.0005, 0.00075, 0.001$ and the distances $L = 20D$ (a) as well as $L = 50D$ (b) according to Fig.4.1-1.

5 Signal Propagation and Detection in Lossy Media

5.1 Planar Wave Solution in Lossy Media

We start with the modified Maxwell equations of Section 1.1. They were written in Section 2.1 for Cartesian coordinates and planar TEM waves. The polarization was extracted by Eqs.(2.1-16), (2.1-17) and Eqs.(2.1-18), (2.1-19) were obtained. We can use these equations here but we leave out the indices i and 1 since we consider waves in one medium rather than at the boundary of two mediums:

$$\frac{\partial E}{\partial y} + \mu \frac{\partial H}{\partial t} + sH = 0 \tag{1}$$

$$\frac{\partial H}{\partial y} + \epsilon \frac{\partial E}{\partial t} + \sigma E = 0 \tag{2}$$

These two equations describe the propagation of a planar TEM wave in a medium characterized by the constant scalars ϵ, μ, σ, and s.

Originally the theory was developed for seawater as medium. Both communication with submarines and detection of submarines by airborne radar were handicapped by the existing theory for the propagation of periodic sinusoidal waves, which applied to the transmission of power rather than signals. The end of the cold war shifted interest from such military applications in seawater to the interstellar *empty space* as a lossy medium. This sounds strange at first but one has to keep in mind that the interstellar space between stars and galaxies contains gas—mostly atomic hydrogen—with a concentration of about 5×10^4 atoms per cubic meter. This gas has little effect for distances of millions of light years. However, over a distance of 1 billion light years the effect is about as great as that of the Earth's atmosphere. For distances of 10 billion light years the losses in interstellar space may dominate the losses in the Earth's atmosphere.

We are currently able to detect electromagnetic waves from sources having a distance of more than 10 billion light years. Nobody expects to receive 'intelligent' signals from such distances, but one does not receive periodic sinusoidal waves as produced by certain radio transmitters or lasers either. What is radiated are always pulses, including sinusoidal pulses. Most of the radiated energy excites first the hydrogen atoms in interstellar space and is then re-radiated. Only a small fraction of the originally radiated energy reaches targets like the Earth where some of the received energy can be stored in chemical form, e.g., by transforming carbon dioxide into carbon and oxygen. Any energy that cannot be stored as non-thermal energy must heat up the Earth and increase re-radiation until an equilibrium is reached.

The re-radiated energy is used mostly to excite further hydrogen atoms that in turn re-radiate it again, etc. This produces in the end very uniform black body radiation. The existence of such radiation is well established but it is usually attributed to the Big Bang. The re-radiation of electromagnetic waves originating from stars by the interstellar hydrogen atoms and objects like Earth provides an either supplementary or alternative explanation. The theory developed here is a first step for a numerical determination of the contribution of re-radiated energy to the observed black body radiation.

One may eliminate H from Eqs.(1) and (2) to get an equation for E alone

$$\frac{\partial^2 E}{\partial y^2} - \mu\epsilon\frac{\partial^2 E}{\partial t^2} - (\mu\sigma + \epsilon s)\frac{\partial E}{\partial t} - s\sigma E = 0 \tag{3}$$

or one may eliminate E to get an equation for H alone:

$$\frac{\partial^2 H}{\partial y^2} - \mu\epsilon\frac{\partial^2 H}{\partial t^2} - (\mu\sigma + \epsilon s)\frac{\partial H}{\partial t} - s\sigma H = 0 \tag{4}$$

If we use Eq.(3) we get $E = E_{\mathrm{E}}$ due to electric excitation and we must calculate the associated magnetic field strength $H = H_{\mathrm{E}}$ from Eqs.(1) or (2). Since both equations must yield the same function $H_{\mathrm{E}}(y,t)$ one obtains a relation that eliminates an integration constant. Alternately we may use Eq.(4) to obtain $H = H_{\mathrm{H}}$ due to magnetic excitation and must then calculate the associated electric field strength $E = E_{\mathrm{H}}$ from Eqs.(1) or (2). Again an integration constant is determined by the requirement that both equations must yield the same function $E_{\mathrm{H}}(y,t)$. If there is both an electric and a magnetic excitation force we obtain from Eqs.(1)–(4) the field strengths $E_{\mathrm{E}}(y,t) + E_{\mathrm{H}}(y,t)$ and $H_{\mathrm{H}}(y,t) + H_{\mathrm{E}}(y,t)$.

If one ignores the existence of electric and magnetic dipole currents one can use $s = 0$ and $\sigma \neq 0$ in Eqs.(1) and (2). But this eliminates the terms $s\sigma E$ and $s\sigma H$ in Eqs.(3) and (4), which leads to different equations. This is the reason why the transition $s \to 0$ generally yields different results when made at the beginning rather than at the end of the calculation .

In addition to the differential equations we need boundary and initial conditions. Furthermore, we must introduce certain physical laws such as the *causality law.* We do not usually hear about the need to introduce the causality law because most of the solutions of Maxwell's equations found in the literature are *steady state* solutions[1] holding for $t \to \infty$. Causality becomes meaningless in this case. Although the causality law is one of the most basic laws of physics and signal transmission one hardly ever finds it stated in other than philosophical books. We have cited it already in Section 1.1, footnote 5, but repeat it here: *Every effect must have a sufficient cause that occurred a finite time earlier.*

There can be no effect before the information about a cause is available. The time difference between cause and effect must be finite and larger than zero since we

[1] The same holds true in quantum mechanics when eigenfunctions are determined.

cannot observe infinite or infinitesimal time differences. The causality law introduces the universally observed distinguished direction of time since the effect comes after the cause. There is no such distinguished direction for a space variable.

Consider first Eq.(3) and an electric force function $E(y = 0, t)$ as boundary condition, having the time variation of a step function:

$$\begin{aligned} E(0,t) = E_0 S(t) &= 0 \quad \text{for } t < 0 \\ &= E_0 \quad \text{for } t \geq 0 \end{aligned} \tag{5}$$

As initial condition $E(y, t = 0)$ we may choose a function that does not depend on the boundary condition, otherwise something that happens at $t > 0$ would have an effect at the time $t = 0$. Of course, satisfying two independent conditions requires a solution of Eq.(3) that is sufficiently general to permit two conditions. The independence of boundary and initial condition introduces the causality law. Let us choose

$$E(y,0) = 0 \quad \text{for } y > 0 \tag{6}$$

as initial condition for the electric field strength. An initial condition $H(y, 0)$ for the magnetic field strength is not needed for the solution of Eq.(3) but it is required for the determination of the associated magnetic field strength from Eqs.(1) or (2). We choose in analogy to Eq.(6):

$$H(y,0) = 0 \quad \text{for } y > 0 \tag{7}$$

It will turn out later on that there are solutions of Eq.(3) that increase exponentially with y. In order to exclude them we need a boundary condition for $y \to \infty$. To obtain such a boundary condition we need a physical law that is routinely and universally used but has no name[2]. We shall call it the *finiteness law*:

All observable physical quantities are finite.

Using this law we can write additional boundary conditions for E and H holding for large values of y:

$$E(y \to \infty, t) = \text{finite}, \quad H(y \to \infty, t) = \text{finite} \tag{8}$$

Anyone with training in physics will accept that only finite or bounded field strengths, currents, voltages, powers, and energies can be observed and are thus part of physics. However, it appears that in Eq.(8) we have introduced an infinite distance y and distances are certainly observable physical quantities. The explanation is that the notation $y \to \infty$ or $t \to \infty$ in physics stand for $y \gg 1\,\text{m}$ and $t \gg 1\,\text{s}$, which are large but finite distances and times. At a 'really infinite' distance y or time t

[2]It is disturbing that there are commonly used laws of nature that have no name and are never listed. We may take comfort from the fact that mathematicians cannot list all their axioms either but can do so for certain parts of mathematics only: Peano's axioms for numbers theory, Euclid's axioms for Euclidean geometry, etc.

we can introduce any assumptions we like without running a risk of being proved wrong by observation. Such non-provable or non-disprovable assumptions are not part of physics.

A further problem of notation is the force function $E(0,t)$ introduced by Eq.(5). One may object that it has infinite duration and energy. We could avoid this objection by using a rectangular pulse of duration T instead of the step function:

$$\begin{aligned} E(0,t) = E_0 R(t) &= 0 \quad \text{for } t < 0 \text{ and } t > T \\ &= E_0 \quad \text{for } 0 \le t \le T \end{aligned} \tag{9}$$

However, $R(t)$ can be represented conveniently by two step functions

$$R(t) = S(t) - S(t-T) \tag{10}$$

and a solution for the boundary condition of Eq.(5) can thus readily be used to write the solution for the boundary condition of Eq.(9) and avoid any infinite energy.

The initial conditions of Eqs.(6) and (7) imply a further set of initial conditions

$$\partial E(y,0)/\partial y = 0, \quad \partial H(y,0)/\partial y = 0 \qquad \text{for } y > 0 \tag{11}$$

while Eqs.(6), (7), and (11) substituted into Eqs.(1) and (2) yield a final set of initial conditions:

$$\partial E(y,t=0)/\partial t = 0, \quad \partial H(y,t=0)/\partial t = 0 \qquad \text{for } y > 0 \tag{12}$$

Using the method of *steady state solution plus deviation from it* we write the expected solution of Eq.(3) in the form

$$E(y,t) = E_{\mathrm{E}}(y,t) = E_0[w(y,t) + F(y)] \tag{13}$$

Substitution of $F(y)$ into Eq.(3) yields an ordinary differential equation

$$d^2F/dy^2 - s\sigma F = 0 \tag{14}$$

with the general solution

$$F(y) = A_{00}e^{-\sqrt{s\sigma}\,y} + A_{01}e^{\sqrt{s\sigma}\,y} \tag{15}$$

The boundary condition of Eq.(8) demands $A_{01} = 0$ while the boundary condition of Eq.(5) calls for $A_{00} = 1$. The function $w(y,t)$—which gives the deviation from the steady state solution $F(y)$—must also satisfy Eq.(3) but the boundary condition of Eq.(5) becomes homogeneous when Eq.(15) is substituted into Eq.(13):

$$\begin{aligned} E_{\mathrm{E}}(0,t) = E_0[w(0,t)+1] &= E_0 \quad \text{for } t \ge 0 \\ w(0,t) &= 0 \quad \text{for } t \ge 0 \end{aligned} \tag{16}$$

The solution for $w(y,t)$ will not be derived here since we want to show in considerable detail how the two physical laws of causality and finiteness are introduced into the mathematical model by means of the boundary and initial conditions. We refer to the literature[3] for the derivation of $w(y,t)$:

$$\begin{aligned} w(y,t) = -\frac{2}{\pi}e^{-at}\Bigg[\int_0^{2\pi K} \left(\mathrm{ch}\left[(a^2 - b^2c^2)^{1/2}t\right] + \frac{a\,\mathrm{sh}\left[(a^2 - b^2c^2)^{1/2}t\right]}{(a^2 - b^2c^2)^{1/2}} \right) \\ \times \frac{\beta \sin \beta y}{b^2}\, d\beta \\ + \int_{2\pi K}^{\infty} \left(\cos\left[(b^2c^2 - a^2)^{1/2}t\right] + \frac{a \sin\left[(b^2c^2 - a^2)^{1/2}t\right]}{(b^2c^2 - a^2)^{1/2}} \right) \\ \times \frac{\beta \sin \beta y}{b^2}\, d\beta \Bigg] \end{aligned}$$

$$\begin{aligned} a &= (\mu\sigma + \epsilon s)c^2/2 = (Z\sigma + s/Z)c/2 = \sigma/2\epsilon + s/2\mu \\ b^2 &= \beta^2 + s\sigma,\ K = (a^2/c^2 - s\sigma)^{1/2}/2\pi \\ c &= 1/\sqrt{\mu\epsilon},\ Z = \sqrt{\mu/\epsilon} \end{aligned} \tag{17}$$

We make the transition $s \to 0$ and introduce the normalized notation

$$\begin{aligned} \alpha &= Zc\sigma/2 = \sigma/2\epsilon \\ \eta &= \beta c/\alpha = 4\pi\kappa/Z\sigma \\ \theta &= \alpha t = Z\sigma ct/2 = \sigma t/2\epsilon \\ \zeta &= \alpha y/c = Z\sigma y/2 = \sqrt{\epsilon/\mu}\,\sigma y/2 \end{aligned} \tag{18}$$

Equations (13) and (17) assume the following form:

$$E_{\mathrm{E}}(\zeta,\theta) = E_0[1 + w(\zeta,\theta)] \tag{19}$$

$$\begin{aligned} w(\zeta,\theta) = -\frac{2}{\pi}e^{-\theta}\Bigg[\int_0^1 \left(\mathrm{ch}\left[(1-\eta^2)^{1/2}\theta\right] + \frac{\mathrm{sh}\left[(1-\eta^2)^{1/2}\theta\right]}{(1-\eta^2)^{1/2}} \right) \frac{\sin \zeta\eta}{\eta}\, d\eta \\ + \int_1^{\infty} \left(\cos\left[(\eta^2-1)^{1/2}\theta\right] + \frac{\sin\left[(\eta^2-1)^{1/2}\theta\right]}{(\eta^2-1)^{1/2}} \right) \frac{\sin \zeta\eta}{\eta}\, d\eta \Bigg] \end{aligned} \tag{20}$$

---

[3] Harmuth 1986, pp. 49–53

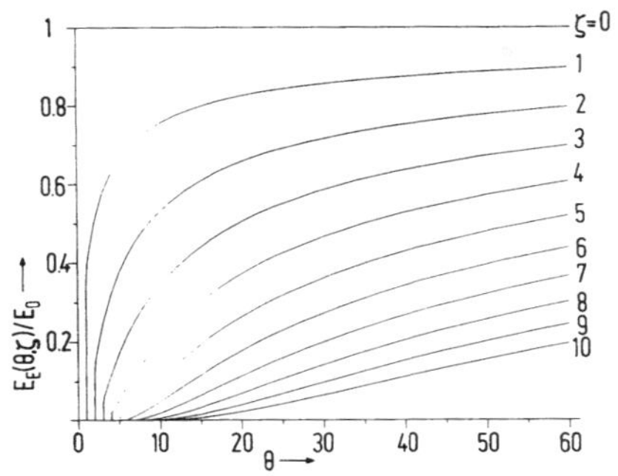

FIG.5.1-1. Plots of $E_E(\zeta,\theta)/E_0$ according to Eq.(19) for $\zeta = 0, 1, \dots, 10$ in the interval $0 \le \theta \le 60$.

To check the existence of the two integrals of Eq.(20) we note that the integrand of the first integral is a continuous, bounded function for all finite values of ζ and $\theta \ge 0$. The integration interval $[0,1]$ is bounded and the integral exists.

The integration interval $1 \le \eta < \infty$ of the second integral may be divided into two subintervals $1 \le \eta < z$ and $z \le \eta < \infty$. The integral over the first interval exists for the reasons stated above. The integral over the second interval can be rewritten for $z \gg 1$ as follows[4]:

$$\int_z^\infty \left(\cos\left[(\eta^2-1)^{1/2}\theta\right] + \frac{\sin\left[(\eta^2-1)^{1/2}\theta\right]}{(\eta^2-1)^{1/2}}\right)\frac{\sin\zeta\eta}{\eta}\,d\eta$$
$$\approx \int_z^\infty \frac{\cos\eta\theta\sin\zeta\eta}{\eta}\,d\eta + \int_z^\infty \frac{\sin\eta\theta\sin\zeta\eta}{\eta^2}\,d\eta, \quad z \gg 1 \qquad (21)$$

The second integral on the right in Eq.(21) cannot be larger than

$$\int_z^\infty \frac{d\eta}{\eta^2}$$

and thus converges uniformly for all $\theta \ge 0$ and $\zeta \ge 0$. The first integral on the right of Eq.(21) can be rewritten:

$$\int_z^\infty \frac{\sin\zeta\eta\cos\eta\theta}{\eta}\,d\eta = \frac{1}{2}\int_z^\infty \frac{\sin[(\theta+\zeta)\eta]}{\eta}\,d\eta + \frac{1}{2}\int_z^\infty \frac{\sin[(\theta-\zeta)\eta]}{\eta}\,d\eta \qquad (22)$$

Both integrals on the right converge conditionally but not absolutely.

Computer plots of $E_E(\zeta,\theta)$ are shown in Fig.5.1-1 for various values of ζ in the time interval $0 \le \theta \le 60$. We note that the plots are curved so that for small values

[4]For more details of the material presented from here to the end of Section 5.5 see Boules 1989a.

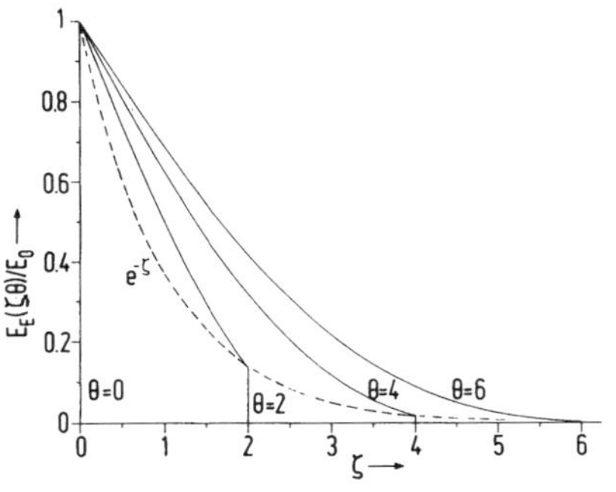

FIG.5.1-2. Plots of $E_{\mathrm{E}}(\zeta,\theta)/E_0$ according to Eq.(19) for $\theta = 0, 2, 4, 6$ in the interval $0 \le \zeta \le 6$. Note the decrease of the jumps according to $e^{-\zeta}$.

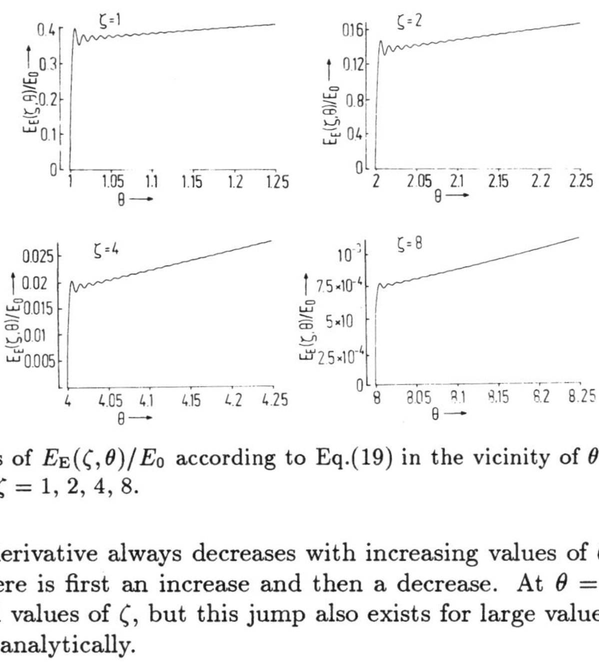

FIG.5.1-3. Plots of $E_{\mathrm{E}}(\zeta,\theta)/E_0$ according to Eq.(19) in the vicinity of $\theta = \zeta$ with a large scale for θ and $\zeta = 1, 2, 4, 8$.

of ζ the first derivative always decreases with increasing values of θ, but for larger values of ζ there is first an increase and then a decrease. At $\theta = \zeta$ we observe a jump for small values of ζ, but this jump also exists for large values of ζ as we are going to show analytically.

Figure 5.1-2 shows again $E_{\mathrm{E}}(\zeta,\theta)$ but with ζ as independent variable and θ as parameter. This illustration emphasizes that $E_{\mathrm{E}}(\zeta,\theta)$ drops to zero at $\zeta = \theta$ and stays zero for $\zeta > \theta$. The illustration also shows the function $e^{-\zeta}$ that determines the height of the jumps as will be shown presently.

Plots of $E_{\mathrm{E}}(\zeta,\theta)$ in the vicinity of $\theta = \zeta$ with a greatly enlarged scale for θ are shown in Fig.5.1-3. We see that there are damped oscillations following the jumps at $\theta = \zeta$ in Fig.5.1-1.

For large values of θ the exponential factor $e^{-\theta}$ in Eq.(20) makes $w(\zeta,\theta)$ approach zero and $E_{\mathrm{E}}(\zeta,\theta)/E_0$ of Eq.(19) approaches 1.

To investigate the jumps of $E_{\mathrm{E}}(\zeta,\theta)$ we recall that the only integral of Eq.(20) that does not converge absolutely is

$$\int_1^\infty \cos\left[(\eta^2-1)^{1/2}\theta\right]\frac{\sin\zeta\eta}{\eta}\,d\eta$$

$$= \int_1^z \cos\left[(\eta^2-1)^{1/2}\theta\right]\frac{\sin\zeta\eta}{\eta}\,d\eta + \int_z^\infty \cos\left[(\eta^2-1)^{1/2}\theta\right]\frac{\sin\zeta\eta}{\eta}\,d\eta \quad (23)$$

We denote the second integral on the right $I_1(\zeta,\theta)$. Equation (19) may then be written as follows:

$$E_{\mathrm{E}}(\zeta,\theta) = E_0\left[f(\zeta,\theta) - \frac{2}{\pi}e^{-\theta}I_1(\zeta,\theta)\right] \quad (24)$$

where $f(\zeta,\theta)$ is a continuous function of both ζ and θ in the intervals $0 \le \zeta < \infty$ and $0 \le \theta < \infty$. If we write $\theta = \zeta^+$ for θ just to the right of $\theta = \zeta$ and $\theta = \zeta^-$ just to the left we get:

$$E_{\mathrm{E}}(\zeta,\zeta^+) - E_{\mathrm{E}}(\zeta,\zeta^-) = -\frac{2}{\pi}E_0 e^{-\zeta}[I_1(\zeta,\zeta^+) - I_1(\zeta,\zeta^-)] \quad (25)$$

For large values of z we may approximate $I_1(\zeta,\theta)$ as follows:

$$I_1(\zeta,\theta) \approx \int_z^\infty \frac{\cos\theta\eta\sin\zeta\eta}{\eta}\,d\eta = \frac{1}{2}\int_z^\infty \left(\frac{\sin[(\theta+\zeta)\eta]}{\eta} - \frac{\sin[(\theta-\zeta)\eta]}{\eta}\right)d\eta$$

$$= \frac{1}{2}\int_0^\infty \left(\frac{\sin[(\theta+\zeta)\eta]}{\eta} - \frac{[(\theta-\zeta)\eta]}{\eta}\right)d\eta$$

$$- \frac{1}{2}\int_0^z \left(\frac{\sin[(\theta+\zeta)\eta]}{\eta} - \frac{\sin[(\theta-\zeta)\eta]}{\eta}\right)d\eta \quad (26)$$

The last integral from 0 to z is a continuous function of ζ and θ. We denote it $g(\zeta,\theta)$ and obtain[5]

$$I_1(\zeta,\theta) = \frac{1}{2}\int_0^\infty \left(\frac{\sin[(\theta+\zeta)\eta]}{\eta} - \frac{\sin[(\theta-\zeta)\eta]}{\eta}\right)d\eta - \frac{1}{2}g(\zeta,\theta)$$

$$= \frac{\pi}{4}[\mathrm{sign}(\theta+\zeta) - \mathrm{sign}(\theta-\zeta)] - \frac{1}{2}g(\zeta,\theta) \quad (27)$$

where $\mathrm{sign}\, x$ is $+1$ for $x > 0$ and -1 for $x < 0$. We get:

[5] Gradshteyn and Ryzhik 1980, p. 405, 3.721/1

$$\begin{aligned} I_1(\zeta,\zeta^+) - I_1(\zeta,\zeta^-) &= \frac{\pi}{4}[\operatorname{sign}(\zeta^+ + \zeta) - \operatorname{sign}(\zeta^+ - \zeta) \\ &\quad - \operatorname{sign}(\zeta^- + \zeta) + \operatorname{sign}(\zeta^- - \zeta)] \\ &= \frac{\pi}{4}[1 - 1 - 1 + (-1)] = -\frac{\pi}{2} \end{aligned} \tag{28}$$

Substitution into Eq.(25) yields

$$E_{\rm E}(\zeta,\zeta^+) - E_{\rm E}(\zeta,\zeta^-) = E_0 e^{-\zeta} \tag{29}$$

Hence, the jumps of $E_{\rm E}(\zeta,\theta)$ in Figs.5.1-1 and 5.1-2 decrease with ζ like $e^{-\zeta}$.

Equation (20) may be rewritten into a form which shows analytically that $E_{\rm E}(\zeta,\theta)$ is zero for $\theta < \zeta$. Using the relations

$$\cos jx = \operatorname{ch} x, \ \sin jx = j \operatorname{sh} x$$

we may rewrite $w(\zeta,\theta)$:

$$w(\zeta,\theta) = -\frac{2}{\pi} e^{-\theta} \int_0^\infty \left(\cos\left[(\eta^2-1)^{1/2}\theta\right] + \frac{\sin\left[(\eta^2-1)^{1/2}\theta\right]}{(\eta^2-1)^{1/2}} \right) \frac{\sin \zeta\eta}{\eta}\, d\eta \tag{30}$$

We further use the relation

$$\cos ax = \frac{d}{dx}\frac{\sin ax}{a} \tag{31}$$

to obtain:

$$w(\zeta,\theta) = -\frac{2}{\pi} e^{-\theta} \left(\frac{\partial}{\partial\theta} + 1\right) \int_0^\infty \frac{\sin\left[(\eta^2-1)^{1/2}\theta\right]}{(\eta^2-1)^{1/2}} \frac{\sin\zeta\eta}{\eta}\, d\eta \tag{32}$$

Using once more Eq.(31) we get

$$\frac{\partial w}{\partial \zeta} = -\frac{2}{\pi} e^{-\theta} \left(\frac{\partial}{\partial\theta} + 1\right) \int_0^\infty \frac{\sin\left[(\eta^2-1)^{1/2}\theta\right]}{(\eta^2-1)^{1/2}} \cos\zeta\eta\, d\eta \tag{33}$$

This integral is tabulated[6]:

$$\begin{aligned} &\int_0^\infty \frac{\sin\left[(\eta^2-1)^{1/2}\theta\right]}{(\eta^2-1)^{1/2}} \cos\zeta\eta\, d\eta \\ &\qquad = \frac{\pi}{2} J_0\left(j\sqrt{\theta^2-\zeta^2}\right) = \frac{\pi}{2} I_0\left(\sqrt{\theta^2-\zeta^2}\right) \quad \text{for } 0<\zeta<\theta \\ &\qquad = 0 \qquad\qquad\qquad\qquad\qquad\qquad\qquad\quad \text{for } 0<\theta<\zeta \end{aligned} \tag{34}$$

[6] Gradshteyn and Ryzhik 1980, p. 472, 3.876/1

In this equation $J_0\left(j\sqrt{\theta^2-\zeta^2}\right)$ is the Bessel function of first kind of order zero and $I_0\left(\sqrt{\theta^2-\zeta^2}\right)$ the modified Bessel function (of first kind) of order zero. Using Eq.(31) we may rewrite Eq.(34):

$$\begin{aligned}\frac{\partial}{\partial\zeta}\int_0^\infty \frac{\sin\left[(\eta^2-1)^{1/2}\theta\right]}{(\eta^2-1)^{1/2}}\frac{\sin\zeta\eta}{\eta}\,d\eta &= \frac{\pi}{2}I_0\left(\sqrt{\theta^2-\zeta^2}\right) \quad \text{for } 0<\zeta<\theta\\ &= 0 \qquad\qquad\qquad\qquad\quad \text{for } 0<\theta<\zeta \qquad (35)\end{aligned}$$

Integration of Eq.(35) introduces integration constants that are arbitrary. The ability to choose these constants reflects the fact that there are infinitely many ways to write Eq.(30). Our goal is to obtain 0 for $0<\theta<\zeta$ in Eq.(38) below. By trial and error one finds that this calls for the integration constant $(\pi/2)\operatorname{sh}\theta$ for $0<\theta<\zeta$ in Eq.(36). The integration constant for $0<\zeta<\theta$ is of lesser importance. Hence, we write

$$\begin{aligned}\int_0^\infty \frac{\sin\left[(\eta^2-1)^{1/2}\theta\right]}{(\eta^2-1)^{1/2}}\frac{\sin\zeta\eta}{\eta}\,d\eta &= \frac{\pi}{2}\int_0^\zeta I_0\left(\sqrt{\theta^2-\xi^2}\right)d\xi \quad \text{for } 0<\zeta<\theta\\ &= \frac{\pi}{2}\operatorname{sh}\theta \qquad\qquad\qquad\qquad \text{for } 0<\theta<\zeta \qquad (36)\end{aligned}$$

Equation (32) may now be written as follows:

$$\begin{aligned}w(\zeta,\theta) &= -e^{-\theta}\left(\frac{\partial}{\partial\theta}+1\right)\int_0^\zeta I_0\left(\sqrt{\theta^2-\xi^2}\right)d\xi \qquad \text{for } 0<\zeta<\theta\\ &= -e^{-\theta}\left(\frac{\partial}{\partial\theta}+1\right)\operatorname{sh}\theta = -e^{-\theta}(\operatorname{ch}\theta+\operatorname{sh}\theta) = -1\\ &\qquad\qquad\qquad\qquad\qquad\qquad\qquad\qquad \text{for } 0<\theta<\zeta \qquad (37)\end{aligned}$$

Substitution into Eq.(19) brings the desired result

$$\begin{aligned}E_{\mathrm{E}}(\zeta,\theta) &= E_0\left\{1-e^{-\theta}\int_0^\zeta\left[\frac{\theta I_1\left(\sqrt{\theta^2-\xi^2}\right)}{(\theta^2-\xi^2)^{1/2}}+I_0\left(\sqrt{\theta^2-\xi^2}\right)\right]d\xi\right\}\\ &\qquad\qquad\qquad\qquad\qquad\qquad\qquad\qquad \text{for } 0<\zeta<\theta\\ &= 0 \quad \text{for } 0<\theta<\zeta \qquad\qquad\qquad\qquad\qquad\qquad (38)\end{aligned}$$

where $I_1\left(\sqrt{\theta^2-\xi^2}\right)$ denotes the modified Bessel function of first kind of order one.

5.2 Signal Distortions

A signal is an electromagnetic wave with a beginning and finite energy. Generally, a signal does not have an end, but this is like saying that a capacitor cannot be discharged completely through a resistor. The rectangular pulse $E_0R(t)$ of Eq.(5.1-9) is an example of a simple signal. Its representation in Eq.(5.1-10) by a superposition of two step functions shows that an electric excitation force $E_0R(t)$ will produce an electric field strength $E_{\mathrm{E,R}}(\zeta,\theta)$ representing the distorted pulse according to Eq.(5.1-19):

$$\begin{aligned} E_{\mathrm{E,R}}(\zeta,\theta) &= S(\theta)E_{\mathrm{E}}(\zeta,\theta) - S(\theta-\Delta\Theta)E_{\mathrm{E}}(\zeta,\theta-\Delta\Theta) \\ &= E_0\{S(\theta)[1-w(\zeta,\theta)] - S(\theta-\Delta\Theta)[1+w(\zeta,\theta-\Delta\Theta]\} \\ &\qquad \Delta\Theta = \alpha T = \sigma T/2\epsilon \end{aligned} \tag{1}$$

Electric field strengths according to this equation are shown in Figs.5.2-1 and 5.2-2 for various values of $\Delta\Theta$ and ζ as functions of θ. All four plots show distorted rectangular pulses at distances $\zeta = 1, 3, 5, 7, 9$ but with various durations $\Delta\Theta = 2, 4, 6, 8$.

Consider a receiver at the distance ζ that produces a voltage $v(\theta)$ proportionate to $E_{\mathrm{E,R}}(\zeta,\theta)$ across a resistance R_{i}. The quantity $v^2(\theta)/R_{\mathrm{i}}$ represents a power and the time integral over this power represents an energy. This justifies to call

$$W_{\mathrm{s}}(\zeta,\Delta\Theta) = \frac{1}{E_0^2}\int_0^\infty E_{\mathrm{E,R}}^2(\zeta,\theta)d\theta \tag{2}$$

the normalized energy at the distance ζ of the signal represented by the electric field strength $E_{\mathrm{E,R}}$ of Eq.(1).

The field strength $E_{\mathrm{E,R}}(\zeta,\theta)$ has an infinitely long trailing edge that decreases slowly according to Figs.5.2-1 and 5.2-2. This suggests to break the integral of Eq.(2) into a first part over the interval $0 \le \theta \le \theta_1$ and a second part over the interval $\theta_1 \le \theta < \infty$. The first part can be integrated numerically while an analytical approximation has to be found for the second part:

$$W_{\mathrm{s}}(\zeta,\Delta\Theta) = W_{\mathrm{s1}}(\zeta,\Delta\Theta) + W_{\mathrm{s2}}(\zeta,\Delta\Theta) \tag{3}$$

$$W_{\mathrm{s1}}(\zeta,\Delta\Theta) = \frac{1}{E_0^2}\int_0^{\theta_1} E_{\mathrm{E,R}}^2(\zeta,\theta)\,d\theta \tag{4}$$

$$W_{\mathrm{s2}}(\zeta,\Delta\Theta) = \frac{1}{E_0^2}\int_{\theta_1}^\infty E_{\mathrm{E,R}}^2(\zeta,\theta)\,d\theta \tag{5}$$

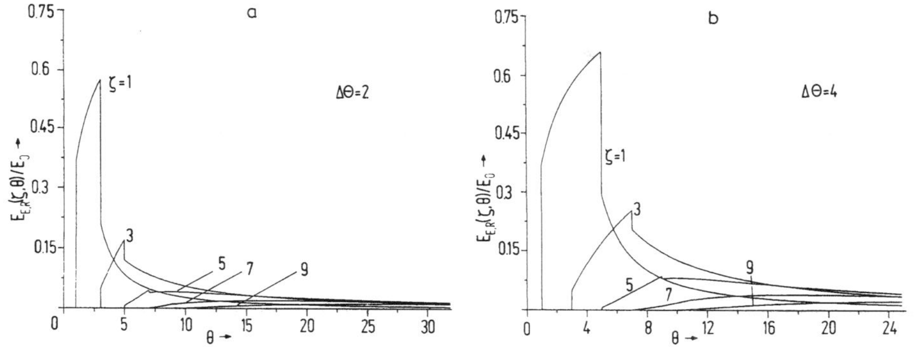

FIG.5.2-1. Electric field strength $E_{\mathrm{E,R}}(\zeta,\theta)/E_0$ according to Eq.(1) for $\zeta = 1, 3, 5, 7, 9$ due to an electric excitation force with the time variation of a rectangular pulse of duration $\Delta\Theta = 2$ in the time interval $0 \leq \theta \leq 30$ (a) and $\Delta\Theta = 4$, $0 \leq \theta \leq 25$ (b).

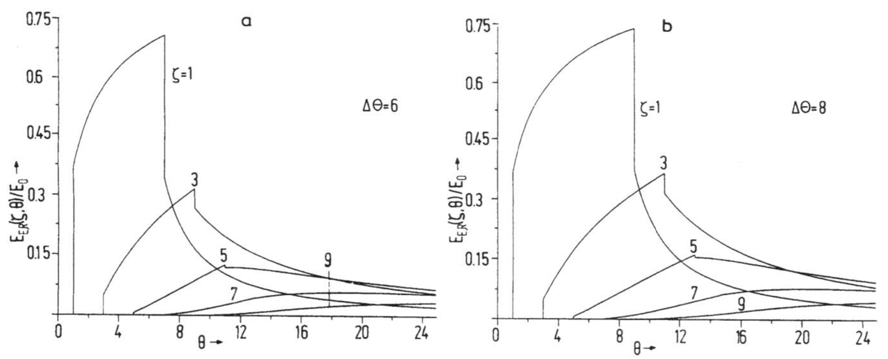

FIG.5.2-2. Electric field strength $E_{\mathrm{E,R}}(\zeta,\theta)/E_0$ according to Eq.(1) for $\zeta = 1, 3, 5, 7, 9$ due to an electric excitation force with the time variation of a rectangular pulse of duration $\Delta\Theta = 6$ (a) and $\Delta\Theta = 8$ (b) in the time interval $0 \leq \theta \leq 25$.

For W_{s2} we use an approximation[1] to $E_{\mathrm{E}}(\zeta,\theta)$ defined by Eqs.(5.1-19), (5.1-20) or (5.1-38) that is simpler and works well everywhere except close to $\theta = \zeta$:

$$\begin{aligned} E_{\mathrm{E}}^{(K)}(\zeta,\theta)/E_0 &= 0 && \text{for } 0 < \theta < \zeta \\ &= e^{-\theta} + \zeta \int_{\zeta}^{\theta} \frac{I_1\left(\sqrt{x^2-\zeta^2}\right)}{(x^2-\zeta^2)^{1/2}} e^{-x}\,dx && \text{for } \zeta \leq \theta < \infty \end{aligned} \tag{6}$$

Figure 5.2-3 shows the approximation. The solid lines represent $E_{\mathrm{E}}(\zeta,\theta)$ according to Eqs.(5.1-19) and (5.1-20) while the stars represent $E_{\mathrm{E}}^{(K)}(\zeta,\theta)$ of Eq.(6).

Substitution of $E_{\mathrm{E}}^{(K)}(\zeta,\theta)$ for $E_{\mathrm{E}}(\zeta,\theta)$ in Eq.(1) yields for $\theta > \zeta + \Delta\Theta$:

[1]Kuester 1987. For illustrations showing the deviations close to $\theta = \zeta$ see Harmuth and Hussain 1994a, pp. 6 and 7.

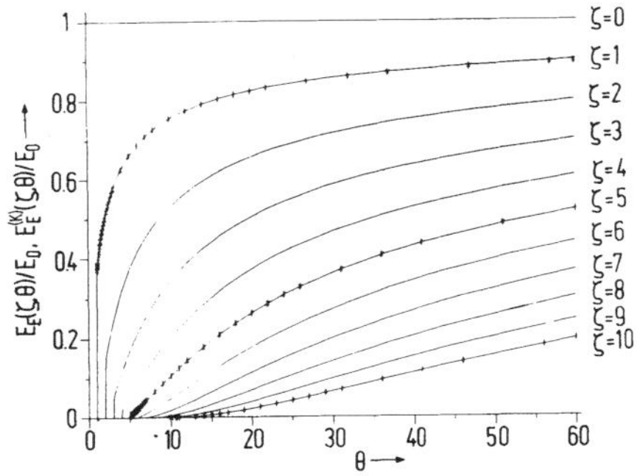

FIG.5.2-3. The function $E_{\mathrm{E}}(\zeta,\theta)/E_0$ according to Eqs.(5.1-19), (5.1-20) shown by solid lines and $E_{\mathrm{E}}^{(K)}(\zeta,\theta)/E_0$ according to Eq.(6) shown by the stars.

$$\begin{aligned} E_{\mathrm{E,R}}(\zeta,\theta)/E_0 &= \zeta\int_{\zeta}^{\theta} \frac{I_1\left(\sqrt{x^2-\zeta^2}\right)}{(x^2-\zeta^2)^{1/2}} e^{-x}dx - \zeta\int_{\zeta}^{\theta-\Delta\Theta} \frac{I_1\left(\sqrt{x^2-\zeta^2}\right)}{(x^2-\zeta^2)^{1/2}} e^{-x}dx \\ &= \zeta\int_{\theta-\Delta\Theta}^{\theta} \frac{I_1\left(\sqrt{x^2-\zeta^2}\right)}{(x^2-\zeta^2)^{1/2}} e^{-x}dx \end{aligned} \tag{7}$$

We may use an asymptotic expansion[2] for $I_1(z)$

$$I_1(z) \approx \frac{e^z}{\sqrt{2\pi z}}\left(1-\frac{3}{8z}\right), \quad z \geq 12 \tag{8}$$

if we choose

$$\begin{aligned} (x^2-\zeta^2)^{1/2} &\geq (x_{\min}^2-\zeta^2)^{1/2} \geq 12 \\ x_{\min} &\geq \zeta\left(\frac{144}{\zeta^2}+1\right)^{1/2} \end{aligned} \tag{9}$$

Substituting the lower integral limit $\theta - \Delta\Theta$ for $x_{\min}$ we get a minimum value of θ_1:

$$\begin{aligned} \theta_1 &\geq \zeta\left(\frac{144}{\zeta^2}+1\right)^{1/2} + \Delta\Theta \geq \zeta(144+1)^{1/2} + \Delta\Theta \\ &\geq 12.05\zeta + \Delta\Theta \quad \text{for } \zeta \geq 1 \end{aligned} \tag{10}$$

With the series expansion

[2] Olver 1964, p. 377, 9.7.1

$$(x^2-\zeta^2)^{-1/2} \approx \frac{1}{x}\left(1+\frac{\zeta^2}{2x^2}\right) \tag{11}$$

we obtain the following approximation for the integrand of Eq.(7):

$$\frac{I_1\left(\sqrt{x^2-\zeta^2}\right)}{(x^2-\zeta^2)^{1/2}}e^{-x} \approx \frac{1}{\sqrt{2\pi}}\left(\frac{1}{x^{3/2}}-\frac{3}{8x^{5/2}}-\frac{\zeta^2}{2x^{5/2}}\right) \tag{12}$$

Equation (7) assumes the form

$$\begin{aligned} E_{\mathrm{E,R}}(\zeta,\theta)/E_0 &\approx \frac{\zeta}{\sqrt{2\pi}}\int\limits_{\theta-\Delta\Theta}^{\theta}\left(\frac{1}{x^{3/2}}-\frac{3}{8x^{5/2}}-\frac{\zeta^2}{2x^{5/2}}\right)dx \\ &\approx \frac{\zeta}{\sqrt{2\pi}}\Big[-2\theta^{-1/2}+2(\theta-\Delta\Theta)^{-1/2} \\ &\quad +\left(\frac{1}{4}+\frac{\zeta^2}{3}\right)\left(\theta^{-3/2}-(\theta-\Delta\Theta)^{-3/2}\right)\Big],\ \theta\geq\theta_1 \end{aligned} \tag{13}$$

We square Eq.(13):

$$\begin{aligned} \left[E_{\mathrm{E,R}}(\zeta,\theta)/E_0\right]^2 &\approx \frac{2\zeta^2}{\pi}\Bigg\{\frac{1}{\theta}+\frac{1}{\theta-\Delta\Theta}-\frac{2}{\theta(\theta-\Delta\Theta)^{1/2}} \\ &\quad +\left(\frac{1}{4}+\frac{\zeta^2}{3}\right)\left[\frac{1}{\theta^2}+\frac{1}{(\theta-\Delta\Theta)^2}-\frac{1}{\theta^{1/2}(\theta-\Delta\Theta)^{3/2}}\right]\Bigg\} \end{aligned} \tag{14}$$

Using the tabulated integrals

$$\begin{gathered} \int\frac{dz}{\sqrt{(z-q)z}} = \ln\left(2\sqrt{z^2-q}+2z-q\right) \\ \int\frac{dz}{z^{3/2}\sqrt{z-q}} = \frac{2}{q}\sqrt{1-\frac{q}{z}},\quad \int\frac{dz}{z^{1/2}(z-q)^{3/2}} = -\frac{2}{q\sqrt{1-q/z}} \end{gathered} \tag{15}$$

we integrate Eq.(14):

$$\begin{aligned} W_{s2}(\zeta,\Delta\Theta) &\approx \frac{2\zeta^2}{\pi}\Bigg\{\ln(\theta^2-\theta\Delta\Theta)-2\ln\left[2(\theta^2-\theta\Delta\Theta)^{1/2}+2\theta-\Delta\Theta\right] \\ &\quad -\left(\frac{1}{4}+\frac{\zeta^2}{3}\right)\Bigg[-\frac{1}{\theta}-\frac{1}{\theta-\Delta\Theta}+\frac{2}{\Delta\Theta\,(1-\Delta\Theta/\theta)^{1/2}} \\ &\quad -\frac{2}{\Delta\Theta}\left(1-\frac{\Delta\Theta}{\theta}\right)^{1/2}\Bigg]\Bigg\}_{\theta_1}^{\infty} \end{aligned} \tag{16}$$

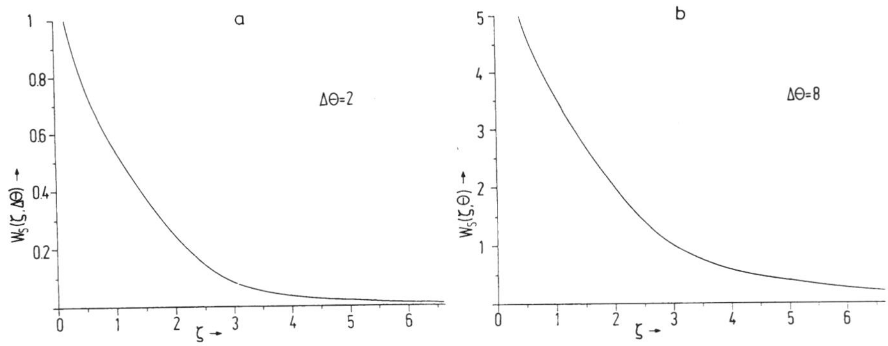

FIG.5.2-4. Normalized signal energy $W_s(\zeta, \Delta\Theta)$ of a distorted rectangular pulse according to Eq.(3) for $\Delta\Theta = 2$ (a) and $\Delta\Theta = 8$ (b) in the interval $0 \le \zeta \le 7$.

To obtain the upper limit $\theta \to \infty$ we calculate the limit of the first two terms of Eq.(16) that contain the logarithmic function:

$$\lim_{\theta\to\infty} \ln \frac{\theta^2 - \theta\Delta\Theta}{\left[2(\theta^2 - \theta\Delta\Theta)^{1/2} + 2\theta - \Delta\Theta\right]^2} = -4\ln 2 \tag{17}$$

The function $W_{s2}(\zeta, \Delta\Theta)$ may then be rewritten into the following rather simple form:

$$W_{s2}(\zeta, \theta) = \frac{2\zeta^2}{\pi}\left[-4\ln 2 - 2\ln\left(\frac{X}{2X + 2\theta_1 - \Delta\Theta^2}\right) + \left(\frac{1}{4} + \frac{\zeta^2}{3}\right)\left(\frac{2}{X} - \frac{2\theta_1 + \Delta\Theta}{\theta_1(\theta_1 - \Delta\Theta)}\right)\right]$$

$$X^2 = \theta_1(\theta_1 - \Delta\Theta),\ \theta_1 \ge 12.05\zeta + \Delta\Theta \tag{18}$$

Using numerical integration for $W_{s1}(\zeta, \Delta\Theta)$ of Eq.(4) as well as for $W_{s2}(\zeta, \Delta\Theta)$ of Eqs.(5) and (18) yields the total energy $W_s(\zeta, \Delta\Theta)$ plotted in Fig.5.2-4 as function of ζ for $\Delta\Theta = 2$ and $\Delta\Theta = 8$.

5.3 DETECTION OF DISTORTED SYNCHRONIZED SIGNALS IN NOISE

Let the input stage of a receiver at the distance ζ from the transmitter[1] transform the normalized electric field strength $E_{E,R}(\zeta, \theta)/E_0$ into a normalized voltage $s(\theta)$

$$s(\theta) = E_{E,R}(\zeta, \theta)/E_0 \tag{1}$$

[1] Distance refers strictly to the distance travelled by the signal in the distorting medium. For a submarine receiving signals from a transmitter above the seawater this distance is the shortest distance from the surface to the receiving antenna.

that represents a received, distorted rectangular pulse. We assume that the receiver is synchronized so that the arrival time of any pulse is known. We further assume that pulses with either positive or negative amplitude but otherwise equal time variation are transmitted. The received signals $s(\theta)$ can than have the two time variations $s_0(\theta)$ or $s_1(\theta) = -s_0(\theta)$. The probability of transmitting $s_1(\theta)$ shall be the same as transmitting $s_0(\theta)$.

Let thermal noise[2] be added to the received signal. We denote with $n(\theta)$ a normalized noise voltage sample. The received signal $r(\theta)$ with superimposed noise is then either

$$r(\theta) = s_0(\theta) + n(\theta) \quad \text{or} \quad r(\theta) = s_1(\theta) + n(\theta) \qquad (2)$$

The receiver must decide whether it is more probable that $s_0(\theta)$ rather than $s_1(\theta)$ was transmitted or vice-versa.

For our assumptions the best method to make a decision is based on the cross-correlation of the received signal with stored sample signals. Refer to Fig.5.3-1a which shows a typical received signal that represents a heavily distorted rectangular pulse. If the distance ζ in the distorting medium between transmitter and receiver is known one can store a sample signal with 'essentially' the same time variation at the receiver. The word 'essentially' is needed since the trailing edge of the signal in Fig.5.3-1a is infinitely long. We can store only a truncated form of the signal with a duration $\Delta\Theta_f$ as shown in Fig.5.3-1b. The larger we make $\Delta\Theta_f$ the better we will discriminate against noise since the truncated sample function of Fig.5.3-1b uses only part of the energy of the signal in Fig.5.3-1a. However, a decision which signal was most probably transmitted cannot be made before the time $\theta = \zeta + \Delta\Theta_f$. In the absence of distortions one would choose $\Delta\Theta_f = \Delta\Theta$, but with distortions there is no obvious upper bound on $\Delta\Theta_f$. The time between the beginning $\theta = 0$ of signal transmission at $\zeta = 0$ and the decision at the receiver what signal was most probably transmitted is $\zeta + \Delta\Theta$ for an undistorted rectangular pulse but $\zeta + \Delta\Theta_f$ for a distorted pulse. The *observed normalized propagation velocity* is $\zeta/(\zeta + \Delta\Theta)$ in the one case and $\zeta/(\zeta + \Delta\Theta_f)$ in the other. The difference is not trivial if we observe the values of ζ, $\Delta\Theta$, and $\Delta\Theta_f$ in Figs.5.3-1a and b. One must always trade noise suppression versus observed propagation velocity.

The plots of Figs.5.3-1a and b start at the same time $\theta = \zeta$ but this does not necessarily yield the largest value of the integral over their product. The peak of their cross-correlation function shown in Fig.5.3-1c may occur at a time $\Theta_{max} > \Delta\Theta_f$.

There is one more difficulty that can be learned from Fig.5.3-1. In principle the signal of Fig.5.3-1a has an infinitely long trailing edge. For a numerical calculation of the cross-correlation function by computer we must truncate the signal. Let us do so at the time $\zeta + \Theta_0$ shown in Fig.5.3-1a. The cross-correlation function in Fig.5.3-1c shows a break at Θ_0. Hence, wherever we see such a break in the plots to be shown we know that only the part for $\theta < \Theta_0$ is correct while the part for $\theta > \Theta_0$ has an error caused by the truncation of the signal $s(\theta)$.

[2] We use the concept of thermal noise (Nyquist 1928, Johnson 1928) rather than that of white, Gaussian noise since we are not restricting the analysis to waves with sinusoidal time variation.

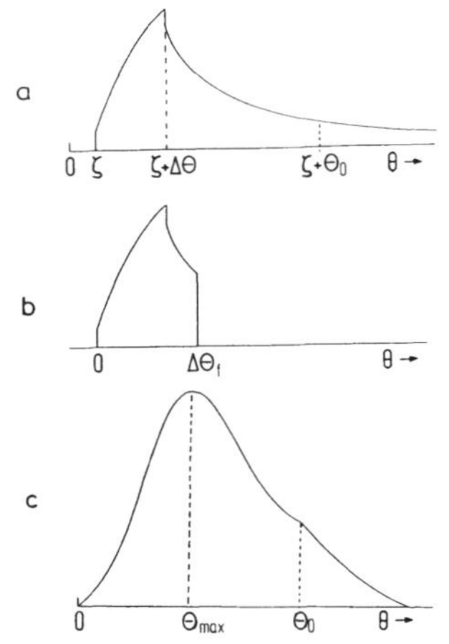

FIG.5.3-1. Distorted rectangular pulse $s(\theta)$ received at the distance ζ from the transmitter (a). Truncated sample signal $h(\Delta\Theta_f, \theta)$ stored at the receiver (b). Cross-correlation function between received distorted rectangular pulse and truncated sample signal (c).

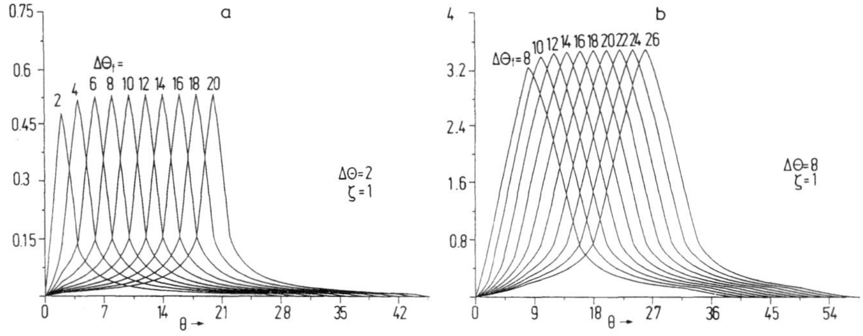

FIG.5.3-2. Cross-correlation functions of the signal $s(\theta)$ defined by $\zeta = 1$ in Fig.5.2-1a with its truncated samples of duration $\Delta\Theta_f = 2, 4, \ldots, 20$, $\Theta_0 = 26$ (a) and corresponding functions for $\zeta = 1$ in Fig.5.2-2b, $\Delta\Theta_f = 8, 10, \ldots, 26$, $\Theta_0 = 32$ (b).

Let the signal $s(\theta)$ in Fig.5.3-1a be the function denoted $\zeta = 1$ in Fig.5.2-1a. Its cross-correlation function with truncated samples $h(\Delta\Theta_f, \theta)$ is shown in Fig.5.3-2a for various values of $\Delta\Theta_f$. We see that the peaks of the functions increase somewhat as we advance from $\Delta\Theta_f = 2$ to 4 and 6 but remain essentially unchanged for larger values of $\Delta\Theta_f$. The time at which a decision about the transmitted signal can be made would be $\zeta + \Delta\Theta = 1 + 2 = 3$ for an undistorted signal but becomes essentially $\zeta + \Delta\Theta_f = 1 + 20 = 21$ for a distorted signal if the truncated sample signal has the duration $\Delta\Theta_f = 20$.

Figures 5.3-3a and 5.3-4a show a repetition of Fig.5.3-2a but the distance is increased to $\zeta = 3$ and $\zeta = 5$. We note that larger values of $\Delta\Theta_f$ now increase the peaks of the cross-correlation functions much more than in Fig.5.3-2a.

Corresponding cross-correlation functions are plotted for the much longer rect-

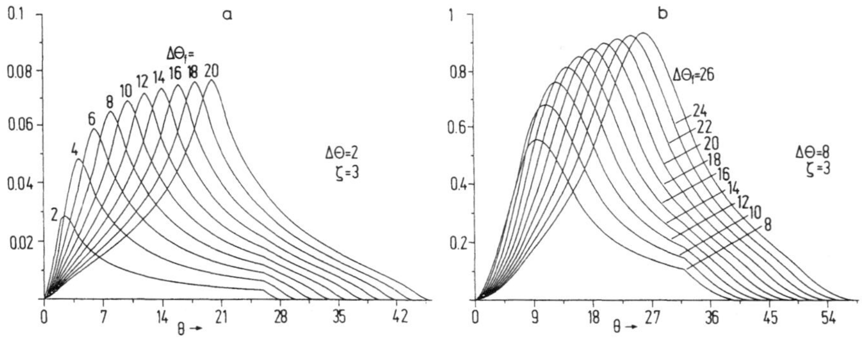

FIG.5.3-3. Cross-correlation functions of the signal $s(\theta)$ defined by $\zeta = 3$ in Fig.5.2-1a with its truncated samples of duration $\Delta\Theta_f = 2, 4, \ldots, 20$, $\Theta_0 = 26$ (a) and corresponding functions for $\zeta = 3$ in Fig.5.2-2b, $\Delta\Theta_f = 8, 10, \ldots, 26$, $\Theta_0 = 32$ (b).

angular pulses with $\Delta\Theta = 8$ of Fig.5.2-2b for $\zeta = 1, 3, 5$ in Figs.5.3-2b to 5.3-4b. The larger value of $\Delta\Theta$ increases the time at which the cross-correlation functions reach their peaks.

We have noted in connection with Fig.5.3-1 that the time $\Theta_{\max}$ for which the peak of the cross-correlation function occurs is not necessarily the same as $\Delta\Theta_f$. Figure 5.3-5a shows $\Theta_{\max}$ as function of $\Delta\Theta_f$ for the signals $\zeta = 1, 3, 5$ of Fig.5.2-1a. Corresponding plots are shown in Fig.5.3-5b for the signals $\zeta = 1, 3, 5$ of Fig.5.2-2b.

We turn to the calculation of the error probability due to additive thermal noise. First we renormalize the time since the calculation is simplified if the signal $s(\theta)$ in Fig.5.3-1a starts at the time $\theta' = \theta - \zeta = 0$ rather than at $\theta = \zeta$; the prime is then dropped and we write θ rather than θ'.

The signal $s_0(\theta)$ of Eq.(2) is shifted by $\Theta_{\max} - \Delta\Theta_f$ in order to make the integral

$$K_{s,h} = \int_0^{\Delta\Theta_f} s_0(\theta + \Theta_{\max} - \Delta\Theta_f) h(\Delta\Theta_f, \theta) d\theta \tag{3}$$

assume its largest possible value according to Fig.5.3-1. Furthermore, we write $h(\Delta\Theta_f, \theta) = h_0(\Delta\Theta_f, \theta)$ and think of this function as one of a system of orthogonal functions $\{h_i(\Delta\Theta_f, \theta)\}$ in the interval $0 \leq \theta \leq \Delta\Theta_f$:

$$\begin{aligned} \int_0^{\Delta\Theta_f} h_i(\Delta\Theta_f, \theta) h_l(\Delta\Theta_f, \theta) d\theta = K_{h,h}\delta_{il} &= K_{h,h} \text{ for } i = l \\ &= 0 \quad \text{ for } i \neq l \end{aligned} \tag{4}$$

A sample λ of thermal noise $n(\theta) = n_\lambda(\theta)$ of Eq.(2) is decomposed by this system of orthogonal functions:

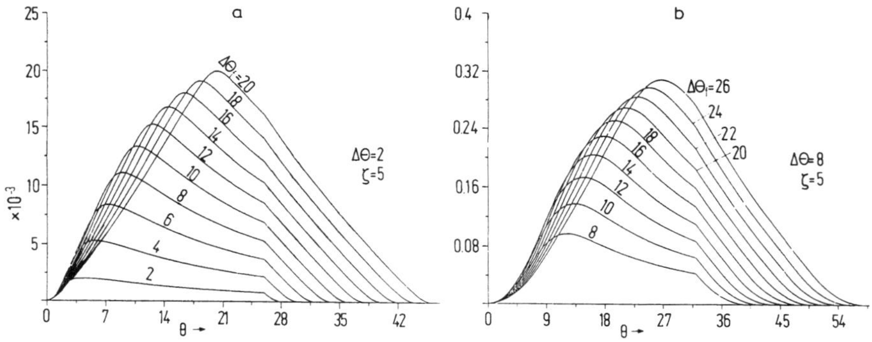

FIG.5.3-4. Cross-correlation functions of the signal $s(\theta)$ defined by $\zeta = 5$ in Fig.5.2-1a with its truncated samples of duration $\Delta\Theta_f = 2, 4, \ldots, 20$, $\Theta_0 = 26$ (a) and corresponding functions for $\zeta = 5$ in Fig.5.2-2b, $\Delta\Theta_f = 8, 10, \ldots, 26$, $\Theta_0 = 32$ (b).

$$n_\lambda(\theta) = \sum_{i=0}^{\infty} a_\lambda(i) h_i(\Delta\Theta_f, \theta), \quad a_\lambda(i) = \int_0^{\Delta\Theta_f} n_\lambda(\theta) h_i(\Delta\Theta_f, \theta) d\theta \tag{5}$$

The distorted, noise-free signal $s_0(\theta + \Theta_{max} - \Delta\Theta_f)$ is also decomposed with the help of the orthogonal system of functions $\{h_i(\Delta\Theta_f, \theta)\}$:

$$\begin{aligned} s_0(\theta + \Delta\Theta_{max} - \Delta\Theta_f) &= \sum_{i=1}^{\infty} a_s(i) h_i(\Delta\Theta_f, \theta) \\ a_s(i) &= \int_0^{\Delta\Theta_f} s_0(\theta + \Theta_{max} - \Delta\Theta_f) h_i(\Delta\Theta_f, \theta) d\theta \end{aligned} \tag{6}$$

For the particular function $h_0(\Delta\Theta_f, \theta) = h(\Delta\Theta_f, \theta)$ we get from Eqs.(3) and (6):

$$a_S(0) = \int_0^{\Delta\Theta_f} s_0(\theta + \Theta_{max} - \Delta\Theta_f) h_0(\Delta\Theta_f, \theta) = K_{s,h} \tag{7}$$

Furhermore, for $\Theta_{max} = \Delta\Theta_f$ the functions $s_0(\theta)$ and $h_0(\Delta\Theta_f, \theta)$ are equal in the interval $0 \leq \theta \leq \Delta\Theta_f$ and we get:

$$a_s(0) = \int_0^{\Delta\Theta_f} h_0^2(\Delta\Theta_f, \theta) d\theta = K_{h,h}, \quad a_s(i) = 0 \text{ for } i \neq 0 \tag{8}$$

The signal $s_0(\theta + \Theta_{max} - \Delta\Theta_f)$ with superimposed noise sample $n_\lambda(\theta)$ is also decomposed:

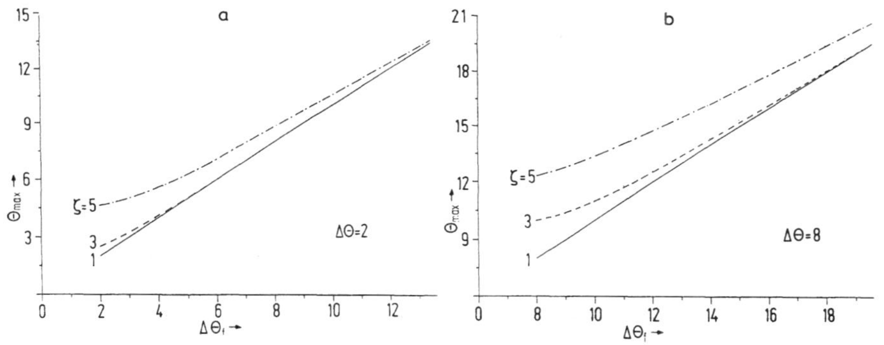

FIG.5.3-5. The time Θ_{max} as function of $\Delta\Theta_f$ for the signals $\zeta = 1, 3, 5$ of Fig.5.2-1a (a) and of Fig.5.2-2b (b).

$$\begin{aligned} r(\theta) &= s_0(\theta + \Theta_{max} - \Delta\Theta_f) + n_\lambda(\theta) \\ &= \sum_{i=0}^{\infty} a(i) h_i(\Delta\Theta_f, \theta), \quad a(i) = a_s(i) + a_\lambda(i) \end{aligned} \tag{9}$$

Since the possible transmitted signals $s_0(\theta)$ and $s_1(\theta) = -s_0(\theta)$ have the same energy one may use cross-correlation to decide whether $s_0(\theta)$ or $s_1(\theta)$ was transmitted:

$$\begin{aligned} \int_0^{\Delta\Theta_f} r(\theta) h_0(\Delta\Theta_f, \theta) d\theta &= \int_0^{\Delta\Theta_f} [s_0(\theta + \Theta_{max} - \Delta\Theta_f) + n_\lambda(\theta)] h_0(\Delta\Theta_f, \theta) d\theta \\ &= K_{s,h} + \int_0^{\Delta\Theta_f} \sum_{i=0}^{\infty} a_\lambda(i) h_i(\Delta\Theta_f, \theta) h_0(\Delta\Theta_f, \theta) d\theta \\ &= K_{s,h} + K_{h,h} a_\lambda(0) \end{aligned} \tag{10}$$

If $s_0(\theta)$ was transmitted, Eq.(10) should be larger than zero, if $s_1(\theta)$ was transmitted it should be smaller than zero. The signals $s_0(\theta)$ and $s_1(\theta)$ were assumed to be transmitted with equal probability 1/2. Two conditions must be met to make Eq.(10) produce a wrong sign and thus an error:

a) $\operatorname{sign} K_{s,h} \neq \operatorname{sign} K_{h,h} a_\lambda(0)$

b) $|K_{h,h} a_\lambda(0)| > |K_{s,h}|$, equivalent $\dfrac{K_{h,h} a_\lambda(0)}{|K_{s,h}|} > 1$ or $\dfrac{K_{h,h} a_\lambda(0)}{|K_{s,h}|} < -1$ (11)

In the case of thermal noise the probability of $a_\lambda(0)$ being positive is 1/2 and the probability of being negative is also 1/2. Hence, the probability of condition (a) being satisfied equals 1/2, regardless of whether $s_0(\theta)$ or $s_1(\theta)$ was transmitted.

The distribution of $x = K_{\mathrm{h,h}} a_\lambda(0)/|K_{\mathrm{s,h}}|$ is needed for the computation of the probability of condition (b) being satisfied. The variable x has the same distribution as $a_\lambda(0)$ since $K_{\mathrm{h,h}}$ and $|K_{\mathrm{s,h}}|$ are constants. Since $a_\lambda(0)$ is a component of a decomposition of a sample of thermal noise by means of a system of orthogonal functions[3], it has the density function $w(x)$:

$$w(x) = \frac{1}{\sqrt{2\pi}\sigma} \exp(-x^2/2\sigma^2), \quad x = K_{\mathrm{h,h}} a_\lambda(0)/|K_{\mathrm{s,h}}| \tag{12}$$

For the determination of the mean-square-deviation σ^2 we must determine the average of $K_{\mathrm{h,h}}^2 a_\lambda^2(0)/K_{\mathrm{s,h}}^2$. Substitution of $K_{\mathrm{s,h}}$ from Eq.(7) yields:

$$\sigma^2 = \langle K_{\mathrm{h,h}}^2 a_\lambda^2(0)/K_{\mathrm{s,h}}^2 \rangle = K_{\mathrm{h,h}}^2 \langle a_\lambda^2(0) \rangle / a_{\mathrm{s}}^2(0) \tag{13}$$

We denote the average energy of the orthogonal component $h_0(\Delta\Theta_{\mathrm{f}}, \theta)$ of thermal noise $W_{\mathrm{N}}(0)$ and the energy of the orthogonal component $h_0(\Delta\Theta_{\mathrm{f}}, \theta)$ of the signal $W_{\mathrm{S}}(0)$. Instead of the energy ratio $W_{\mathrm{S}}(0)/W_{\mathrm{N}}(0)$ we can also write the signal-to-noise power ratio $P_{\mathrm{S}}(0)/P_{\mathrm{N}}(0) = [W_{\mathrm{S}}(0)/\Delta\Theta_{\mathrm{f}}]/[W_{\mathrm{N}}(0)/\Delta\Theta_{\mathrm{f}}]$. The energy $W_{\mathrm{S}}(0)$ is the signal energy used for detection; we write W_{S} and by implication P_{S} rather than $W_{\mathrm{S}}(0)$ and $P_{\mathrm{S}}(0)$. We may also write W_{N} and P_{N} rather than $W_{\mathrm{N}}(0)$ and $P_{\mathrm{N}}(0)$ since the average power $P_{\mathrm{N}}(i)$ of thermal noise is the same for any orthogonal component $h_i(\Delta\Theta_{\mathrm{f}}, \theta)$. If we use sinusoidal functions to represent the noise in a non-normalized orthogonality interval of duration T, the average noise power P_{N} would be the same as that in a frequency interval $\Delta f = 1/2T$; the factor 2 is due to the fact that for any sine function there is an orthogonal cosine function. The mean-square-deviation σ^2 becomes:

$$\sigma^2 = K_{\mathrm{h,h}}^2 W_{\mathrm{N}}/W_{\mathrm{S}} = K_{\mathrm{h,h}}^2 P_{\mathrm{N}}/P_{\mathrm{S}} \tag{14}$$

The probability $p(x > 1) + p(x < -1)$ that x is larger than $+1$ or smaller than -1 follows from Eq.(12) by integration:

$$p_{\mathrm{e}} = p(x>1) + p(x<1) = \frac{1}{\sqrt{2\pi}\sigma} \int_1^\infty \exp\left(-\frac{x^2}{2\sigma^2}\right) dx = \frac{2}{\sqrt{\pi}} \int_{1/\sqrt{2}\sigma}^\infty e^{-y^2} dy$$

$$= 1 - \operatorname{erf}\left(\frac{1}{\sqrt{2}\sigma}\right) = 1 - \operatorname{erf}\left(\frac{P_{\mathrm{S}}}{2K_{\mathrm{h,h}}^2 P_{\mathrm{N}}}\right)^{1/2} \tag{15}$$

Before discussing plots of p_{e} we still have to investigate the effect of thermal noise on the observed propagation velocity of distorted signals. Let a rectangular pulse $s(t)$ be transmitted so that its leading edge occurs at the time $t = 0$ at the location $y = 0$. The receiver shall be at a distance y. According to Figs.5.1-1, 5.2-1, and 5.2-2 the leading edge of the distorted pulse always propagates with the vacuum velocity of light c. The leading edge thus arrives at the distance y at the time t_{d}:

[3] Harmuth 1972, p. 331

TABLE 5.3-1

NORMALIZED OBSERVED PROPAGATION VELOCITY v_S/c ACCORDING TO EQ.(19) FOR TRANSMITTED RECTANGULAR PULSES OF WIDTH $\Delta\Theta = 2$ AND $\Delta\Theta = 8$ FOR DISTANCES $\zeta = 1, 3, 5$, STORED SAMPLE SIGNALS OF WIDTH $\Delta\Theta_f$, AND TIMES Θ_{max} FOR MAXIMUM CORRELATOR OUTPUT.

$\Delta\Theta = 2$

ζ	$\Delta\Theta_f$	Θ_{max}	v_S/c	ζ	$\Delta\Theta_f$	Θ_{max}	v_S/c	ζ	$\Delta\Theta_f$	Θ_{max}	v_S/c
1	2	2.00	0.3333	3	2	2.50	0.5455	5	2	4.50	0.5263
	4	4.00	0.2000		4	4.10	0.4225		4	5.60	0.4717
	6	6.00	0.1429		6	6.00	0.3333		6	7.20	0.4098
	8	8.00	0.1111		8	8.00	0.2727		8	9.00	0.3571
	10	10.00	0.0909		10	10.00	0.2308		10	10.70	0.3185
	12	12.00	0.0769		12	12.00	0.2000		12	12.60	0.2841
	14	14.00	0.0667		14	14.00	0.1765		14	14.50	0.2564
	16	16.00	0.0588		16	16.00	0.1579		16	16.40	0.2336
	18	18.00	0.0526		18	18.00	0.1429		18	18.30	0.2146
	20	20.00	0.0476		20	20.00	0.1304		20	20.20	0.1984

$\Delta\Theta = 8$

ζ	$\Delta\Theta_f$	Θ_{max}	v_S/c	ζ	$\Delta\Theta_f$	Θ_{max}	v_S/c	ζ	$\Delta\Theta_f$	Θ_{max}	v_S/c
1	8	8.00	0.1111	3	8	9.70	0.2362	5	8	12.20	0.2907
	10	10.00	0.0909		10	11.00	0.2143		10	13.40	0.2717
	12	12.00	0.0769		12	12.60	0.1923		12	14.70	0.2538
	14	14.00	0.0667		14	14.40	0.1724		14	16.20	0.2358
	16	16.00	0.0588		16	16.20	0.1563		16	17.80	0.2193
	18	18.00	0.0526		18	18.10	0.1422		18	19.50	0.2041
	20	20.00	0.0476		20	20.10	0.1299		20	21.20	0.1908
	22	22.00	0.0435		22	22.00	0.1200		22	23.00	0.1786
	24	24.00	0.0400		24	24.00	0.1111		24	24.90	0.1672
	26	26.00	0.0370		26	26.00	0.1034		26	26.80	0.1572

$$t_d = y/c \tag{16}$$

Due to the presence of noise we make the decision whether the signal $s_0(t)$ or $s_1(t)$ has arrived at the later time $t_d + T_{max}$. The observed signal propagation velocity v_S becomes:

$$v_S = \frac{y}{t_d + T_{max}} = \frac{y/t_d}{1 + T_{max}/t_d} = \frac{c}{1 + T_{max}/t_d} \tag{17}$$

In normalized notation we get

$$T_{max}/t_d = \Theta_{max}/\zeta \tag{18}$$

and

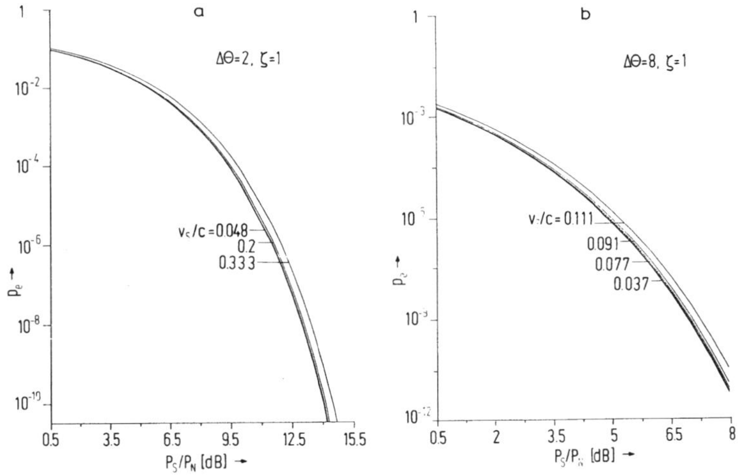

FIG.5.3-6. Error probability p_e as function of the signal-to-noise ratio P_S/P_N according to Eq.(15) for the distance $\zeta = 1$ and various values of the normalized observable propagation velocity v_S/c for the pulse durations $\Delta\Theta = 2$ (a) and $\Delta\Theta = 8$ (b).

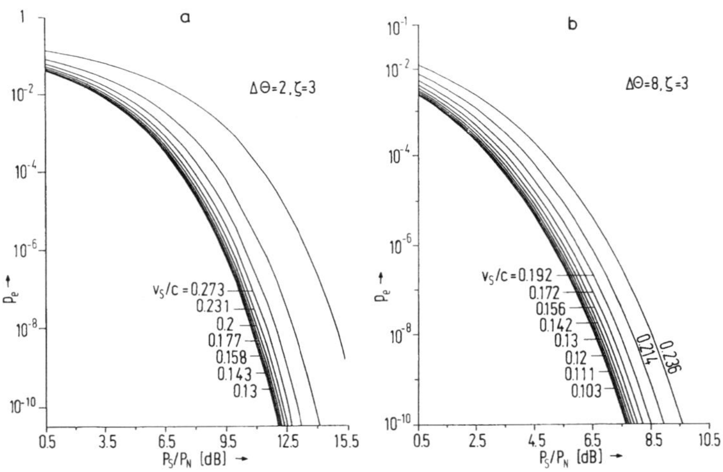

FIG.5.3-7. Error probability p_e as function of the signal-to-noise ratio P_S/P_N according to Eq.(15) for the distance $\zeta = 3$ and various values of the normalized observable propagation velocity v_S/c for the pulse durations $\Delta\Theta = 2$ (a) and $\Delta\Theta = 8$ (b).

$$\frac{v_S}{c} = \frac{1}{1 + \Theta_{max}/\zeta} \tag{19}$$

The time θ_{max} at which the correlator output reaches its maximum value is itself a function of the width $\Delta\Theta$ of the transmitted signal and the width $\Delta\Theta_f$ of the stored sample signal as well as of ζ. Values of v_S/c for various values of $\Delta\Theta$, ζ, $\Delta\Theta_f$, and Θ_{max} are listed in Table 5.3-1. Since the propagation velocity v_S/c

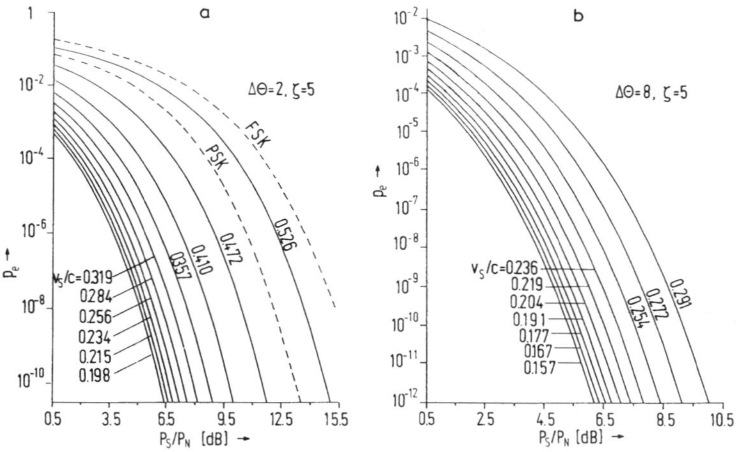

FIG.5.3-8. Error probability p_e as function of the signal-to-noise ratio P_S/P_N according to Eq.(15) for the distance $\zeta = 5$ and various values of the normalized observable propagation velocity v_S/c for the pulse durations $\Delta\Theta = 2$ (a) and $\Delta\Theta = 8$ (b). The dashed plots in (a) show for comparison the error probability for conventional coherent frequency shift keying (FSK) and coherent phase shift keying (PSK).

depends on $\Delta\Theta_f$ it is a function of the error probability p_e and the signal-to-noise ratio P_S/P_N. Figures 5.3-6 to 5.3-8 present the error probability p_e as function of the signal-to-noise ratio for various values of the parameters $\Delta\Theta$, ζ, and v_S/c.

A study of these illustrations shows that the error probability decreases with the signal-to-noise ratio P_S/P_N and the velocity v_S/c for fixed values of $\Delta\Theta$ and ζ. If the duration $\Delta\Theta_f$ of the stored sample signal is increased, the velocity of v_S/c decreases and the error probability for a fixed ratio P_S/P_N decreases. The reduction of the error probability becomes more conspicuous with increasing values of ζ. The reason is that Θ_{max}—when the correlator output reaches its maximum—deviates from $\Delta\Theta_f$ more and more as ζ increases as shown in Fig.5.3-5.

To provide a comparison we show in Fig.5.3-8a plots of the error probability versus the signal-to-noise ratio for the conventional coherent phase and frequency shift keying PSK and FSK. These two plots assume sinusoidal pulses propagating in a loss-free medium and additive thermal noise.

5.4 Detection of Distorted Radar Signals in Noise

In the case of the detection of communication signals we know that a signal has arrived but must decide which one out of at least two possible signals it was. In the radar case we must decide whether at any time only noise or noise plus a signal was received. The received signal may differ from the radiated one. A signal returned from a scattering point will only have the distortions caused by the medium, while a larger target will add its 'target signature' to the distorted signal. We will restrict ourselves to the case of a scattering point.

A radar in a non-distorting medium like the atmosphere determines the distance

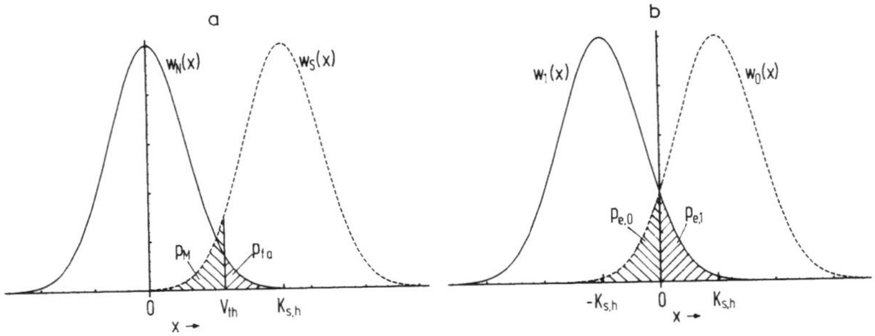

FIG.5.4-1. Density functions $w_N(x)$ and $w_S(x)$ according to Eqs.(2) and (5) for the case of radar detection (a). For comparison the density functions $w_0(x)$ and $w_1(x)$ are shown for two signals $s_0(\theta)$ and $s_1(\theta) = -s_0(\theta)$ used for the case of communications in Section 5.3 (b).

to a target by measuring the time between radiation and reception of a signal. This principle can also be used in distorting media. However, there is a second principle that can be exploited. A look at Figs.5.2-1 and 5.2-2 shows that the time variation of a distorted signal is a function of the distance ζ traveled in the medium. Hence, the distortion is distance information.

We use the same system of orthogonal functions $\{h_i(\Delta\Theta_f, \theta)\}$ as in Section 5.3. If only thermal noise $n_\lambda(\theta)$ is received we get instead of Eq.(5.3-10):

$$\int_0^{\Delta\Theta_f} n_\lambda(\theta) h_0(\Delta\Theta_f, \theta) d\theta = K_{h,h} a_\lambda(0) \tag{1}$$

Since $K_{h,h}$ is a constant we have the density function $w_N(x)$ for $K_{h,h}a_\lambda(0)$:

$$w_N(x) = \frac{1}{\sqrt{2\pi}\sigma} \exp(-x^2/2\sigma^2), \quad x = K_{h,h} a_\lambda(0) \tag{2}$$

The mean-square-deviation σ^2 is the same as in Eq.(5.3-14):

$$\sigma^2 = K_{h,h}^2 P_N / P_S \tag{3}$$

If a signal $s_0(\theta)$ plus noise $n_\lambda(\theta)$ is received we get according to Eq.(5.3-10):

$$\int_0^{\Delta\Theta_f} [s_0(\theta + \Theta_{max} - \Delta\Theta_f) + n_\lambda(\theta)] h_0(\Delta\Theta_f, \theta) d\theta = K_{s,h} + K_{h,h} a_\lambda(0) \tag{4}$$

The density function $w_S(x)$ is the same as in Eq.(2) but shifted from the average value $x = 0$ to $x = K_{s,h}$:

$$w_S(x) = \frac{1}{\sqrt{2\pi}\sigma} \exp\left[-(x - K_{s,h})^2/2\sigma^2\right], \quad x = K_{h,h} a_\lambda(0) \tag{5}$$

The density functions $w_{\rm N}(x)$ and $w_{\rm S}(x)$ are shown in Fig.5.4-1a. We introduce a threshold (voltage) $V_{\rm th}$ as shown. If[1] noise only was received and we decide that a signal was received, we obtain a *false alarm probability* represented by the area denoted $p_{\rm fa}$. If a signal plus noise was received and we decide that noise only was received, we have an error probability or a *miss probability* represented by the area denoted $p_{\rm M}$.

Let us turn to the false alarm probability $p_{\rm fa}$. Taking the integral over $w_{\rm N}(x)$ in Eq.(2) we obtain:

$$\begin{aligned} p_{\rm fa} &= \int_{V_{\rm th}}^{\infty} w_{\rm N}(x)dx = \frac{1}{\sqrt{2\pi}\sigma} \int_{V_{\rm th}}^{\infty} \exp(-x^2/2\sigma^2)dx \\ &= \frac{1}{\sqrt{\pi}} \int_{V_{\rm th}/\sqrt{2}\sigma}^{\infty} e^{-y^2} dy = \frac{1}{2}\left(1 - \operatorname{erf} \frac{V_{\rm th}}{\sqrt{2}\sigma}\right) \end{aligned} \tag{6}$$

The miss probability $p_{\rm M}$ or the detection probability $p_{\rm D} = 1 - p_{\rm M}$ follows from Eq.(5):

$$\begin{aligned} 1 - p_{\rm D} = p_{\rm M} &= \int_{-\infty}^{V_{\rm th}} w_{\rm S}(x)dx = \frac{1}{\sqrt{2\pi}\sigma} \int_{-\infty}^{V_{\rm th}} \exp\left[-(x - K_{\rm s,h})^2/2\sigma^2\right] dx \\ &= \frac{1}{2}\left(1 - \operatorname{erf} \frac{K_{\rm s,h} - V_{\rm th}}{\sqrt{2}\sigma}\right) \end{aligned} \tag{7}$$

For a chosen false alarm probability $p_{\rm fa}$ and a signal-to-noise ratio $P_{\rm S}/P_{\rm N}$ one obtains from Eqs.(6) and (3) a threshold $V_{\rm th}$. Substitution of this threshold into Eq.(7) then yields the detection probability $p_{\rm D}$ as function of the false alarm probability $p_{\rm fa}$ and the signal-to-noise ratio $P_{\rm S}/P_{\rm N}$.

Figure 5.4-2a shows the detection probability $p_{\rm D}$ as function of the signal-to-noise ratio $P_{\rm S}/P_{\rm N}$ with the false alarm probability $p_{\rm fa}$ as parameter for a pulse of original duration $\Delta\Theta = 2$, a propagation distance $\zeta = 5$, and a value $\Delta\Theta_{\rm f} = 2$ according to Fig.5.3-1. Corresponding plots for $\Delta\Theta = \Delta\Theta_{\rm f} = 8$ are shown in Fig.5.4-2b. As one would expect a larger signal-to-noise ratio reduces the false alarm probability for a fixed detection probability or increases the detection probability for a fixed false alarm probability. Furthermore, for a fixed signal-to-noise ratio the detection probability increases with the false alarm probability.

We turn to the computation of the observed propagation velocity of the detected signal. Figure 5.4-3 shows the cross-correlation function of the noise-free received signal $s(\theta) = s_0(\theta)$ and the stored sample signal $h(\Delta\Theta_{\rm f}, \theta) = h_0(\Delta\Theta_{\rm f}, \theta)$ as used in Eq.(4). We set the threshold $V_{\rm th}$ of the false alarm probability in Eq.(6) to

[1] For details see Van Trees 1968, Weber 1987, or many other textbooks.

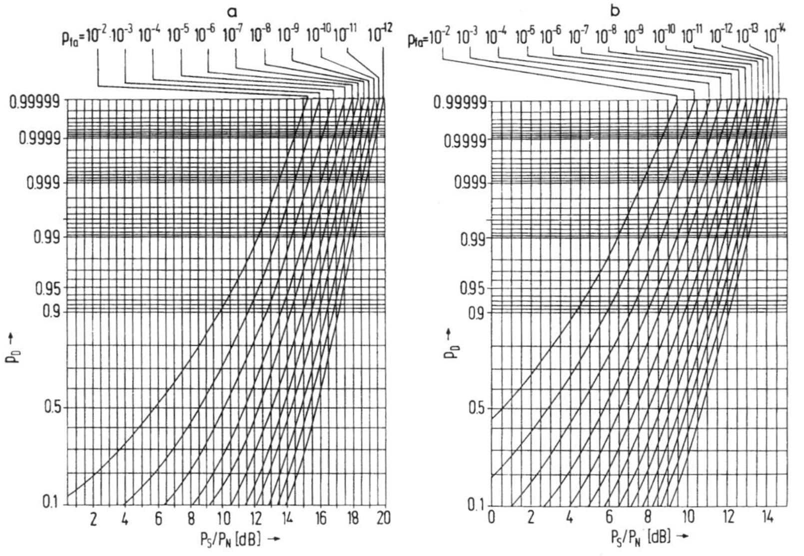

FIG.5.4-2. Detection probability p_D versus signal-to-noise ratio P_S/P_N with the false alarm probability p_{fa} for a propagation distance $\zeta = 5$ and $\Delta\Theta = \Delta\Theta_f = 2$ (a) as well as $\Delta\Theta = \Delta\Theta_f = 8$ (b).

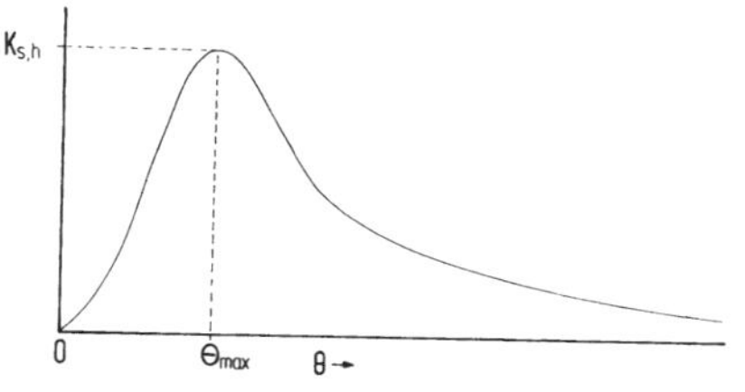

FIG.5.4-3. Cross-correlation function of the noise-free signal $s(\theta) = s_0(\theta)$ and the stored sample signal $h(\Delta\Theta_f, \theta) = h_0(\Delta\Theta_f, \theta)$. In the absence of noise, a decision that a signal was received is made in the time interval $0 \leq \theta \leq \Theta_{max}$.

$$V_{th} = \gamma K_{s,h}, \quad 0 < \gamma \leq 1 \tag{8}$$

and solve Eq.(6) for γ as function of p_{fa}. The plots of Fig.5.4-4 are obtained.

For the false alarm and miss probabilities of practical interest the detection decision will always be made in the rising part of the cross-correlation function in Fig.5.4-3 during the time interval $0 < \theta < \Theta_{max}$. This rising part of the cross-correlation function divided by the maximum value $K_{s,h}$ is shown in Fig.5.4-5. It shows the relation between $\gamma = V_{th}/K_{s,h}$ and the time Θ_{th} when the threshold V_{th} is reached. This decision time $\Theta_{th} > 0$ represents a delay relative to the time $\theta = \zeta$ when the beginning of the signal $s(\theta)$ arrived at the receiver.

We use again Eqs.(5.3-16)–(5.3-19), but replace the time Θ_{max} by the threshold

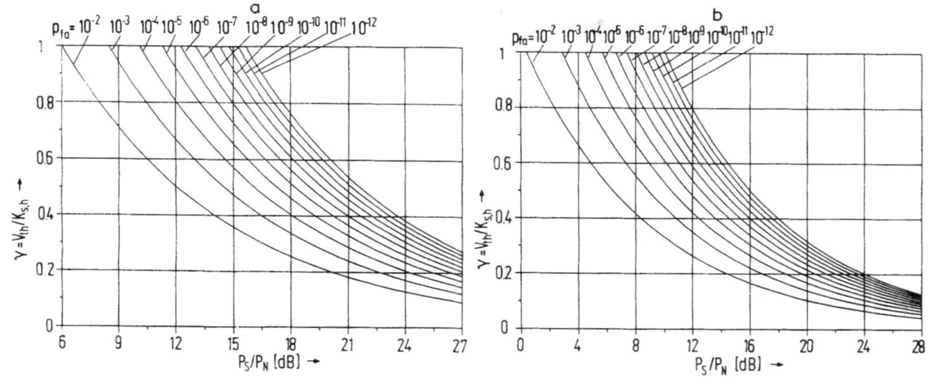

FIG.5.4-4. Relative threshold $\gamma = V_{th}/K_{s,h}$ as function of the signal-to-noise ratio P_S/P_N and the false alarm probability p_{fa} as parameter for $\zeta = 5$ and $\Delta\Theta = \Delta\Theta_f = 2$ (a) as well as $\Delta\Theta = \Delta\Theta_f = 8$ (b).

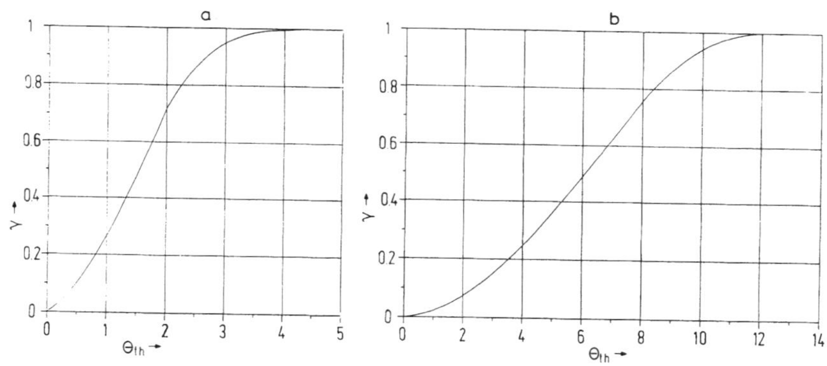

FIG.5.4-5. The rising part of the cross-correlation function of Fig.5.4-3 divided by $K_{s,h}$ for $\zeta = 5$ and $\Delta\Theta = \Delta\Theta_f = 2$ (a) as well as $\Delta\Theta = \Delta\Theta_f = 8$ (b).

time Θ_{th}:

$$\frac{v_S}{c} = \frac{1}{1 + \Theta_{th}/\zeta}, \quad 0 < \Theta_{th} \leq \Theta_{max} \tag{9}$$

The normalized velocity v_S/c is plotted in Fig.5.4-6.

In order to see how the plots of Figs.5.4-4 to 5.4-6 can be used to determine the observed signal propagation time we choose a false alarm probability $p_{fa} = 10^{-10}$ and a detection probability $p_D = 0.998$. From Fig.5.4-2a we get the signal-to-noise ratio $P_S/P_N = 18$ dB, holding for $\zeta = 5$ and $\Delta\Theta = \Delta\Theta_f = 2$.

We combine Figs.5.4-4a, 5.4-5a, and 5.4-6a, which all hold for $\zeta = 5$ and $\Delta\Theta = \Delta\Theta_f = 2$, as shown in Fig.5.4-7. The plot on the left (a) is entered with the values $P_S/P_N = 18$ dB—obtained from Fig.5.4-2a—and $p_{fa} = 10^{-10}$—chosen originally—and the value $\gamma = 0.68$ is found. For this value of γ the center plot (b) yields

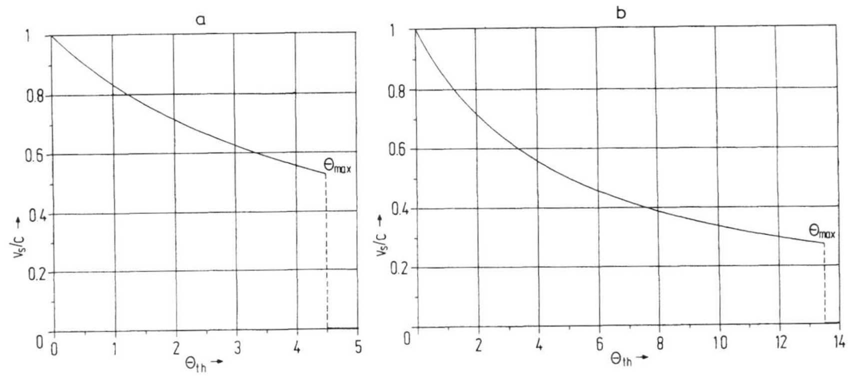

FIG.5.4-6. Normalized signal velocity v_S/c as function of the threshold time Θ_{th} according to Eq.(9) for $\zeta = 5$ and $\Delta\Theta = \Delta\Theta_f = 2$ (a) as well as $\Delta\Theta = \Delta\Theta_f = 8$ (b).

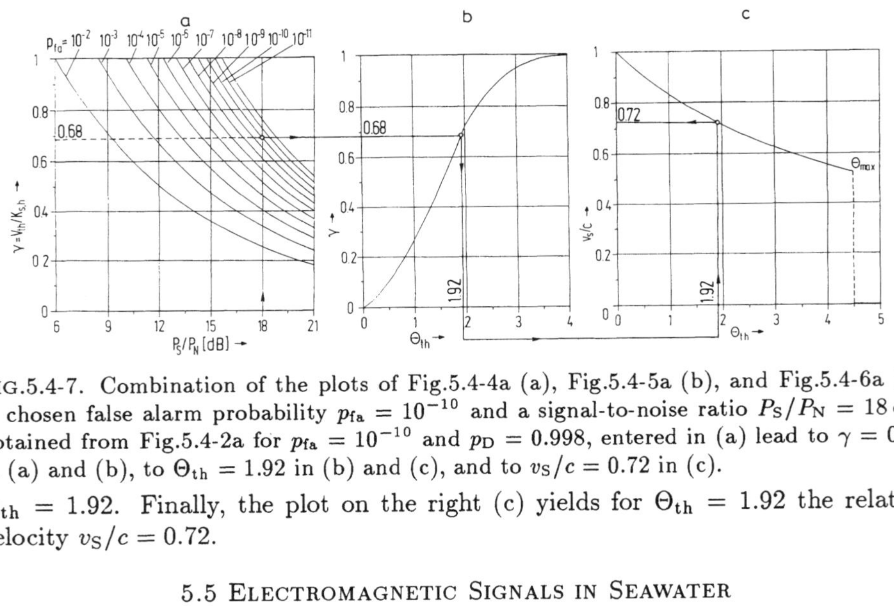

FIG.5.4-7. Combination of the plots of Fig.5.4-4a (a), Fig.5.4-5a (b), and Fig.5.4-6a (c). A chosen false alarm probability $p_{fa} = 10^{-10}$ and a signal-to-noise ratio $P_S/P_N = 18\,\text{dB}$, obtained from Fig.5.4-2a for $p_{fa} = 10^{-10}$ and $p_D = 0.998$, entered in (a) lead to $\gamma = 0.68$ in (a) and (b), to $\Theta_{th} = 1.92$ in (b) and (c), and to $v_S/c = 0.72$ in (c).

$\Theta_{th} = 1.92$. Finally, the plot on the right (c) yields for $\Theta_{th} = 1.92$ the relative velocity $v_S/c = 0.72$.

5.5 Electromagnetic Signals in Seawater

For seawater the constants μ. ϵ, and σ describing the medium of propagation have the following typical values:

$$\mu \approx \mu_0 = 4\pi \times 10^{-7}\,\text{Vs/Am}, \quad \epsilon \approx 80\epsilon_0 \approx 7.1 \times 10^{-10}\,\text{As/Vm}, \quad \sigma \approx 4\,\text{A/Vm} \tag{1}$$

We obtain with the help of Eq.(5.1-18):

$$\theta \approx 2.8 \times 10^9 t, \quad \zeta \approx 9.1 y \tag{2}$$

if t is measured in seconds and y in meter. For distances of practical interest in seawater, e.g., 50 m to 1500 m, the normalized distance ζ varies from 455 to 13650.

This implies that θ must be at least 455 since the electric and magnetic field strengths are zero for $\theta < \zeta$. We will soon see that θ will generally be larger than 1000. It is next to impossible to evaluate Eq.(5.1-20) by computer for such large values of ζ and θ; an analytical approximation of this equation is needed.

As starting point for such an analytical approximation we observe that the following relations are satisfied for $\eta \geq 1$

$$(\eta^2-1)^{1/2} \geq 0, \quad \left|\cos\left[(\eta^2-1)^{1/2}\theta\right]\right| \leq 1, \quad \left|\frac{\sin\left[(\eta^2-1)^{1/2}\theta\right]}{(\eta^2-1)^{1/2}}\right| \leq \theta \tag{3}$$

and we may rewrite the second integral of Eq.(5.1-20) as follows:

$$\begin{aligned}|I_{\mathrm{E}2}| &= \left|e^{-\theta}\int_1^\infty\left(\cos\left[(\eta^2-1)^{1/2}\theta\right]+\frac{\sin\left[(\eta^2-1)^{1/2}\theta\right]}{(\eta^2-1)^{1/2}}\right)\frac{\sin\zeta\eta}{\eta}\,d\eta\right| \\ &\leq e^{-\theta}(1+\theta)\left|\int_1^\infty\frac{\sin\zeta\eta}{\eta}\,d\eta\right| = (1+\theta)e^{-\theta}\int_{\zeta>0}^\infty\frac{\sin x}{x}\,dx \\ &\leq (1+\theta)e^{-\theta}\int_0^\infty\frac{\sin x}{x}\,dx = \frac{\pi}{2}(1+\theta)e^{-\theta}\end{aligned} \tag{4}$$

The last term on the right of Eq.(4) equals 5.90×10^{-42} for $\theta = 100$ and is negligible compared with 1 in Eq.(5.1-19). We rewrite Eq.(5.1-19):

$$E_{\mathrm{E}}(\zeta,\theta)/E_0 = 1-\frac{2}{\pi}I_{\mathrm{E}1} \tag{5}$$

$$I_{\mathrm{E}1} = -e^{-\theta}\int_0^1\left(\mathrm{ch}\left[(1-\eta^2)^{1/2}\theta\right]+\frac{\mathrm{sh}\left[(1-\eta^2)^{1/2}\theta\right]}{(1-\eta^2)^{1/2}}\right)\frac{\sin\zeta\eta}{\eta}\,d\eta \tag{6}$$

The integral $I_{\mathrm{E}1}$ has nonnegligible value only if the hyperbolic sine and cosine functions are of the order of e^θ. This is the case if η is neither equal nor close to 1. In this case we may write

$$\mathrm{ch}\left[(1-\eta^2)^{1/2}\theta\right] \approx \mathrm{sh}\left[(1+\eta^2)^{1/2}\theta\right] \approx \frac{1}{2}\exp\left[(1-\eta^2)^{1/2}\theta\right], \quad \eta \neq 1,\ \theta \gg 1$$

and $I_{\mathrm{E}1}$ becomes:

$$I_{\mathrm{E}1} \approx \frac{1}{2}\int_0^\delta \exp\left\{\left[(1-\eta^2)^{1/2}-1\right]\theta\right\}\left(1+\frac{1}{(1-\eta^2)^{1/2}}\right)\frac{\sin\zeta\eta}{\eta}\,d\eta, \quad \delta < 1 \tag{7}$$

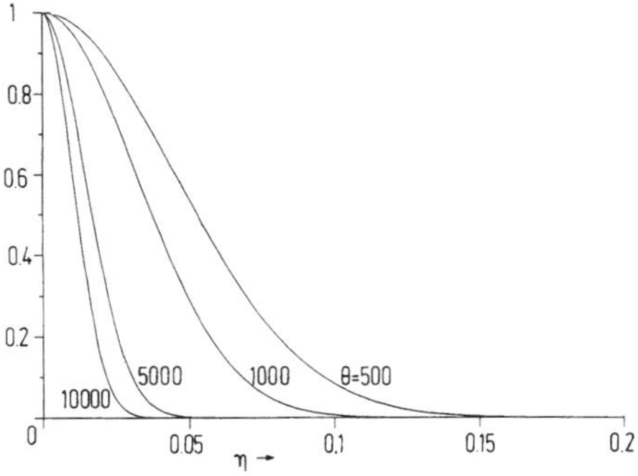

FIG.5.5-1. The exponential factor $\exp\left\{\left[(1-\eta^2)^{1/2}-1\right]\theta\right\}$ of Eq.(8) for $\theta = 500$, 1000, 5000, 10000 as function of η.

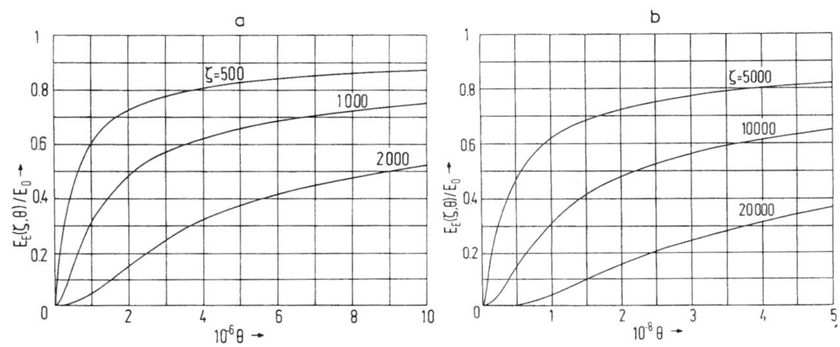

FIG.5.5-2. Electric field strength $E_{\mathrm{E}}(\zeta,\theta)/E_0$ according to Eq.(8) for electric step excitation as function of θ for $\zeta = 500$, 1000, 2000 (a) and for $\zeta = 5000$, 10000, 20000 (b).

The electric field strength of Eq.(5) becomes:

$$E_{\mathrm{E}}(\zeta,\theta)/E_0 \approx$$

$$1-\frac{1}{\pi}\int_0^{\delta} \exp\left\{\left[(1-\eta^2)^{1/2}-1\right]\theta\right\}\left(1+\frac{1}{(1-\eta^2)^{1/2}}\right)\frac{\sin\zeta\eta}{\eta}\,d\eta \quad (8)$$

The factor $\exp\left\{\left[(1-\eta^2)^{1/2}-1\right]\theta\right\}$ is plotted in Fig.5.5-1 for η in the interval $0 \le \eta \le 0.2$ with $\theta = 500$, 1000, 5000, 10000 as parameter. It is evident that this factor is practically zero over most of the integration interval $0 \le \eta < \delta < 1$ and is nonnegligible only if η is close to zero. We conclude from this illustration that $\delta = 0.5$ will yield a satisfactory accuracy for $\theta \ge 500$.

The electric field strength $E_{\mathrm{E}}(\zeta,\theta)/E_0$ of Eq.(8) is plotted in Fig.5.5-2a for $\zeta = 500$, 1000, 2000 as function of θ in the range $0 \le \theta \le 10^7$. Figure 5.5-2b shows corresponding plots for $\zeta = 5000$, 10000, 20000 and θ in the range $0 \le \theta \le 5 \times 10^8$. The beginnings of some of the plots are shown with larger scales for θ in Figs.5.5-3 and 5.5-4.

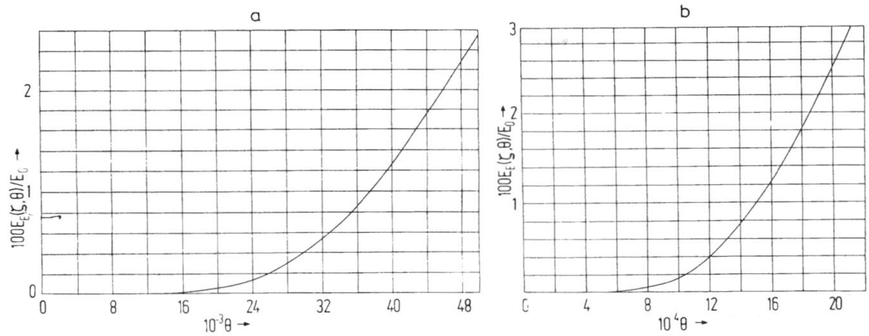

FIG.5.5-3. Beginning part of the electric field strength $E_E(\zeta,\theta)$ according to Eq.(8) for $\zeta = 500$ or $y = 55$ m (a) and for $\zeta = 1000$ or $y = 110$ m (b).

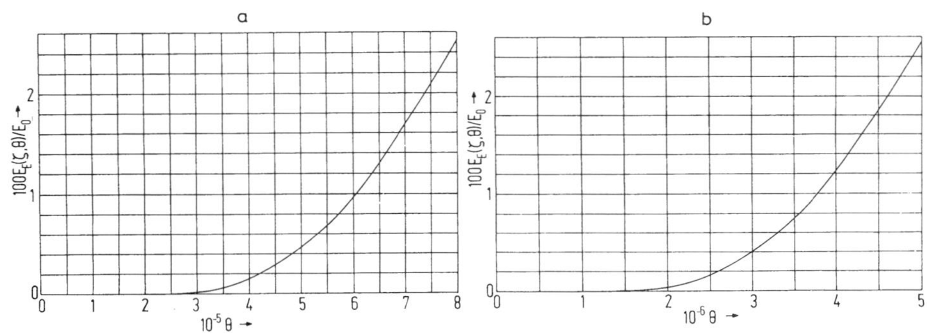

FIG.5.5-4. Beginning part of the electric field strength $E_E(\zeta,\theta)$ according to Eq.(8) for $\zeta = 2000$ or $y = 220$ m (a) and for $\zeta = 5000$ or $y = 550$ m (b).

Equation (8) can be further simplified so that the integral can be replaced by a tabulated function. We recognize that $E_E(\zeta,\theta)/E_0$ in Figs.5.5-3 and 5.5-4 is almost zero until very large values of θ are reached. For example, in Fig.5.5-3b for $\zeta = 1000$ the plot is visible for $\theta > 5 \times 10^4$. For such large values of θ the integrand of Eq.(8) is almost zero according to Fig.5.5-1 unless η is very close to zero. In this case we may use the following approximations:

$$1 + \frac{1}{(1-\eta^2)^{1/2}} \approx 2, \quad \exp\left\{\left[(1-\eta^2)^{1/2} - 1\right]\theta\right\} \approx \exp\left(-\frac{\eta^2\theta}{2}\right) \tag{9}$$

Equation (8) is rewritten:

$$E_E(\zeta,\theta)/E_0 \approx 1 - \frac{2}{\pi}\int_0^\delta e^{-\eta^2\theta/2}\,\frac{\sin\zeta\eta}{\eta}\,d\eta \tag{10}$$

Let us write I_E for the integral:

$$I_{\mathrm{E}} = \int_0^{\delta} e^{-\eta^2\theta/2} \frac{\sin\zeta\eta}{\eta}\,d\eta \tag{11}$$

Since this integral is very close to zero for values of η that are *not* close to zero, we may write

$$\int_{\delta}^{\infty} e^{-\eta^2\theta/2} \frac{\sin\zeta\eta}{\eta}\,d\eta = 0 \tag{12}$$

and I_{E} is approximated by

$$I_{\mathrm{E}} \approx \int_0^{\infty} e^{-\eta^2\theta/2} \frac{\sin\zeta\eta}{\eta}\,d\eta \tag{13}$$

Using the substitution $\eta^2 = u$ we get:

$$I_{\mathrm{E}} \approx \int_0^{\infty} e^{-\theta u/2} \frac{\sin(\zeta\sqrt{u})}{2u}\,du = \frac{1}{2}\int_0^{\infty} e^{-pu} \frac{\sin(2\sqrt{au})}{u}\,du$$
$$p = \theta/2,\ a = \zeta^2/4 \tag{14}$$

The last integral is found in a table of Laplace transforms[1]:

$$\int_0^{\infty} e^{-pu} \frac{\sin(2\sqrt{au})}{u}\,du = 2\sqrt{\pi}\int_0^{\sqrt{a/p}} e^{-x^2}\,dx = \pi\,\mathrm{erf}\left(\sqrt{\frac{a}{p}}\right) \tag{15}$$

Equation (14) becomes

$$I_{\mathrm{E}} \approx \frac{\pi}{2}\,\mathrm{erf}\left(\frac{\zeta}{\sqrt{2\theta}}\right) \tag{16}$$

and the electric field strength of Eq.(5) assumes the following form:

$$E_{\mathrm{E}}(\zeta,\theta) = E_0\left[1 - \mathrm{erf}\left(\frac{\zeta}{\sqrt{2\theta}}\right)\right] \tag{17}$$

A comparison of $E_{\mathrm{E}}(\zeta,\theta)/E_0 = E_{\mathrm{E}}^{(1)}(\zeta,\theta)$ according to Eq.(8) and $E_{\mathrm{E}}(\zeta,\theta)/E_0 = E_{\mathrm{E}}^{(2)}(\zeta,\theta)$ according to Eq.(17) is provided by Table 5.5-1. This table lists $E_{\mathrm{E}}^{(1)}(\zeta,\theta)$

[1] Bateman 1954, vol. 1, p. 154(34)

TABLE 5.5-1

A COMPARISON OF THE ELECTRIC FIELD STRENGTH $E^{(1)}(\zeta,\theta)$ ACCORDING TO EQ.(8) WITH THE ELECTRIC FIELD STRENGTH $E^{(2)}(\zeta,\theta)$ ACCORDING TO EQ.(17) FOR $\zeta = 2000$ AND θ IN THE RANGE $5\times 10^5 \le \theta \le 10^7$.

θ	$E^{(1)}(\zeta,\theta)$	$E^{(2)}(\zeta,\theta)$	θ	$E^{(1)}(\zeta,\theta)$	$E^{(2)}(\zeta,\theta)$
5.0×10^5	0.00467775	0.00467777	5.5×10^6	0.39376863	0.39376867
1.0×10^6	0.04550025	0.04550028	6.0×10^6	0.41421618	0.41421622
1.5×10^6	0.10247042	0.10247046	6.5×10^6	0.43276758	0.43276763
2.0×10^6	0.15757570	0.15729922	7.0×10^6	0.44969207	0.44969183
2.5×10^6	0.20590321	0.20590323	7.5×10^6	0.46520882	0.46520883
3.0×10^6	0.24821306	0.24821311	8.0×10^6	0.47950012	0.47950017
3.5×10^6	0.28504940	0.28504938	8.5×10^6	0.49271668	0.49271667
4.0×10^6	0.31731050	0.31731057	9.0×10^6	0.50498508	0.50498509
4.5×10^6	0.34577858	0.34577858	9.5×10^6	0.51641227	0.51641232
5.0×10^6	0.37109336	0.37109339	1.0×10^7	0.52708925	0.52708924

and $E_{\rm E}^{(2)}(\zeta,\theta)$ for $\zeta = 2000$ and θ in the range $5\times 10^5 \le \theta \le 10^7$. The values according to Eqs.(8) or Eq.(17) agree to at least five significant decimal digits.

We may use the simplified formula of Eq.(17) to determine the attenuation of a rectangular pulse $E_{\rm E,R}(\zeta,\theta)$ according to Eq.(5.2-1) in seawater. At a distance $\zeta > 0$ from the transmitter or the surface of the sea the non-distorted pulse has the time variation

$$\begin{aligned} E_{\rm E,R}(\zeta,\theta) &= E_{\rm E}(\zeta,\theta) && \text{for } \zeta < \theta \le \zeta + \Delta\Theta \\ &= E_{\rm E}(\zeta,\theta) - E_{\rm E}(\zeta,\theta-\Delta\Theta) && \text{for } \theta > \zeta + \Delta\Theta \end{aligned} \tag{18}$$

where $E_{\rm E}(\zeta,\theta)$ is defined by Eq.(17). We rewrite Eq.(18) as follows:

$$\begin{aligned} E_{\rm E,R}(\zeta,\theta) &= 1 - \operatorname{erf}\left(\zeta/\sqrt{2\theta}\right) && \text{for } \zeta < \theta \le \zeta + \Delta\Theta \\ &= -\operatorname{erf}\left(\zeta/\sqrt{2\theta}\right) + \operatorname{erf}\left(\zeta/\sqrt{2(\theta-\Delta\Theta)}\right) && \text{for } \theta > \zeta + \Delta\Theta \end{aligned} \tag{19}$$

Plots of Eq.(19) are shown for $\Delta\Theta = 2\times 10^7$ and 20×10^7 in Fig.5.5-5. These plots represent pulses of duration $\Delta T = 7.14\,\text{ms}$ and $\Delta T = 71.4\,\text{ms}$ in non-normalized notation according to Eq.(2); the propagation distance $\zeta = 10^4$ corresponds to a distance $y = 1100\,\text{m}$.

At the receiver the signal starts at the time $\theta = \zeta$ and is truncated at the time $\theta = \zeta + \Delta\Theta_{\rm f}$. We assume $\Delta\Theta_{\rm f} > \Delta\Theta$. If we represent the transmitted energy W_0 at $\zeta = 0$ by

$$W(0,\Delta\Theta) = E_0^2\Delta\Theta \tag{20}$$

we may represent the received energy at ζ by

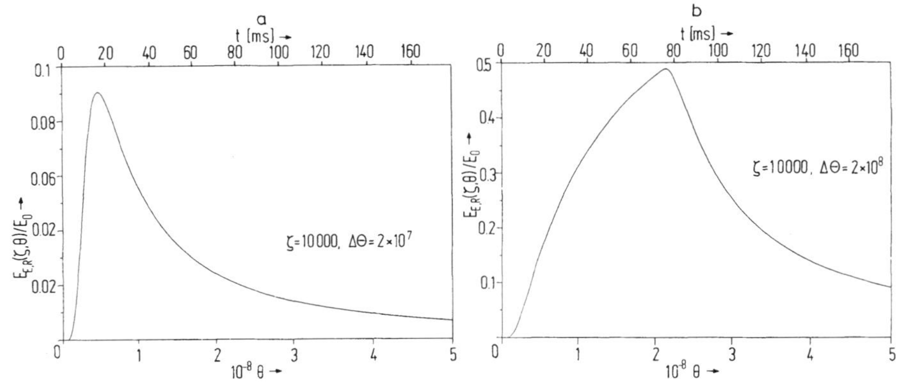

FIG.5.5-5. The pulses $E_{\mathrm{E,R}}(\zeta,\theta)/E_0$ according to Eq.(19) for $\Delta\Theta = 2\times 10^7$ (a) and $\Delta\Theta = 2\times 10^8$ (b). The non-normalized values are in (a) pulse duration $\Delta\Theta \to \Delta T = 7.14\,\mathrm{ms}$, distance $\zeta \to y = 1100\,\mathrm{m}$, time interval $0 \le \theta \le 5\times 10^8 \to 0 \le t \le 179\,\mathrm{ms}$, and in (b) $\Delta T = 71.4\,\mathrm{ms}$, $y = 1100\,\mathrm{m}$, $0 \le t \le 179\,\mathrm{ms}$.

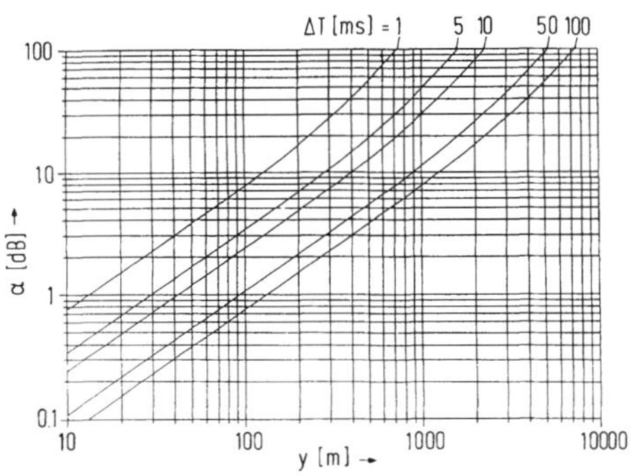

FIG.5.5-6. Attenuation $\alpha(y,\Delta T,\Delta T)$ as function of the propagation distance y in seawater for an originally rectangular pulse of duration ΔT according to Eq.(22). Note that electromagnetic pulses with a duration of 10 to 100 ms may be transmitted over distances of several kilometer without undue attenuation by absorption.

$$W(\zeta,\Delta\Theta,\Delta\Theta_{\mathrm{f}}) = \int_{\zeta}^{\zeta+\Delta\Theta_{\mathrm{f}}} E^2_{\mathrm{E,R}}(\zeta,\theta)d\theta = \int_0^{\Delta\Theta_{\mathrm{f}}} E^2_{\mathrm{E,R}}(\zeta,u)du \qquad (21)$$

The energy attenuation by absorption $\alpha(\zeta,\Delta\Theta,\Delta\Theta_{\mathrm{f}})$ is the ratio of Eqs.(21) and (20):

$$\alpha(\zeta,\Delta\Theta,\Delta\Theta_{\mathrm{f}}) = \frac{W(\zeta,\Delta\Theta,\Delta\Theta_{\mathrm{f}})}{W(0,\Delta\Theta)}$$

$$= \int_{\zeta}^{\Delta\Theta} \left[1 - \operatorname{erf}\left(\frac{\zeta}{\sqrt{2u}}\right)\right]^2 du \qquad \text{for } \zeta < u \leq \Delta\Theta$$

$$= \int_{\Delta\Theta}^{\Delta\Theta_f} \left[\operatorname{erf}\left(\frac{\zeta}{\sqrt{2(u-\Delta\Theta)}}\right) - \operatorname{erf}\left(\frac{\zeta}{\sqrt{2u}}\right)\right]^2 du \quad \text{for } \Delta\Theta \leq u \leq \Delta\Theta_f \tag{22}$$

Figure 5.5-6 shows plots of $\alpha(\zeta, \Delta\Theta, \Delta\Theta_f) = \alpha(y, \Delta T, \Delta T_f)$ for $\Delta\Theta_f = \Delta\Theta$, using the non-normalized notation $\Delta T = 1$, 5, 10, 50, 100 ms and distances in the interval $10 \leq y \leq 10000$ m. This plot shows that, for instance, a propagation distance $y = 500$ m and an acceptable absorption attenuation $\alpha = 20$ dB requires a pulse duration $\Delta T = 5$ ms or longer. We caution that in addition to the absorption attenuation one may have a boundary attenuation at the air-seawater boundary and a geometric propagation attenuation.

5.6 Distance Information from Distortions

The possibility of an airborne anti-submarine radar has been of interest for a long time. The classical radar using pulses modulated unto a sinusoidal carrier and measuring the round-trip propagation time of such pulses returned by a scatterer cannot be used for this task, but the solution of Maxwell's equations for step functions opens new applications. Let us investigate what one can expect to achieve.

Seawater has a dc conductivity of about 4 S/m. As a result, time varying electromagnetic waves can generally not propagate very far in seawater. Exceptions in terms of sinusoidal time variation are waves with frequencies below about 100 Hz or around 10^{15} Hz, which is the visible light region. In terms of pulses—which for simplicity one may think of to be approximately rectangular—one needs pulses of at least millisecond or tens of milliseconds duration according to Fig.5.5-6, or at most picosecond duration. The short pulses and the high frequencies are possible because the conductance of seawater is due mainly to sodium and chlorine ions with a mass equal to about 40000 and 60000 electron masses. It takes some time before such heavy particles get moving and the dc conductivity of 4 S/m is established.

We will study the use of long pulses, typically in the order of tens of milliseconds. This solves the problem of attenuation in seawater. But three major problems come immediately to ones mind: a) Sufficient energy is needed to overcome the reflection losses at the air-seawater boundary. b) The duration of the pulses precludes distance measurements based on the round-trip propagation time of signals as used by the classical radar since its very beginning; a new principle for distance measurement applicable to long, distorted pulses must be used. c) A new radiator is needed that produces powerful pulses with high efficiency, but is small enough to be carried by an airplane.

Consider the reflection losses first. An electromagnetic wave with perpendicular incidence going through the air-seawater boundary suffers a reflection loss of about 40 – 50 dB (Horvat 1969). This transition has not been studied satisfactorily for pulses—we know nothing about the distortions—but enough theory has been developed in Chapters 2 and 3 to work it out.

Since a radar signal has to go twice through this boundary we have to accept a loss of 80 – 100 dB. However, most of this loss is made up by the *geometric advantage* of the anti-submarine radar. If such a radar is mounted on an airplane flying about 500 m above the sea surface and a submarine is about 500 m below the sea surface—the airplane flying in a search pattern and always looking essentially straight down since the attenuation is too great in any other direction—one needs a radar with a range of 1 km. This is an unusually short range. Airborne search radars operating in the atmosphere routinely look for targets at distances of 100 km or more. Since the energy returned by a target varies like R^{-4}, where R is the distance between radar and target, the short distance implies an enormous saving of power or energy. This geometric advantage is a main reason why we can accept an attenuation of 80 – 100 dB caused by the signal having to go twice through the air-seawater boundary.

To obtain a first estimate of the required transmitter power of an anti-submarine radar we start with Eq.(39) on page 19 of Harmuth (1990), which gives the minimum energy W_{min} of a received pulse for a detection probability of 0.99, a false alarm probability of 10^{-8}, and a noise temperature of 450 K of the input amplifier:

$$W_{\mathrm{min}} = 2.40 \times 10^{-19}\,[\mathrm{J}] \tag{1}$$

The received energy W_{r} is given by Eq.(4) on page 5 of the same reference,

$$W_{\mathrm{r}} = \frac{W_{\mathrm{t}} G A_{\mathrm{e}} \sigma_{\mathrm{r}}}{(4\pi)^2 R^4} \tag{2}$$

where W_{t} is the transmitted energy, G the radiating antenna gain, A_{e} the effective area of the receiving antenna, σ_{r} the radar cross-section of the target, and R the distance to the target. The radiating antenna gain G is 2 for the large-current radiator to be discussed in Section 5.7. Beam forming by means of radiator arrays is currently impossible due to the long pulses used.

We substitute W_{min} for W_{r} and solve Eq.(2) for W_{t}:

$$W_{\mathrm{t}} = \frac{(4\pi)^2 W_{\mathrm{min}} R^4}{G A_{\mathrm{e}} \sigma_{\mathrm{r}}} \tag{3}$$

We choose $R = 1000$ m (airplane at altitude 500 m, submarine at depth 500 m), take W_{min} from Eq.(1), and choose $A_{\mathrm{e}} = 10\,\mathrm{m}^2$, $\sigma_{\mathrm{r}} = 10\,\mathrm{m}^2$. We obtain:

$$W_{\mathrm{t}} = \frac{(4\pi)^2 \times 2.40 \times 10^{-19} \times 10^{12}}{2 \times 10 \times 10} = 1.9 \times 10^{-7}\,[\mathrm{J}] \tag{4}$$

This is the minimum transmitted energy required if the target is in the air. The air-seawater boundary attenuation causes an energy loss of 80 – 100 dB. Taking the median 90 dB, we get the required energy:

$$W_{\mathrm{t}}' = 190\,[\mathrm{J}] \tag{5}$$

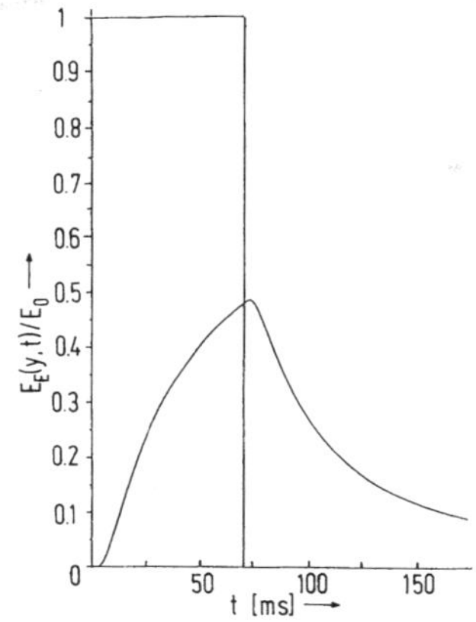

FIG.5.6-1. Rectangular pulse of duration 71 ms and the resulting distorted pulse produced by propagating 1100 m through seawater.

We still have to add the absorption loss in the seawater. From Fig.5.5-6 we get for a distance $y = 1000$ m and a pulse duration $\Delta T = 50$ ms the approximate attenuation $\alpha(y, \Delta T, \Delta T) = 10$ dB. Hence, Eq.(5) becomes:

$$W_t'' = 1900\,[\mathrm{J}] \tag{6}$$

This is about 40 times the energy transmitted with a typical shipboard radar signal. However, we must keep in mind that our pulse is $\Delta T = 50$ ms long rather than a few microseconds. The transmitted peak power is thus not large at all,

$$P_{\mathrm{peak}} = \frac{W_t''}{\Delta T} = \frac{1900}{50 \times 10^{-3}} = 38\,[\mathrm{kW}] \tag{7}$$

while the average power is about what a microwave oven delivers,

$$P_{\mathrm{av}} = 1.9\,[\mathrm{kW}] \tag{8}$$

if one pulse is transmitted per second. These are rather small values for radar.

Our first estimate of energy and power will be refined as time goes on, just as it was done for the classical radar operating in the atmosphere over the last 60 years. At the present we can say that the required transmitter power is well within our means.

The use of pulses with a duration in the order of 50 ms raises the question of range determination. One cannot expect to measure a distance of 1000 m by observing the round-trip propagation time of such a long pulse. We must find a new principle.

Figure 5.6-1 shows a radiated rectangular pulse and the distorted pulse obtained after the pulse has propagated 1100 m through seawater. This corresponds to a submarine at a depth of 550 m since the pulse must propagate twice the depth. There is a time delay between the two pulses which implies that in principle one could derive the distance of 550 m from an observation of the round-trip propagation time, but this is not a practical proposition. Measuring the arrival time of the returned pulse by observing when it or its cross-correlation function with a sample pulse exceeds a certain threshold voltage is not realistic since a small change of the threshold results in an enormous change of distance due to the large times shown in Fig.5.6-1. Furthermore, the amplitude of the returned pulse depends on the radar cross-section of the target; a changed amplitude of the returned pulse of Fig.5.6-1 and a fixed threshold would again give enormous variations of the distance.

Let us observe that for a pulse with a duration of about 50 ms any submarine is a point scatterer. Hence, the specific structure of a submarine has essentially no effect on the returned signal. In the language of radar, there will be no observable target signature. We can calculate the time variation of distorted pulses for n round-trip propagation distances in the water of 100 m, 200 m, ... , 1100 m as discussed in Sections 5.1 to 5.5. The received distorted pulse can then be correlated with these sample pulses in n correlators. Whenever the output voltage of one of these n correlators exceeds a threshold determined by detection and false alarm probability as discussed in Section 5.5, we assume that a target was detected. If the threshold was exceeded by the correlator output for the sample pulse calculated for a round-trip distance of 500 m, and the outputs of all the other correlators are less, we know that there is a target at a depth of 250 ± 25 m.

Due to the long duration of the pulses in Fig.5.6-1 the correlations can be done by an integrate-and-sample circuit, an analog-to-digital converter, and a computer. One does not even have to do the correlations simultaneously or in real time. If the received, distorted signal is sampled, transformed into a series of binary numbers, and stored one can do the correlations sequentially, at a time of ones choosing and with the speed of ones choosing. This is completely different from the classical radar[1].

One more concept is needed. The returned, distorted pulse in Fig.5.6-1 must be received while the radiator is still transmitting the rectangular pulse. This is not possible with any available receiver. But we can replace the one rectangular pulse in Fig.5.6-1 with a series of much shorter rectangular pulses separated by a time gap during which there is no transmission. Figure 5.6-2 shows such a sequence of pulses with duration of 2 ms and gaps with duration of 41 μs. Consider an airplane at an altitude of $R_{\rm air} = 500$ m. After the end of radiation a return will be received from the surface of the sea for a time

$$\Delta t = \frac{2R_{\rm air}}{c} = \frac{2 \times 500}{3 \times 10^8} = 3.33\,[\mu{\rm s}]$$

Some more time is required for the input amplifier to recover. Then one can receive the returned signal by integrating the output voltage of the input amplifier for about 35 μs, sampling this voltage, transforming it with an analog-to-digital converter to a binary number, and feeding it to the computer that does the correlations. The integration over 35 μs provides an initial filtering. The final filtering is done during the cross-correlation with the calculated, distorted sample pulse.

[1] For more details see Harmuth 1992b.

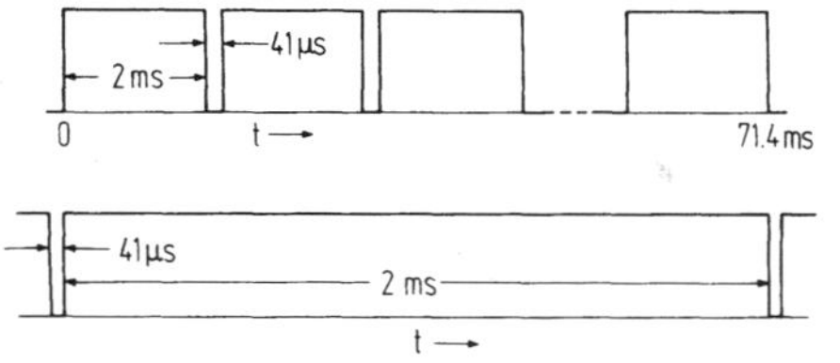

FIG.5.6-2. A pulse of duration 71.4 ms is composed of 35 pulses of 2 ms duration with gaps of 41 μs duration between them.

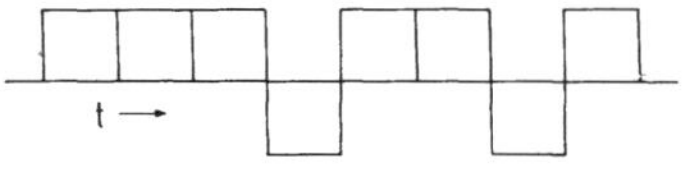

FIG.5.6-3. A sequence of pulses with positive or negative amplitude makes selective reception possible without using a sinusoidal carrier.

It is evident that the choice of gaps permitting reception in Fig.5.6-2 is not optimal but is used strictly to show the principle. Optimizing these gaps with respect to detection and false alarm probability requires knowledge of the radiator that will be discussed in Section 5.7.

The gaps of 41 μs duration in Fig.5.6-2 occur only in the radiated signal but not in the received one. Seawater has for electric and magnetic field strengths a similar effect as a low-pass filter has for voltages and currents. Narrow gaps are smoothed out as one may readily deduce from Sections 5.1 to 5.5.

We have so far assumed that a pulse with a duration in the order of tens of milliseconds is to be radiated. Figure 5.6-1 in particular holds for a pulse duration of 71 ms. The optimal pulse duration is determined by two contradicting requirements: 1. The shorter the pulse the more drastic is the distortion and the better the range resolution, if the signal-to-noise ratio is no problem. 2. The longer the pulse the more energy it can contain when radiated and the loss of energy in the seawater becomes less important, which means it will produce a good signal-to-noise ratio at the threshold detector. For a given noise temperature—either due to received noise or due to noise produced in the input amplifier—there is an optimal pulse duration for a certain probing depth.

As long as there is only one anti-submarine radar operating one may use a single, nominally rectangular pulse as shown in Fig.5.6-1. If more than one radar is operating one must mark their signals in some way to permit selective reception. Conventionally, the marking in radar transmission is provided by a sinusoidal carrier with certain frequency. Such an approach to the selective reception problem would increase the absorption losses of an anti-submarine radar too much to be acceptable. In the transmission of digital signals over wire lines we have used for more than a century pulse sequences as shown in Fig.5.6-3 to produce signals that can be separated from others and thus be received selectively. This technique is applicable to radio transmission too[2].

[2]For more details see Harmuth 1991.

Signals like that of Fig.5.6-3 are sometimes objected to because they have a dc component and it is widely believed that electromagnetic waves not guided by wires cannot have a dc component. Textbooks deriving the one-dimensional wave equation from Maxwell's equations and then showing its general solution by d'Alembert demonstrate that this is not in line with the accepted theory of electromagnetic waves[3].

5.7 Radiation of Slowly Varying EM Waves

Efficient radiators for sinusoidal waves can be implemented by using the resonance effect of time-invariant rods or wires with sinusoidal currents. The many variations of the resonating dipole offer examples. The principle is sometimes obscured by using terms like *standing waves* rather than resonance, but a standing wave is a resonance effect. The use of resonance is restricted to waves with frequencies that permit resonating structures with acceptable size.

For 30 years the radiation of very slowly varying electromagnetic waves has attracted attention in connection with communication to submarines (Merrill 1974). Success was always limited by the fact that the known radiators required immense areas, had efficiencies below 1%, and could radiate only a few watt of power. Great efforts went into finding better radiators. The lack of success of these efforts created the general assumption that nature did not permit such radiators. There is no theoretical basis for this assumption and there were always attempts to find better radiators not based on resonance (Corum 1986, 1988).

To elucidate the matter let us look at the radiation produced by a Hertzian electric dipole. This is not a resonating dipole but an inherently small radiator that will radiate waves with any time variation. Such a dipole with length s in the direction $\mathbf{s}$ is shown in Fig.5.7-1. The current $i(t)$ flowing in the dipole produces the field strengths $\mathbf{E}$ and $\mathbf{H}$ at the location $\mathbf{r}$ (Harmuth 1984, pp. 49, 52):

$$\mathbf{E} = \frac{Z_0 s}{4\pi c}\left[\frac{1}{r}\frac{di}{dt}\frac{\mathbf{r}\times(\mathbf{r}\times\mathbf{s})}{sr^2} + \left(\frac{c}{r^2}i + \frac{c^2}{r^3}\int i\,dt\right)\left(\frac{\mathbf{r}\times(\mathbf{r}\times\mathbf{s})}{sr^2} + 2\frac{(\mathbf{s}\cdot\mathbf{r})\mathbf{r}}{sr^2}\right)\right] \tag{1}$$

$$\mathbf{H} = \frac{s}{4\pi c}\left(\frac{1}{r}\frac{di}{dt} + \frac{c}{r^2}i\right)\frac{\mathbf{s}\times\mathbf{r}}{sr} \tag{2}$$

where $Z_0 = \sqrt{\mu/\epsilon} \approx 377\,\Omega$, $c = 1/\sqrt{\mu\epsilon} \approx 3\times 10^8\,\mathrm{m/s}$, and $r = |\mathbf{r}|$. We note that both $\mathbf{E}$ and $\mathbf{H}$ have a far zone term that decreases like $1/r$. In addition there are near zone terms decreasing like $1/r^2$, or $1/r^3$. The vector terms on the right of Eqs.(1) and (2) give the directional dependence of $\mathbf{E}$ and $\mathbf{H}$ or the *radiation diagrams* of $\mathbf{E}$ and $\mathbf{H}$.

---

[3]See pages 145 – 147 of Harmuth (1984) for oscillograms of radiated pules that clearly have a dc component. Note the emphasis on 'dc component'. Steady state dc lasting from minus infinity to plus infinity can be radiated no more than steady state, periodic sinusoidal waves since the energy would be infinite. Such steady state waves are outside the conservation law of energy as well as outside the causality law, which explains why they are perfectly useful in some applications but fail in others.

FIG.5.7-1. Hertzian electric dipole represented by a dipole vector **s** with the current $i(t)$ flowing in the direction of **s**. The field strengths $\mathbf{E} = \mathbf{E}(\mathbf{r}, t)$ and $\mathbf{H} = \mathbf{H}(\mathbf{r}, t)$, as well as Poynting's vector $\mathbf{P} = \mathbf{P}(\mathbf{r}, t) = \mathbf{E} \times \mathbf{H}$ are produced at the location **r**.

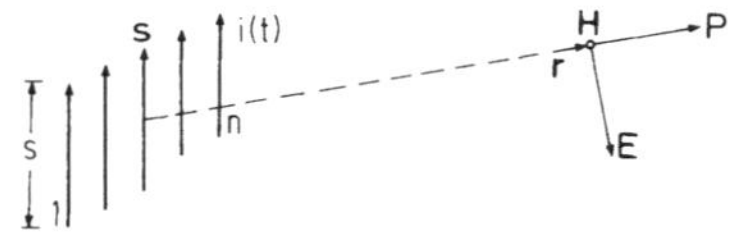

FIG.5.7-2. Array of n radiators of length s, with current $i(t)$ flowing in each radiator in the direction of **s**. Field strengths **E** and **H** as well as Poynting's vector $\mathbf{P} = \mathbf{E} \times \mathbf{H}$ are shown at the location **r**.

The vector product $\mathbf{E} \times \mathbf{H}$ gives Poynting's vector **P** and the integral of **P** over the surface of a sphere with radius r gives the power flowing through this surface (Harmuth 1984, pp. 52, 53):

$$P(t) = \frac{Z_0 s^2}{6\pi c^2} \left[\left(\frac{di}{dt} \right)^2 + \frac{2c}{r} i \frac{di}{dt} + \frac{c^2}{r^2} \left(i^2 + \frac{di}{dt} \int i\, dt \right) + \frac{c^3}{r^3} i \int i\, dt \right] \quad (3)$$

The term $(di/dt)^2$ does not contain r; this power is radiated to infinity. The power of the other terms decreases with increasing radius r of the integration sphere, which means that energy is stored in space.

In order to generalize from one to n Hertzian electric dipoles as shown in Fig.5.7-2 we replace $i(t)$ by $ni(t)$. This approximation assumes that the interaction between the dipoles has a negligible delay, which will be correct if the distance between the dipoles divided by c is small compared with the duration of rectangular pulses that we want to radiate. The current $i(t)$ is separated into the current amplitude I and its time variation $f(t)$:

$$i(t) = If(t), \quad -1 \le f(t) \le +1 \quad (4)$$

where $f(t) = \sin \omega t$ holds in the case of a sinusoidal current. The far zone term $P_{\mathrm{rad}}(t)$ of $P(t)$ assumes the form:

$$P_{\mathrm{rad}} = \frac{Z_0}{6\pi} \left(\frac{nsI}{c} \right)^2 \left(\frac{df}{dt} \right)^2 \quad (5)$$

If we want to make the radiated power large we can increase the length s and the number n of the radiators—both having a practical limit particularly if the radiator is to be transportable. Furthermore, one can increase the current amplitude I. All three possibilities are used by long wave radiators. The radiated power and

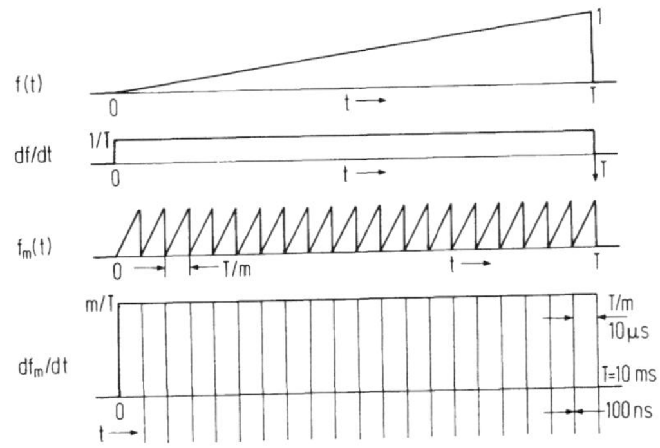

FIG.5.7-3. A time function $f(t)$ rising from 0 to 1 during the time T yields as derivative df/dt a rectangular pulse with amplitude $1/T$ and duration T. The saw-tooth function $f_m(t)$ yields as derivative $df_m(t)/dt$ a rectangular pulse with amplitude m/T and duration T.

the efficiency of the radiators used by the projects Sanguine and Seafarer[1] are representative for what can be achieved in this way. A transportable radiator is not possible if one relies on large values of s, n, and I.

The factor df/dt in Eq.(5) has the same effect as the factors n, s, or I. For instance, for a sinusoidal time variation $f(t) = \sin\omega t$ we get:

$$\frac{df}{dt} = \omega\cos\omega t \tag{6}$$

A frequency $\omega/2\pi = 100\,\text{Hz}$ yields

$$\left(\frac{df}{dt}\right)^2 = 10^4(2\pi)^2\cos^2\omega t \tag{7}$$

while a frequency $\omega/2\pi = 1\,\text{MHz}$ yields:

$$\left(\frac{df}{dt}\right)^2 = 10^{12}(2\pi)^2\cos^2\omega t \tag{8}$$

The difference of the amplitudes of Eqs.(7) and (8)—and thus of the peak and average radiated power—is a factor 10^8. This is the reason why powerful, high-efficiency radiators can be built readily for sinusoidal waves with high frequency but not with low frequency.

Is it possible to make df/dt large but still have a wave with slow time variation? As long as we stay with sinusoidal waves and the small-relative-bandwidth technology we cannot do so since df/dt and ω are linked by Eq.(6). However, pulses and the large-relative-bandwidth technology make it possible to have large values of df/dt but slowly varying waves.

---

[1]See the Special Issue on Extremely Low-Frequency Communications of IEEE Trans. Communications, COM-22, 1974.

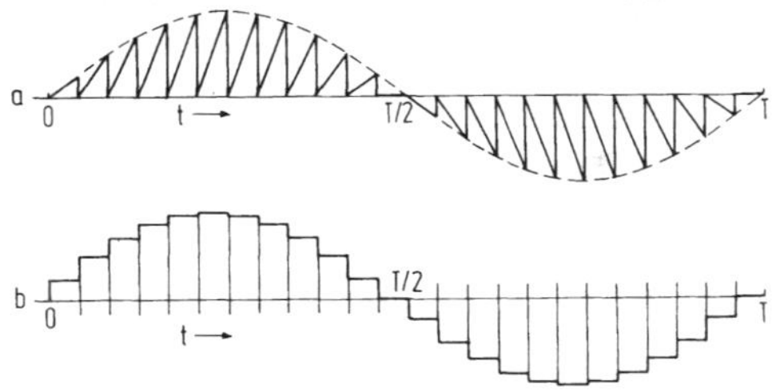

FIG.5.7-4. Ramp function $f_{\mathrm{m}}(t)$ with amplitudes determined by the sinusoidal function $\sin 2\pi t/T$ rather than a rectangular pulse (a) and the derivative of this function (b) with undetermined spikes added at the discontinuities.

For an explanation refer to Fig.5.7-3. On top is shown a function $f(t)$ that rises linearly from 0 to 1 during the time T. The derivative is a rectangular pulse of duration T and amplitude $1/T$. At the end of the pulse, $t = T$, a negative spike is shown with unspecified amplitude or time variation. Mathematically, this spike does not exist, at least not within the theory of differential calculus based on Leibnitz and Newton which yields no defined derivative at the time $t = T$. But the spikes exist for practical antennas due to the near field terms in Eq.(3) in addition to the far field terms[2].

Consider next the function $f_{\mathrm{m}}(t)$ in Fig.5.7-3 that rises from 0 to 1 in the time $\Delta T = T/m$, drops to zero, and does so m times during the time T. The derivative df_{m}/dt is a rectangular pulse with amplitude m/T and duration T. The negative spikes at multiples of T/m occur due to the near zone terms of $P(t)$ in Eq.(3).

The important result is that m narrow pulses with duration T/m and amplitude m/T can be superimposed to yield a long pulse with duration T and amplitude m/T. Consider a pulse of duration $t = 10\,\mathrm{ms}$ and let it be composed of pulses with duration $\Delta T = T/m = 10\,\mu\mathrm{s}$. We obtain $m = 10^{-2}/10^{-5} = 1000$. The factor $(df/dt)^2$ in Eq.(5) is increased by a factor $m^2 = 10^6$. Hence, the term $(nsI)^2$ can be decreased by a factor 10^{-6} without changing the radiated power. Those knowledgeable with radiators for very slowly varying waves will recognize that the problem of ohmic losses is overcome. A reduction of the product ns, which represents the length of a wire, or of the current amplitude I implies a reduction of the ohmic losses and thus an increase in efficiency, while a reduction of s implies a reduction of the physical size of the radiator[3]

It is worth reflecting that the superposition of $m = 1000$ time shifted sinusoidal functions with a frequency of 100 kHz would not produce a sinusoidal function with a frequency of 100 Hz! Our result depends essentially on the use of nonsinusoidal radiator currents. However, one can use the method to produce sinusoidal waves as is made clear by Fig.5.7-4.

---

[2] Differential calculus according to Leibnitz and Newton assumes infinitely short distances dx and dt in space and time, which is already an abstraction since we can observe only finite distances. A further abstraction leads to *distributions* and within this theory the negative spikes can exist mathematically but we must provide additional information for their definition. There is no need to use this abstraction.

[3] The transition from $f(t)$ to $f_{\mathrm{m}}(t)$ in Fig.5.7-3 is closely related to the cyclotron principle used in particle accelerators.

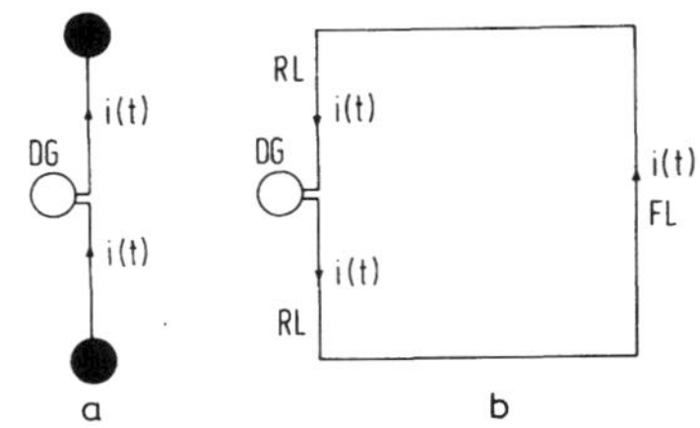

FIG.5.7-5. Hertzian electric dipole (a) and Hertzian magnetic dipole (b); FL forward loop, RL return loop.

The spikes shown in Figs.5.7-3 and 5.7-4 will readily be suppressed by seawater which acts like a low-pass filter for field strengths rather than voltages and currents. One can also put the radiator into a box and wrap aluminum foil or space cloth around it to achieve the same filtering effect.

Radiators for short, nominally rectangular pulses have been built for at least 20 years. A particular type is discussed in detail in the literature under the name *large-current radiator*. The use of light-activated semiconductor switches has advanced these radiators recently from the academic to the practical level (Lukin, Masalov, and Pochanin 1996). The emphasis was in the past on short pulses of about 1 ns duration rather than a time variation approximating a rectangular pulse. Here we shall think of pulse durations ΔT of the order of 100 ns to 10 μs. Such long pulses make it easier to control their time variation.

For a short explanation of the radiator consider the Hertzian electric dipole shown in Fig.5.7-5a. It consists of a short rod with metallic spheres at the ends. A current driver DG pushes a current $i(t)$ through the rod. In principle one can substitute the Hertzian electric dipole for the radiating rods in Fig.5.7-2. However, it is evident that enormous driving voltages would be required to drive a current of tens of amperes through a Hertzian electric dipole, even for a time of the order of 1 μs. The problem of enormous driving voltages is overcome by the Hertzian magnetic dipole of Fig.5.7-5b. This radiator will permit large currents without excessively large voltages. But a new problem arises. The current in the *forward loop* FL flows 'up' in Fig.5.7-5b but it flows in the opposite direction in the *return loop* RL. As a result the radiation from the two sections of the Hertzian magnetic dipole essentially cancel; more precisely, low-power quadrupole rather than high-power dipole radiation is produced.

To make the Hertzian magnetic dipole a useful radiator one can do two things. 1. The forward and the return loop can be made physically very different to prevent total cancellation of the dipole radiation. This principle is used in long-wave radiators with a wire for the forward loop and a ground return for the return loop. 2. One can put a radiation shield around the return loop that prevents radiation. This method is generally restricted to radiators of small size and its implementation is very much at the cutting edge of technology.

A practical method to implement a different forward and return loop for a radiator is shown in Fig.5.7-6. The forward loop is a metal plate, the return loop a metal rod or wire. The two triangular pieces of metal connecting the plate with

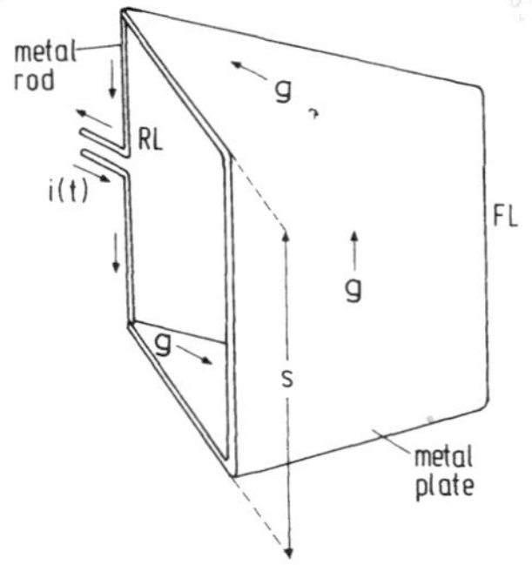

FIG.5.7-6. A large-current radiator uses a metal plate in the forward loop FL and a metal rod or wire in the return loop RL to make them physically very different. The two triangular sections connecting the metal plate with the metal rod are made essentially equal to make their radiation cancel.

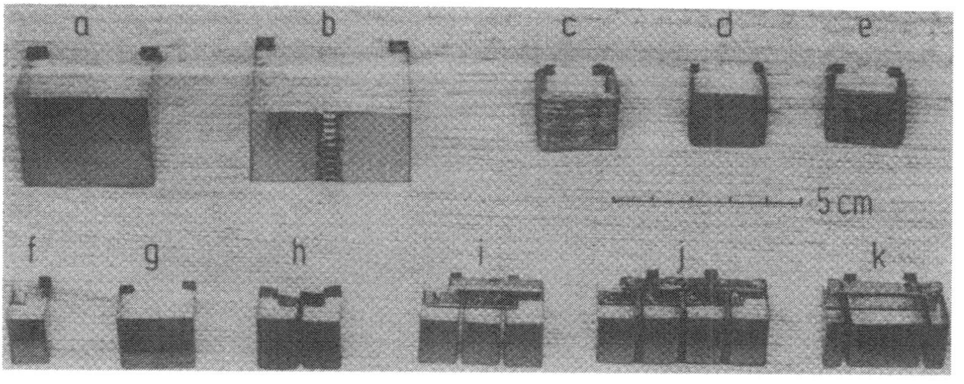

FIG.5.7-7. Experimental development of the large-current radiator. (a) radiator according to Fig.5.7-6; (b) radiator with resistors inserted in the middle; (c)—(e) radiators with metal plate replaced by many wires with one layer (c), two layers (b), and three layers (e); (f)—(h) very small radiators for pulses with a duration of less than 1 ns; (i)—(j) line arrays of large-current radiators to increase the radiated energy and improve the pulse shape (Courtesy C.Kutchera and R.Fleming, Aether Wire & Location Inc., Nicasio, California).

the rod are made practically equal to make their radiation cancel. Radiation from the return loop of this radiator can be suppressed relatively efficiently. Figure 5.7-7 shows several stages in the experimental developement of large-current radiators with small size. A larger radiator for currents of 10 A and up is shown in Fig.5.7-8.

A further advance is shown in Fig.5.7-9. The metal plate is replaced by $n = 4$ wires spaced apart while the wires are bundled together in the return loop. The similarity of this radiating structure with the radiators in Fig.5.7-2 is evident. It is also evident from Fig.5.7-9 that one needs only one current driver for the current $i(t)$ to feed n radiators. This makes it possible to trade voltage for current and helps matching the radiators to the available power source.

The radiation shield in Fig.5.7-9 is currently the technical challenge. Good results have been achieved for pulses with duration of about 1 ns by using a metal tube covered with a thick layer of ferrite and carbon powder. The ferrite, having

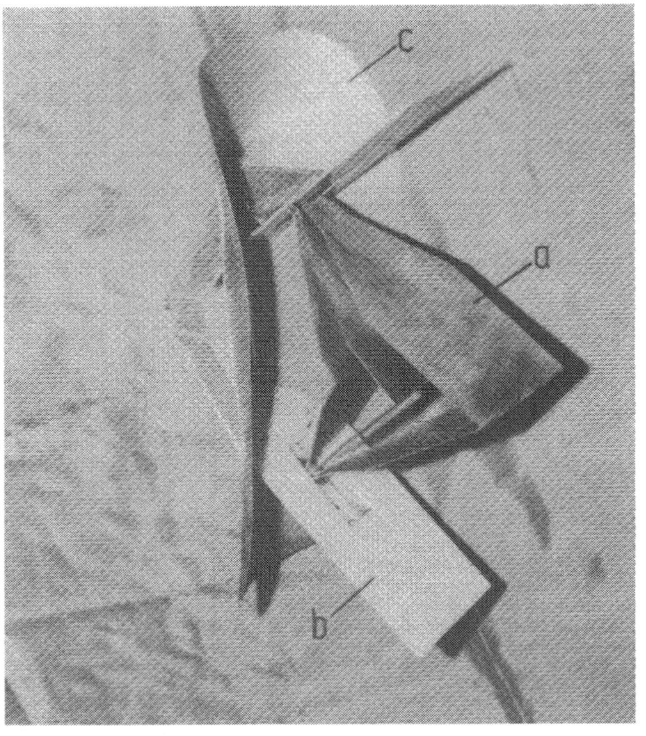

FIG.5.7-8. Large-current radiator according to Fig.5.7-6 for antenna currents of 10 A and up. (a) assembly of three radiating "plates"; (b) metal shield between power source and radiating plates; (c) feed "horns" for the transition from the two-wire coaxial conductor of the power source to the single-wire conductor of the radiating plates (Courtesy S.Masalov and G.Pochanin, Institute for Radiophysics and Electronics, Academy of Science of Ukraine, Kharkov).

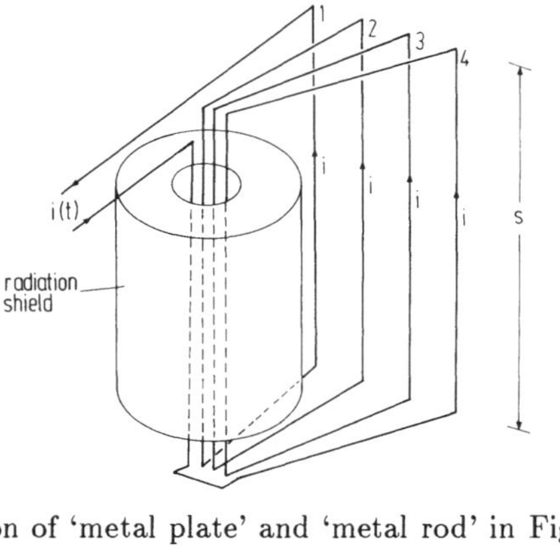

FIG.5.7-9. Implementation of 'metal plate' and 'metal rod' in Fig.5.7-6 by widely spaced and closely bundled wires similar to those in Figs.5.7-7c to d.

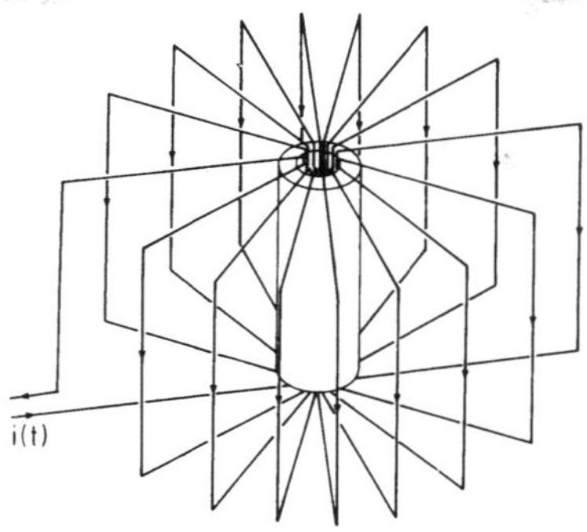

FIG.5.7-10. Further development of the radiator according to Fig.5.7-9.

a large relative permeability and permittivity, reduces the velocity $c = 1/\sqrt{\mu\epsilon}$ and thus the propagation velocity of any wave, while the carbon powder absorbs the wave. One would expect that for pulses with a duration of 1 μs or longer the ferrite material could be replaced by permalloy and barium-titanate, but this has not been investigated experimentally yet.

A further readily understandable advancement of the radiator of Fig.5.7-9 is shown in Fig.5.7-10.

We derive some characteristic numerical values in order to see where the difficulties of a design according to Fig.5.7-9 are—apart from the problem of a good radiation shield—and what one can expect from its further development. Let us assume that each wire loop is square, which means the height of the radiator is s, its diameter $2s$, and the length of one loop is $4s$. The total length of the wire through which the current $i(t)$ in Fig.5.7-9 flows is thus $4ns$. The duration $\Delta T = T/m$ of the pulses of $f_{\rm m}(t)$ in Fig.5.7-3 multiplied by c must be large compared with $4ns$ to make our assumption valid that an equal current flows in each one of the n loops of the radiator. We choose $c\Delta T$ ten times as large as $4ns$:

$$c\Delta T = 40ns, \quad \Delta T = 40ns/c \tag{9}$$

The radiated power according to Eq.(5) becomes with $df/dt = 1/\Delta T$:

$$\begin{aligned} P_{\rm rad}(t) &= \frac{Z_0}{6\pi}\left(\frac{nsI}{40ns}\right)^2 = \frac{Z_0}{6\pi}\frac{I^2}{1600} \\ &= 0.0125 I^2 \quad [P_{\rm rad} \text{ in watt, } I \text{ in ampere}] \end{aligned} \tag{10}$$

The required driving voltage v for the radiated power becomes:

$$\begin{aligned} v &= P_{\rm rad}/I \\ &= 0.0125 I \quad [I \text{ in ampere, } v \text{ in volt}] \end{aligned} \tag{11}$$

A current of 100 A will require a voltage of 1.25 V and radiate a power of 125 W according to Eq.(10), while a current of 1000 A will require 12.5 V and radiate 12.5 kW. These are reasonable values for semiconductor driving circuits.

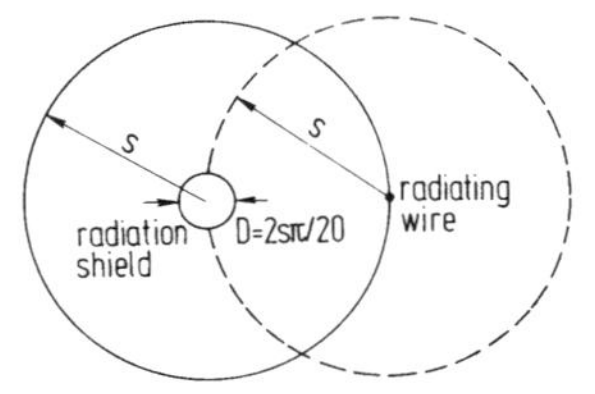

FIG.5.7-11. Choosing the radius and the height of the radiator so that 5% of the radiation hits the radiation shield with a diameter D.

Consider the resistance R_{ohm} of the wire of length $4ns$,

$$R_{\text{ohm}} = 4ns\rho/a \tag{12}$$

where a is the area of the cross-section of the wire and $\rho = 1.7 \times 10^{-8}\,\Omega\text{m}$ is the specific resistance of copper. A saw-tooth current proportionate to $f_{\text{m}}(t)$ in Fig.5.7-3 produces in a resistance R_{ohm} the average loss:

$$\frac{R_{\text{ohm}}}{\Delta T}\int_0^{\Delta T}\left(I\frac{t}{\Delta T}\right)^2 dt = \frac{1}{3}R_{\text{ohm}}I^2 \tag{13}$$

If we want the radiated power to be q times as large as the power lost due to the resistance of the wire loops, we must make Eq.(10) q times as large as Eq.(13):

$$\frac{q}{3}\frac{4ns\rho}{a}I^2 = P_{\text{rad}}(t) = \frac{Z_0}{6\pi}\frac{I^2}{1600}$$

$$ns = \frac{Z_0 a}{12800\pi\rho q} \tag{14}$$

Let us assume we want an *ohmic efficiency* of 90%, which means 10% of the power delivered to the radiator goes for ohmic losses. This yields $q = 9$. We further assume that the 'wire' loops in Fig.5.7-9 consist of 10 HF stranded cables in parallel, each with a diameter of 6 mm. We obtain:

$$a = 10d^2\pi/4 = 283\,[\text{mm}^2] = 2.83 \times 10^{-4}\,[\text{m}^2] \tag{15}$$

Equations (14) and (9) yield:

$$ns = 17.34\,[\text{m}] \tag{16}$$

$$\Delta T = 2.3\,[\mu\text{s}] \tag{17}$$

We still have to separate the product ns into the length s and the number n of loops. This is done with the help of Fig.5.7-11 which shows a top view of the radiator. If we want that only 5% of the power radiated by one radiating wire hits the radiation shield we must make the diameter of this shield $D = 2s\pi/20$.

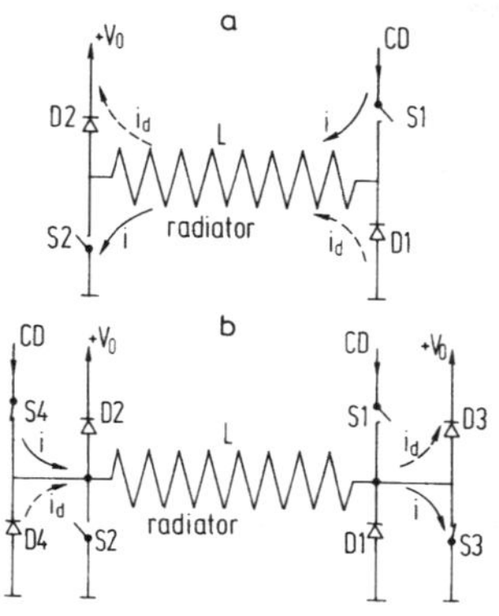

FIG.5.7-12. Current driver for unipolar (a) and bipolar (b) operation. The current i flows when the switches are closed, the current i_{d} when they are opened. CD current driver.

Neglecting the wall thickness of this cylinder and assuming perfect filling of the available space by the n loops of HF stranded cable we get with $D^2\pi/4 = na$:

$$\begin{aligned} D &= \sqrt{4na/\pi} = 2s\pi/20 \\ s^2 &= 400na/\pi^3 \end{aligned} \tag{18}$$

The substitution of n from Eq.(16) yields:

$$\begin{aligned} s^2 &= 400 \times 17.34 \times 2.83 \times 10^{-4}/\pi^3 = 0.0633\,[\mathrm{m}^3] \\ s &\approx 0.4\,[\mathrm{m}] \\ n &= 17.34/0.4 \approx 43 \end{aligned} \tag{19}$$

Hence, the radiator is $s = 40$ cm high, has a diameter of $2s = 80$ cm, and $10 \times 43 = 430$ turns of HF cable with a diameter of 6 mm are wound around it. Ten current drivers pushing 10 A each with a voltage of 1.25 V in 2.3 μs through it are required to radiate somewhat more than 100 W.

We have made many simplifications to derive these numbers but the main point was to show that one obtains reasonable values[4].

We turn to the current drivers for the discussed radiators. The principle is shown in Fig.5.7-12. A current source CD is connected via switch S1 to the radiator and via switch S2 to ground. A ramp current i is driven through the radiator during the time $0 \leq t \leq \Delta T$. At $t = \Delta T$ the switches S1 and S2 are opened. If all the power fed to the radiator had been radiated there would be no electromagnetic energy in the vicinity of the radiator. However, Eq.(3) shows that in

[4] Large-current radiators according to Fig.5.7-6 using an avalanche transistor switch can currently switch 0.7 A in 1 ns (Lukin, Masalov, and Pochanin 1997) while light activated semiconductor switches - referred to as s-diodes in Russian - have achieved 10–20 A in 1 ns.

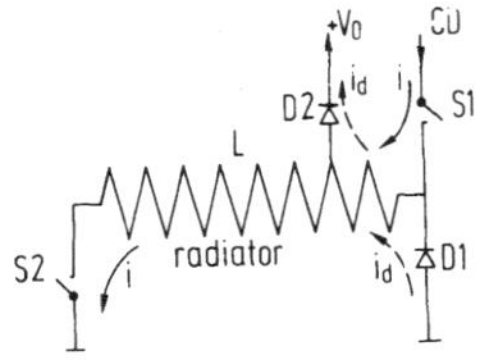

FIG.5.7-13. Modification of the circuit of Fig.5.7-12a to use only part of the windings for the discharge current i_d.

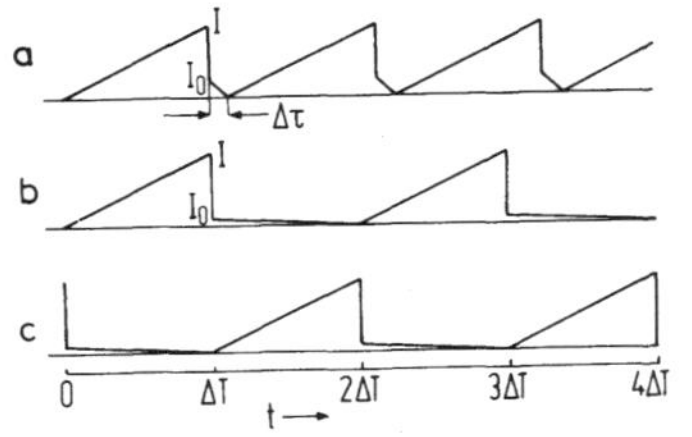

FIG.5.7-14. More detailed possible time variations of an antenna current $If_\mathrm{m}(t)$ according to Fig.5.7-3.

addition to the radiated power term $(di/dt)^2$ there are near zone terms that imply energy stored in the space around the radiator. We can use Eq.(3) only for a qualitative discussion of this stored energy since it is based on the assumption that the radiator has a negligible diameter. All terms in Eq.(3) except the radiation term diverge for $r \to 0$. Better approximations of the large-current radiator have been worked out but they are too complicated for use here (Harmuth 1990). We see from Eq.(3) that an increase of di/dt without change of i or $\int i\,dt$ will favor the radiation term $(di/dt)^2$ over the others which contain di/dt in first order or not at all. The transition from $i(t) = If(t)$ to $i(t) = If_\mathrm{m}(t)$ does not affect the average value of $i(t)$ or $\int i\,dt$ according to Fig.5.7-3. Hence, our method of operation produces inherently small near field components compared with the far field component.

The term $i\,di/dt$ in Eq.(3) is the most important near zone component. It has the form of the terms describing the power fed into an inductive coil due to the current i and the voltage $L\,di/dt$. We will neglect all the other near zone terms for the further discussion and assign the inductivity L to the radiator in Fig.5.7-12a. When the switches S1 and S2 open, the energy stored around the radiator will drive a current $i_\mathrm{d}(t)$ through the path provided by the diodes D1 and D2. This current does not need to flow through all the windings of the radiator. A variation of Fig.5.7-12a using only part of the windings for the discharge of the near zone energy is shown in Fig.5.7-13.

Let the switches S1 and S2 in Fig.5.7-12a open at the time $t = \Delta T$. The following equation applies:

$$L\frac{di_\mathrm{d}}{dt} + V_0 = 0 \tag{20}$$

It has the solution

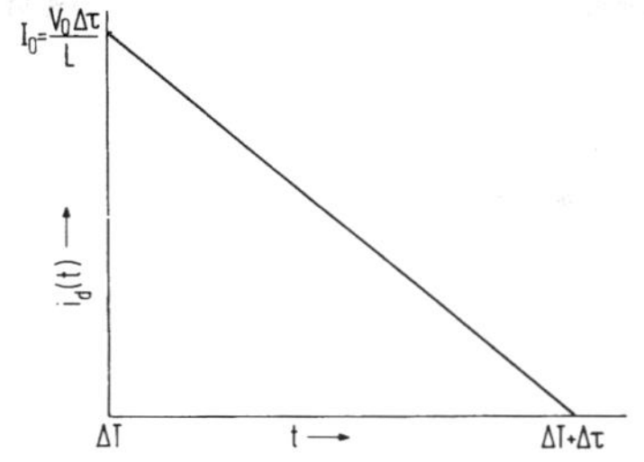

FIG.5.7-15. Time variation of the discharge current $i_d(t)$.

$$i_d(t) = I_0 - \frac{V_0}{L}(t - \Delta T), \quad t \geq \Delta T \tag{21}$$

where I_0 is a constant. We may choose to have the current $i_d(t)$ become zero at the time $t = \Delta T + \Delta\tau$ as shown in Fig.5.7-14a. Substitution of this value of t into Eq.(21) determines I_0:

$$I_0 = \frac{V_0 \Delta\tau}{L}, \quad V_0 = \frac{L I_0}{\Delta\tau}$$
$$i_d(t) = \frac{V_0 \Delta\tau}{L}\left(1 - \frac{t - \Delta T}{\Delta\tau}\right) \tag{22}$$

The time variation of $i_d(t)$ is shown in Fig.5.7-15. The energy in the inductance L is $LI^2/2$ when the switches are opened at the time $t = \Delta T$ according to Fig.5.7-14a. This energy is either returned to the power supply or radiated, if the ohmic losses of the radiator are ignored. The energy returned to the power source is defined by

$$W_{\text{ret}} = \int_{\Delta T}^{\Delta T+\Delta\tau} V_0 i_d \, dt = \frac{V_0^2 \Delta\tau}{L} \int_{\Delta T}^{\Delta T+\Delta\tau} \left(1 - \frac{t - \Delta T}{\Delta\tau}\right) dt$$
$$= \frac{(V_0 \Delta\tau)^2}{2L} = \frac{1}{2} L I_0^2 \tag{23}$$

while the energy according to Eq.(5) radiated as a spike becomes:

$$W_{\text{spike}} = \int_{\Delta T}^{\Delta T+\Delta\tau} \frac{Z_0}{6\pi}\left(\frac{ns}{c}\right)^2 \left(\frac{di_d}{dt}\right)^2 dt = \frac{Z_0}{6\pi}\left(\frac{nsV_0}{Lc}\right)^2 \Delta\tau$$
$$= \frac{Z_0}{6\pi\Delta\tau}\left(\frac{nsI_0}{c}\right)^2 \tag{24}$$

The returned energy and the energy radiated as a spike must equal the stored energy $LI^2/2$:

$$\frac{1}{2}LI_0^2 + \frac{Z_0}{6\pi\Delta\tau}\left(\frac{ns}{c}\right)^2 I_0^2 = \frac{1}{2}LI^2$$

$$I_0 = I\left[1 + \frac{Z_0}{3\pi L\Delta\tau}\left(\frac{ns}{c}\right)^2\right]^{-1/2} \tag{25}$$

The returned energy W_{ret} and the radiated energy W_{spike} are obtained by substituting Eq.(25) into Eqs.(23) and (24):

$$W_{\text{ret}} = \frac{3\pi}{2}\frac{I^2L^2c^2\Delta\tau}{Z_0n^2s^2 + 3\pi Lc^2\Delta\tau} \tag{26}$$

$$W_{\text{spike}} = \frac{1}{2}\frac{I^2LZ_0n^2s^2}{Z_0n^2s^2 + 3\pi Lc^2\Delta\tau} \tag{27}$$

For $L \to 0$ both W_{ret} and W_{spike} vanish. Hence, one will try to design the radiator so that as little energy as possible is stored in the near zone. Equation (3) shows that a large derivative di/dt is one way to approach this goal. For the design of the current driver we can see that one can choose $\Delta\tau$; current I_0 and voltage V_0 are then determined by Eqs.(25) and (22). In the limit $\Delta\tau \to 0$ the returned energy in Eq.(26) vanishes and the radiated energy of the spike becomes $LI^2/2$. The voltage V_0 in Eq.(22) approaches infinity and I_0 in Eq.(25) approaches zero. The wanted radiated energy follows from Eq.(5) for a linear ramp function with $df/dt = 1/\Delta T$:

$$W_{\text{rad}} = \int_0^{\Delta T} P_{\text{rad}}(t)dt = \frac{Z_0}{6\pi\Delta T}\left(\frac{nsI}{c}\right)^2 \tag{28}$$

Hence, a decrease of ΔT and a proportionate decrease of I^2 will leave W_{rad} unchanged but decrease the unwanted radiated energy W_{spike} of Eq.(27).

A second interesting case is $\Delta\tau = \Delta T$. According to Fig.5.7-14b, c one needs two radiators and two drivers that work alternately. According to Eqs.(26) and (27) a large value of $\Delta\tau$ increases the returned energy and decreases the unwanted radiated energy in the spikes.

References and Bibliography

Akers, D. (1994). Further evidence for magnetic charge from hadronic spectra, *Int. J. Theoretical Physics*, vol. 33 (9), 1817–1829.

Anderson, F., Fullerton, L., Christensen, W., and Kortegaard, B. (1991). Wideband beam patterns from sparce arrays, in *Ultra-Wideband Radar: Proceedings of the First Los Alamos Symposium*, B.Noel ed., 273–286, CRC Press, Boca Raton.

Barrett, T.W. (1993). Electromagnetic phenomena not explained by Maxwell's equations; in *Essays on the Formal Aspects of Electromagnetic Theory*, A.Lakhtakia ed., 6–86. World Scientific, Singapore.

Barrett, T.W. (1995). Sagnac effect: A consequence of conservation of action due to gauge field global conformal invariance in a multiple joined topology of coherent fields; in *Advanced Electromagnetism: Foundations, Theory, and Applications*, T.W.Barrett and D.M.Grimes eds., 278–313. World Scientific, Singapore.

Bateman, H. (1954). *Tables of Integral Transforms*. McGraw-Hill: New York.

Becker, R. (1964). *Theorie der Elektriziät*, 18th ed. B.G.Teubner: Stuttgart, Germany.

Becker, R. (1982). *Electromagnetic Fields and Interactions*, (transl. of 16th ed. of *Theorie der Elektrizität* by A.W.Knudsem). Reprinted Dover: New York.

Borisov, V.V. (1996). *Electromagnetic Fields of Transient Currents*, (in Russian). St.Petersburg University Press, Russia.

Boules, R. N. (1989a). Propagation velocity of electromagnetic signals in lossy media in the presence of noise, *PhD Thesis*, Department of Electrical Engineering, The Catholic University of America, Washington, DC.

Boules, R. N. (1989b). Adaptive filtering using the fast Walsh-Hadamard transformation, *IEEE Trans. Electromagn. Compat.*, vol. EMC-31, 125–128.

Boules, R. N. (1991). Parabolid array beam forming for nonsinusoidal waves, *IEEE Trans. Electromagn. Compat.*, vol. EMC-33, 35–41.

Boules, R. N. (1991). Synchronous communication in a lossy medium using transient waves, *IEEE Trans. Electromagn. Compat.*, vol. EMC-33, 295–303.

Brillouin, L. (1914). Über die Fortpflanzung des Lichtes in dispergierenden Medien, *Ann. Phys.*, vol. 4, 203–240.

Brillouin, L. (1946). *Wave Propagation in Periodic Structures*. McGraw-Hill: New York.

Brillouin, L. (1960). *Wave Propagation and Group Velocity*. Academic Press: New York.

Brinson, A., Min, K., and Willis Jr., M. (1996). Eglin designed radar system for explosives (B. Rolfsen, reporter). *Northwest Florida Daily News, Ft.Walton-Beach, FL*, 9 June, p. B1 (A layer of fence wire over an air raid shelter neutralizes the ground probing radar).

Brittingham, J.N. (1983). Focus wave modes in homogeneous Maxwell's equations: Transverse electric mode, *J. Appl. Phys.*, vol. 54, 1179–1189.

Carin, L. and Agi, K. (1993). Ultra-wideband transient microwave scattering measurements using opto-electronically switched antennas, *IEEE Trans. Microwave Theory & Techn.*, vol. MTT-41, 250–254.

Cook, C.E. (1960). Proposed monocycle-pulse very high frequency radar for airborne ice and snow measurements. *Trans. AIEE Commun. Electron.*, vol. 79, 588–594.

Corum, J.F. (1986). Toroidal antenna. US Patent 4,622,558.

Corum, J.F. (1988). Electromagnetic structure and method. US Patent 4,751,515.

Ehrenfest, P. (1910). Mißt der Aberrationswinkel im Falle einer Dispersion des Äthers die Wellengeschwindigkeit? *Ann. Phys.*, 4. Folge, vol. 33, 1571–1576.

Fushchich, W.I. and Nikitin, A.G. (1983). *Symmetries of Maxwell's Equations*. Reidel Publishing Co.: Dordrecht.

Gang Wang, WenBing Wang, and Chang Hong Liang (1998). Further modification of the radar equation due to the slow decay effect in electromagnetic radiation, *IEEE Trans. Electromagn. Compat.*, vol. EMC-40, in press.

Geyi, W., Chengli, R., and Weigan, L. (1991a). Backscattering of electromagnetic missile by a perfectly conducting elliptical cylinder, *J. Appl. Phys.*, 70(1), 1-3.

Geyi, W., Chengli, R., and Weigan, L. (1991b). Unified theory of the backscattering of electromagnetic missiles by a perfectly conducting sphere, *J. Appl. Phys.*, 70(8), 4053–4056.

Geyi, W., Chengli, R., and Weigan, L. (1992a). Unified theory of the backscattering of electromagnetic missiles by a perfectly conducting target, *J. Appl. Phys.*, 71(7), 3103–3106.

Geyi, W., Chengli, R., and Weigan, L. (1992b). A unified theory of electromagnetic missiles generated by an arbitrary plane current source, *Microwave & Optical Letters*, 5(7), 337–340.

Geyi, W., Chengli, R., and Weigan, L. (1992c). Study of backscattering of electromagnetic missiles, *Acta Electronica Sinica (Beijing Dianzi Xuebao)*, 20(6), 26–35.

Gradshteyn, I.S. and Ryzhik, I.M. (1980). *Tables of Integrals, Series, and Products*, (English edition by A.Jeffrey). Academic Press, New York.

Harmuth, H.F. (1960). Radio communication with orthogonal time functions. *Trans. AIEE Commun. Electron.*, vol. 79, 221–228; On the transmission of information by orthogonal time functions. *Trans. AIEE Commun. Electron.*, vol. 79, 248–255.

Harmuth, H.F. (1972). *Transmission of Information by Orthogonal Functions.* Springer-Verlag, Berlin.

Harmuth, H.F. (1977). *Sequency Theory—Foundations and Applications.* Academic Press, New York.

Harmuth, H.F. (1981). *Nonsinusoidal Waves for Radar and Radio Communication.* Academic Press, New York.

Harmuth, H.F. (1984). *Antennas and Waveguides for Nonsinusoidal Waves.* Academic Press, New York.

Harmuth, H.F. (1985). Frequency independent shielded loop antenna. US Patent 4,506,267.

Harmuth, H.F. (1986a). Correction of Maxwell's equations for signals I, II. *IEEE Trans. Electromagn. Compat.*, vol. EMC-28, 250–258, 259–266.

Harmuth, H.F. (1986b). Propagation velocity of electromagnetic signals. *IEEE Trans. Electromagn. Compat.*, vol. EMC-28, 267–272.

Harmuth, H.F. (1986c). *Propagation of Nonsinusoidal Electromagnetic Waves.* Academic Press, New York.

Harmuth, H.F. (1989a). *Information Theory Applied to Space-Time Physics* (in Russian). MIR, Moscow. English edition: World Scientific Publishing Co., Singapore, 1993.

Harmuth, H.F. (1989b). Radar equation for nonsinusoideal waves. *IEEE Trans. Electromagn. Compat.*, vol. EMC-31, 138–147.

Harmuth, H.F. (1990). *Radiation of Nonsinusoidal Electromagnetic Waves.* Academic Press, New York.

Harmuth, H.F. (1991). Efficient operation of probing radar in absorbing media. US Patent 5,057,846.

Harmuth, H.F. (1992a). Response to a letter by J.R.Wait on magnetic dipole currents. *IEEE Trans. Electromagn. Compat.*, vol. EMC-34, 374–375.

Harmuth, H.F. (1992b). Range information from signal distortions. US Patent 5,159,343.

Harmuth, H.F. (1993a). Electromagnetic transients not explained by Maxwell's equations; in *Essays on the Formal Aspects of Electromagnetic Theory*; A.Lakhtakia ed., 87–126 ; World Scientific Publishing Co., Singapore.

Harmuth, H.F. (1993b). Transient solutions of Maxwell's equations for lossy media. *Proceedings of the X School-Seminar on Wave Diffraction and Propagation*, Moscow, 7–15 February, 256–267.

Harmuth, H.F. (1994a). Comments on "General Fourier solution and causality in attenuating media" by R.E.Duren. *IEEE Trans. Electromagn. Compat.*, vol. EMC-36, 156–157.

Harmuth, H.F. (1994b). Response to letters 6 and 7 by Professor J.R.Wait. *IEEE Trans. Electromagn. Compat.*, vol. EMC-36, 410–411.

Harmuth, H.F. (1994c). Transmitter signature and target signature of radar signals, in *Introduction to Ultrawideband Radar Technology* (J.D.Taylor ed.), 435–456. CRC Press, Boca Raton, FL.

Harmuth, H.F. (1994d). Radiator for slowly varying electromagnetic waves. US Patent 5,307,081.

Harmuth, H.F. (1994e). *Propagation of Electromagnetic Signals.* World Scientific, Singapore.

Harmuth, H.F. (1995). Extension of Ohm's law to electric and magnetic dipole currents, in *Advanced Electromagnetism: Foundations, Theory & Applications*, T.W.Barrett and D.M.Grimes, eds., 506–540. World Scientific, Singapore.

Harmuth, H.F. (1996). Sliding correlator for nanosecond pulses. US Patent 5,523,758.

Harmuth, H.F., Boules, R.N., and Hussain, M.G. (1987). Reply to Kuester's comments on the use of a magnetic conductivity. *IEEE Trans. Electromagn. Compat.*, vol. EMC-29, 318–320.

Harmuth, H.F., Boules, R.N., and Hussain, M.G.M. (1988). Response to comments by M.J.Neatrour on Maxwell's equations. *IEEE Trans. Electromagn. Compat.*, vol. EMC-30, 90–91.

Harmuth, H.F. and Hussain, M.G.M. (1994a). *Propagation of Electromagnetic Signals*. World Scientific, Singapore.

Harmuth, H.F. and Hussain, M.G.M. (1994b). Signal solutions of Maxwell's equations for charge carriers with non-negligible mass. *IEEE Trans. Electromagn. Compat.*, vol. EMC-36, 411–412.

Heyman, E. and Melamed, T. (1994). Certain considerations in aperture synthesis of ultrawideband/short pulse radiation, *IEEE Trans. Antennas and Propagation*, vol. 42, 518–525.

Hillion, P. (1987). Spinor focus wave modes, *J. Math. Phys.*, vol. 28, 1743–1748.

Hillion, P. (1990). Boundary value problems for the wave equation. *Rev. math. Phys.*, vol. 2, 177–191.

Hillion, P. (1991). Remarks on Harmuth's 'Correction of Maxwell's equations for signals I'. *IEEE Trans. Electromagn. Compat.*, vol. EMC-33, 144.

Hillion, P. (1992a). Response to "The magnetic conductivity and wave propagation". *IEEE Trans. Electromagn. Compat.*, vol. EMC-34, 376–377.

Hillion, P. (1992b). A further remark on Harmuth's Problem. *IEEE Trans. Electromagn. Compat.*, vol. EMC-34, 377–378..

Hillion, P. (1993). Some comments on electromagnetic signals; in *Essays on the Formal Aspects of Electromagnetic Theory*, A.Lakhtakia ed., 127–137. World Scientific Publishing Co., Singapore.

Hillion, P. and Lakhtakia. A. (1993). On an initial boundary value problem involving Beltrami-Moses fields in electromagnetic theory. *Phil. Trans. R. Soc. Lond.*, vol. A344, 235–248.

Hondzoumis, V.A., Wu, T.T., and Myers, J.M. (1994). Backscattering of an electromagnetic missile by a metal cylinder of degree higher than two, *J. Appl. Phys.*, vol. 80, 14–24.

Horvat, V. (1969). Underwater radio wave transmission; in *Handbook of Ocean and Underwater Engineering*, J.J.Myers, ed. McGraw-Hill, New York.

Hussain, M.G.M. (1988a). General solution of Maxwell's equations for signals in a lossy medium: I. Electric and magnetic field strengths due to electric exponential ramp function excitation. *IEEE Trans. Electromagn. Compat.*, vol. EMC-30, 29–36.

Hussain, M.G.M. (1988b). General solutions of Maxwell's equations for signals in a lossy medium: II. Electric and magnetic field strengths due to magnetic exponential ramp function excitation. *IEEE Trans. Electromagn. Compat.*, vol. EMC-30, 37–40.

Hussain, M.G.M. (1988c). General Solutions of Maxwell's equations for signals in a lossy medium: III. Electric and magnetic field strengths due to electric and magnetic sinusoidal pulse excitations. *IEEE Trans. Electromagn. Compat.*, vol. EMC-30, 41–47.

Hussain, M.G.M. (1988d). Antenna patterns of nonsinusoidal waves with the time variation of a Gaussian pulse, Part I. *IEEE Trans. Electromagn. Compat.*, vol. EMC-30, 504–512.

Hussain, M.G.M. (1988e). Antenna patterns of nonsinusoidal waves with the time variation of a Gaussian pulse, Part II. *IEEE Trans. Electromagn. Compat.*, vol. EMC-30, 513–522.

Hussain, M.G.M. (1989a). Antenna patterns of nonsinusoidal waves with the time variation of a Gaussian pulse, Part IV. *IEEE Trans. Electromagn. Compat.*, vol. EMC-31, 49–54.

Hussain, M.G.M. (1989b). Principles of high-resolution radar based on nonsinusoidal waves, Part I: Signal representation and pulse compression. *IEEE Trans. Electromagn. Compat.*, vol. EMC-31, 359–368.

Hussain, M.G.M. (1989c). Principles of high-resolution radar based on nonsinusoidal waves, Part II: Generalized ambiguity function. *IEEE Trans. Electromagn. Compat.*, vol. EMC-31, 369–375.

Hussain, M.G.M. (1990). Principles of high-resolution radar based on nonsinusoidal waves, Part III: Radar target reflectivity model. *IEEE Trans. Electromagn. Compat.*, vol. EMC-32, 144–152.

Hussain., M.G.M. (1993). A comparison of transient solutions of Maxwell's equations to that of modified Maxwell's equations. *IEEE Trans. Electromagn. Compat.*, vol. EMC-34, 482–486.

Hussain, M.G.M., Al-Habib, M.M., and Omar, A.A. (1989). Antenna patterns of nonsinusoidal waves with the time variation of a Gaussian pulse, Part III. *IEEE Trans. Electromagn. Compat.*, vol. EMC-31, 34–48.

Johnson, J.B. (1928). Thermal agitation of electricity in conductors. *Phys. Rev.*, vol. 32, 97–109.

Kaplan, A.E. and Shkolnikov (1995). Electromagnetic "bubbles" and shock waves: unipolar, nonoscillating EM solutions. *Phys. Rev. Let.*, vol. 75, 2316–2227.

King, R.W.P. (1993). The propagation of a Gaussian pulse in sea water and its application to remote sensing. *IEEE Trans. Geoscience and Remote Sensing*, vol. 31, 595–605.

Kuester, E.F. (1987). Comments on 'Correction of Maxwell's equations for signals I', 'Correction of Maxwell's equations for signals II', and 'Propagation of electromagnetic signals'. *IEEE Trans. Electromagn. Compat.*, vol. EMC-29, 187–190.

Laue, M. (1905). Die Fortpflanzung der Strahlung in dispergierenden Medien. *Ann. Phys.*, vol. 18, 523–566.

Liang, C.H. and Wang, G. (1997). Appropriate short pulses to generate the desired slow decay effect at designated large distance. *Chinese J. Electron.*, vol. 6, no. 2, 80–85.

Lorber, H.W., A time domain radar range equation, in *Ultra-Wideband, Short-Pulse Electromagnetics,* L.Carin and L.B.Felsen eds., Plenum, New York.

LoVetri, J. and Ehrman, J.B. (1994). Time domain electromagnetic plane waves in static and dynamic conducting media, I. *IEEE Trans. Electromagn. Compat.*, vol. EMC-36, 221–228.

Lukin, K.A., Masalov, S.A., and Pochanin, G.P. (1997). Large-current radiator with avalanche transistor switch. *IEEE Trans. Electromagn. Compat.*, vol. EMC-39, in press.

Maindardi, F. (1983). Signal velocity for transient waves in linear dissipative media. *Wave Motion*, vol. 5, 33–41.

Maxwell, J.C. (1891). *A Treatise on Electricity and Magnetism.* Reprinted by Dover, New York 1954.

Merril, J. (1974). Some early historical aspects of project Sanguine. *IEEE Trans. Communication*, vol. COM-22, 359–363.

Mohamed, N.J. (1995). Carrier-free signal design for look-down radar. *IEEE Trans. Electromagn. Compat.*, vol. EMC-37, 51–61.

Morey, R.M. (1974). Geophysical survey system employing electromagnetic impulses. US Patent 3,806,795.

Müller, C. (1967). *Foundations of the Mathematical Theory of Electromagnetic Waves.* Grundlagen der Mathematischen Wissenschaften in Einzeldarstellung, vol. 155. Springer-Verlag, Berlin.

Natio, Yoshiyuki and Takahashi, Michiharu (1989). Electromagnetic wave absorber. US Patent 4,862,174.

Nielsen, N. (1904). *Handbuch der Theorie der Cylinderfunktionen*, Teubner, Leipzig.

Nyquist, H. (1928). Thermal agitation of electric charges in conductors. *Phys. Rev.*, vol. 32, 110–113.

Olver, F.W.J. (1964). Bessel functions of integer order; in *Handbook of Mathematical Functions*, Abramovitz M. and Stegun I.A. eds. US Government Printing Office, Washington, DC.

Oughstum, K.E. and Sherman, G.C. (1994). *Electromagnetic Pulse Propagation in Causal Dielectrics.* Springer-Verlag, Berlin.

Pan, W.Y. (1985). An experimental investigation of the distribution of current and charge induced in a tubular conducting cylinder by an electromagnetic pulse, *IEEE Trans. Electromagn. Compat.*, vol. EMC-27, 88–95.

Pan Zhongying, Yang Qiji, Jiang Changyin, Lin Pingshi, and Guo Huamin (1994). Measurment of the impulse responses and the wideband RCS of targets using transient electromagnetic fields. *Chinese J. of Electronics*, vol. 3, no. 1, 1–9.

Pan Zhongying, Yang Qiji, and Jiang Changyin (1994). Modified steepest descent method for inverting the impulse response of targets. *Chinese J. of Electronics*, vol. 3, no. 4, 14–20.

Papazoglou, T. M. (1975). Transmission of a transient electromagnetic plane wave into a lossy half-space. *J. Appl. Phys.*, vol. 46, 3333–3341.

Reitz, J.R., Milford, F.J., and Christy, R.W. (1980). *Foundations of Electromagnetic Theory*, 3rd ed. Addison-Wesley: Reading, MA.

Ruan, C. and Wan, C. (1989). Choice of excitation in EM missiles, *Electron. Lett.*, vol. 25, 1321–1323.

Samaddar, S.N. (1993). Behavior of the energy density in the Fresnel region of a conducting circular disk excited by single cycle sinusoidas, *J. Appl. Phys.*, vol. 74, 3013–3023.

Sezginer, A. (1985). A general formulation of focus wave modes. *J. Appl. Phys.* **57** (3), 678–683.

Shao Dingrong, Li Shujian, and Zhou Bin (1993). A research of an acousto-optic correlator for a new telecommunication receiver system. *J. SPIE, OIptical Information Processing*, vol. 2051, 992–1002.

Shvartsburg, A.B. (1996). *Time Domain Optics of Ultrashort Waveforms.* Oxford University Press: Oxford, England.

Sommerfeld, A. (1907). Ein Einwand gegen die Relativitätstheorie der Elektrodynamic und seine Beseitigung. *Physik Z.*, vol. 8, 841.

Sommerfeld, A. (1914). Über die Fortpflanzung des Lichtes in dispergierenden Medien. *Ann. Phys.*, vol. 44, 177–202.

Van Trees, H.L. (1968). *Detection, Estimation, and Modulation Theory, Part I.* Wiley: New York.

Voigt, W. (1899). Über die Änderung der Schwingungsform des Lichtes beim Fortschreiten in einem dispergierenden oder absorbierende Mittel. *Ann. Phys.*, Neue Folge, vol. 68, 598–603.

Wang, G., Wang, W.B., and Liang.C.H. (1998). Further modification of the radar range equation due to slow decay effect in electromagnetic radiation. *IEEE Trans. Electromagn. Compat.*, *vol. EMC-40*, 77–82.

Watson, G.N. (1966). *A Treatise on the Theory of Bessel Functions*, 2nd ed., Cambridge University Press, London.

Weber, C. (1987). *Elements of Detection and Signal Design*, 2nd ed. Springer-Verlag, Berlin.

Whittaker, E.T. and Watson, G.N. (1952). *A Course of Modern Analysis*, Cambridge University Press, London.

Withington II, P. and Fullerton, L.W. (1993). An impulse radio communication system, in *Ultra-Wideband, Short-Pulse Electromagnetics*, H.L.Bertoni, L.Carin, and L.B.Felsen eds., 113–120. Plenum Press, New York.

Wu, T.T. (1985). Electromagnetic missiles, *J. Appl. Phys.*, 57(7), 2370–2373.

Wu, T.T. and Shen, H.M. (1988). Generalized analysis of the spherical lens as launcher of electromagnetic missiles, *J. Appl. Phys.*, 63(12), 5647–5653.

Wu, T.T. and Shen, H.M. (1989). Circular cylindrical lens as a line-source electromagnetic-missile launcher, *IEEE Trans. Antenna & Propag.*, vol. AP-37, 39–44.

Wu, T.T. and Shen, H.M. (1990). Fun with pulses, *Physics World*, Nov. 39–42.

Yingzheng, R. and Weigan, L. (1988). Prospect for stealth, counter-stealth technologies, *Electronic Science and Technology (Beijing Dianzi Kexue Jishu)*, 18(11), 2–4.

Zaiping, N. (1983). Radiation characteristics of travelling-wave antennas excited by nonsinusoidal current, *IEEE Trans. Electromagn. Compat.*, vol. EMC-25, 24–31.

Ziolkowski, R.W. (1989). Localized transmission of electromagnetic energy, *Phys. Rev. A*, vol. 39, 2005-20033.

Ziolkowski, R.W. (1992). Properties of electromagnetic beams generated by ultra-wide bandwidth pulse-driven arrays, *IEEE Trans. Antennas and Propagation*, vol. 40, 888-905.

Index

SPRINGER NATURE

GPSR Compliance

The European Union's (EU) General Product Safety Regulation (GPSR) is a set of rules that requires consumer products to be safe and our obligations to ensure this.

If you have any concerns about our products, you can contact us on ProductSafety@springernature.com

In case Publisher is established outside the EU, the EU authorized representative is:

Springer Nature Customer Service Center GmbH
Europaplatz 3
69115 Heidelberg, Germany

FSC
www.fsc.org
MIX
Papier | Fördert gute Waldnutzung
FSC® C083411

Kolibri 360 GmbH
Ferdinand-Jühlke-Straße 7
99095 Erfurt, Deutschland
produktsicherheit@kolibri360.de